Contemporary Urban Design Thinking

Series Editor
Rob Roggema
Research Centre for the Built Environment NoorderRuimte
Hanze University of Applied Sciences
Groningen, The Netherlands

CITTA IDEALE
Office for Adaptive Planning
Wageningen, The Netherlands

This series will investigate contemporary insights in urban design theory and practice. Urbanism has considerably changed and developed over the years and is about to undergo a transformation moving into a new era.

In the 1990's and early 2000's economic driven urban design was prevalent in many countries around the world. Moving forward it is no longer feasible to continue to develop in the same way and new ideas for creating urbanism are urgently required.

This series will publish titles dealing with innovative methods of urbanism including, sustainability driven urbanism, smart urbanism, population driven urbanism, and landscape based urban design.

The series will include books by top researchers and leaders in the fields of urban design, city development and landscape urbanism. The books will contain the most recent insights into urbanism and will provide actual and timely reports filling a gap in the current literature.

The series will appeal to urbanists, landscape architects, architects, policy makers, city/urban planners, urban designers/researchers, and to all of those interested in a wide-ranging overview of contemporary urban design innovations in the field.

More information about this series at http://www.springer.com/series/15794

Rob Roggema
Editor

Nature Driven Urbanism

Springer

Editor
Rob Roggema
Research Centre for the Built Environment NoorderRuimte
Hanze University of Applied Sciences
Groningen, The Netherlands

CITTA IDEALE
Office for Adaptive Planning
Wageningen, The Netherlands

ISSN 2522-8404 ISSN 2522-8412 (electronic)
Contemporary Urban Design Thinking
ISBN 978-3-030-26716-2 ISBN 978-3-030-26717-9 (eBook)
https://doi.org/10.1007/978-3-030-26717-9

This Springer imprint is published by the registered company Springer Nature Switzerland AG.
The registered company address is: Gewerbestrasse 11, 6330 Cham, Switzerland

Contents

Chapter 1
Nature-Driven Urbanism

Rob Roggema

Abstract The city is nature. In many ways this bold statement can be contested, but at the same time wildlife is so abundant Rotterdam is called a wilderness park (Reumer, Wildpark Rotterdam. De stad als natuurgebied. Historische Uitgeverij, Groningen, 2014). One can discuss whether this is true or not, but more interesting is to see the city as a piece of nature, and as such undertake the actions to develop it further. In a city nature should not be treated as something worth to preserve, after all such unique nature can hardly be found inside urban contexts, rather something to increase, enrich and make more resilient.

Keywords Nature · Urban ecology · Urban nature · Nature-driven · Nature-based solutions

The city is nature. In many ways this bold statement can be contested, but at the same time wildlife is so abundant Rotterdam is called a wilderness park (Reumer 2014). One can discuss whether this is true or not, but more interesting is to see the city as a piece of nature, and as such undertake the actions to develop it further. In a city nature should not be treated as something worth to preserve, after all such unique nature can hardly be found inside urban contexts, rather something to increase, enrich and make more resilient.

Nature in cities is discovered. In many studies 'green Infrastructure' is declared to be beneficial to decrease concentration disorder with kids, reduce violence in house, minimize obesity, reduce the recovery time after illness and improve the exercise rate of people living close to green spaces (see Chap. 5).

Green cover, trees, green roofs and facades have the ability to reduce the Urban Heat Island (UHI) effect, a phenomenon that, with climate change, will increase in the (near) future. Recently an Urban heat island is no longer defined as an island in

R. Roggema (✉)
Research Centre for the Built Environment NoorderRuimte,
Hanze University of Applied Sciences, Groningen, The Netherlands

CITTA IDEALE, Office for Adaptive Planning, Wageningen, The Netherlands
e-mail: r.e.roggema@pl.hanze.nl; rob@cittaideale.eu

R. Roggema (ed.), *Nature Driven Urbanism*, Contemporary Urban Design Thinking, https://doi.org/10.1007/978-3-030-26717-9_1

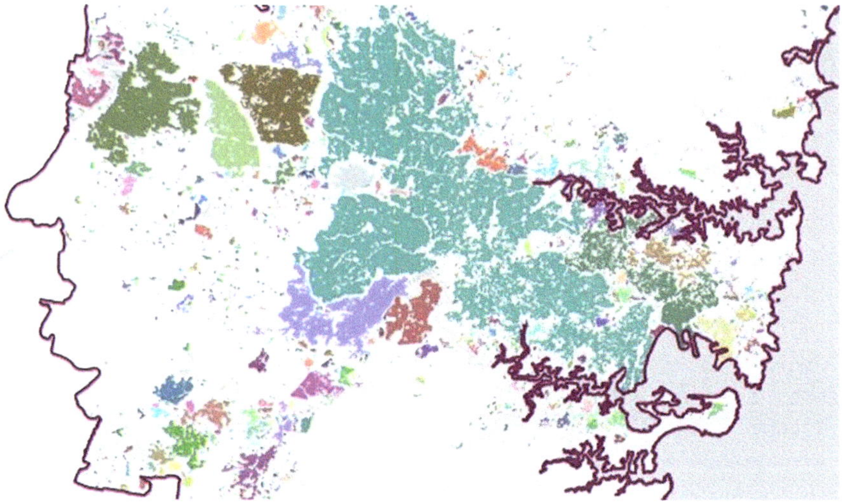

Fig. 1.1 Urban heat Continents in Sydney (Brook 2019)

a broader urban context, but the first 'urban continents of heat (Fig. 1.1), a conglomerate of linked islands, have been identified (Brook 2019). Health of human beings, but also of the ecosystems themselves are at stake, and a profound effort to increase the amount and quality of green spaces is needed to create livable urban areas. And with green spaces close to home being beneficial for your health, it also increases the real estate value of your house (Swinbourne and Rozenwax 2018) hence there is no reason not to pay more attention to green in the urban environment.

Even in highly technological contexts the value of green and nature is increasingly used as an important factor in city planning, as is proven in ten cities in the US (Ahuja 2016). The starting point of designing the city driven by nature, is to analyze the existing pockets of nature, such as the Nature map of Rotterdam (LOLA 2016) illustrates (Fig. 1.2). The map also illustrates the fragmented nature of the existing patches of ecological valuable space.

If we take a look at how the population values green and landscape space it becomes clear the landscapes close to urban or industrialized land use are valued significantly lower than green and undisrupted areas (Buijs et al. 2019). So, what can be concluded is that on the one hand side people value green and nature in their vicinity, it also has benefits for their wellbeing, but on the other side green spaces are diminishing, fragmented and losing a critical size. Therefore, a fundamental choice has to be made: we no longer pursue better and more nature in the urban environment, or we take it seriously hence pay more attention. The latter seems to be the preferred choice for all the good reasons, however existing habits of urbanization prevent this from happening (Fig. 1.3).

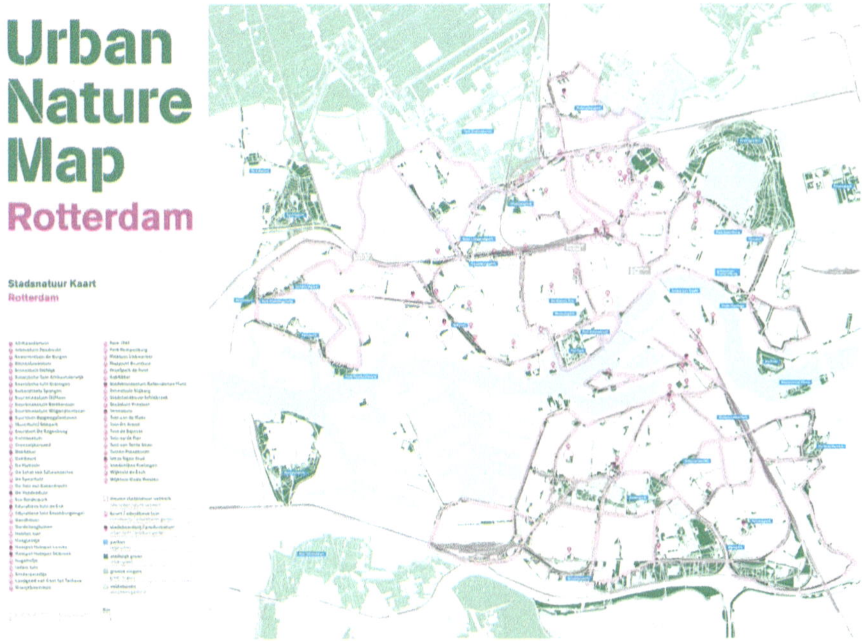

Fig. 1.2 Urban nature map Rotterdam (LOLA 2016)

In order to comply with the desired role of nature and green a new approach to urban design is needed (Garrard et al. 2017). The following objectives can be defined:

1. To protect and create habitat, extend the size of small areas;
2. To help species disperse, establish connections between ecological cores;
3. To minimize anthropogenic threats, minimize environmental impacts and disturbance;
4. To promote ecological processes, allow the space for ecology to emerge, decline, grow, and follow the resilience cycle (Holling and Gunderson 2002);
5. To encourage positive human-nature interactions through arranging green in close proximity and improving the accessibility of green spaces.

A range of design solutions are possible (Roijackers 2017), such as green roofs and facades, climate trees and water to deal with heat stress, planting trees, wadi's, infiltration gardens and water squares to minimize urban floods, using helophytes to improve the water quality, integrate sufficient parks, trees and green to improve the air quality and reduce sealed surfaces, enlarge the area of parks and green spaces and increase the number of street trees to contribute to biodiversity.

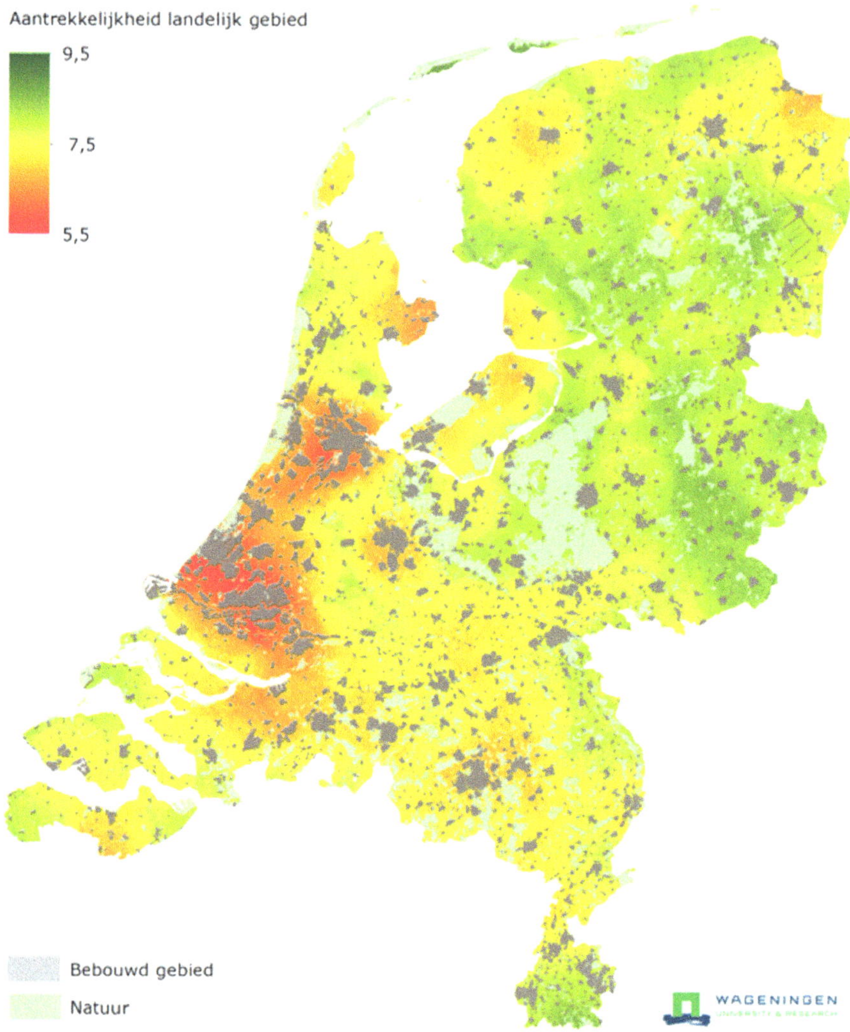

Fig. 1.3 Average attractiveness of the Dutch landscape (Buijs et al. 2019)

Besides new approaches to urban design, and the accompanying solutions, which are readily available, a mental transformation may have more impact. This mental switch should place nature, green and landscape upfront in every planning process. Every step, from the regional planning of land-use until the detailed design of public spaces, should start with the spatial analysis of the ecological system, followed by an iterative process of designing green and ecological options trough the scales, evaluating their benefits as a ground layer for programmatic demands of other land uses. Where current practice often starts with the, economically driven, program to be realized, after which green spaces are fitted in, the new mental model will start with the socio-ecological system and design a robust foundation, within which other functions can be embedded.

In this book the authors discuss their perspectives and show examples with this main frame in mind: an urbanism which is driven by nature and green. In Chap. 2 Dirk Sijmons discusses in his foundational article how the role of nature in urban development is moving from being a contrasting element, distinct from the city, towards making contact with the urban environment, ending in establishing its role as a contract, crystal clear and self-conscious (Sijmons this volume). As Veldman calls it (see Chap. 13), this requires a bold vision, using green spaces and ecology as the condition for livability (Veldman this volume). One of the first examples in the Netherlands applying this 'contract' principle is described in Chap. 8, where the historic creek system is used to determine the urban design of a modern sustainable neighborhood (Chap. 8, Roggema this volume-a), or design road infrastructure (Chap. 14, Collins this volume). In Chap. 10, Phillip Birtles uses South Creek in Western Sydney as the new focus point for urban development, similar to the way in Breda Westerpark was approached (Birtles this volume). Elaborating on this the Third City in Sydney's West can entirely be based on the landscape principles and provide cooling conditions through placing vegetation and water from the beginning at the core of the design for the regional plan (Chap. 5, Roggema this volume-b). Similarly, the water system of Rotterdam is used to define the spatial conditions for a flourishing urban ecology (Chap. 6, Tillie this volume) and, in the case of Doha, water is the main focus of urbanism (Chap. 9, Grichting this volume). Using the basic grids of water and green as the starting point of planning is important, however, the ochre grid, representing the aboriginal heritage and their connection to the land, deserves a similar role in the design process (Chap. 12, Tyrrell this volume). Additionally, nature can play a temporary role, being accepted as a place where ecological values are built up but at the same time moving spatially to elsewhere after a certain period. Backes et al., coin the concept of temporary nature in Chap. 3 (Backes et al. this volume), while Qiao Han and Greg Keeffe design the ecological movement of urban forestry through the city of Manchester, leaving specific areas to temporality (Chap. 4, Han and Keeffe this volume). Within the urban context ecological incentives and interventions can be projected to initiate and form the starting point of eco-urbanism (Chap. 7, Casagrande this volume), which increases the anti-fragility of the entire urban precinct (Chap. 11, Monti this volume). The same principle can be applied in architecture as the work of LAVA illustrates (Chap. 15, Bosse this volume), using the principles, systems and structures found in nature to construct and design contemporary buildings.

Nature and water can play a thriving role at different scales (Fig. 1.4). Birtles, Tyrrell, Roggema (3rd City) and Collins take the regional scale as the starting point of their thinking, whereas Veldman, Qiyao/Keeffe, and Tillie take the city scale as the point of entry. While Roggema (Breda) takes the neighborhood scale as a focus, Backes et al., Monti and Grichting look at smaller spaces and locations within the city, and Casagrande and Bosse see the building as the scale to give nature and ecology a profound role. In his chapter Sijmons links the scales from built elements to the city-region.

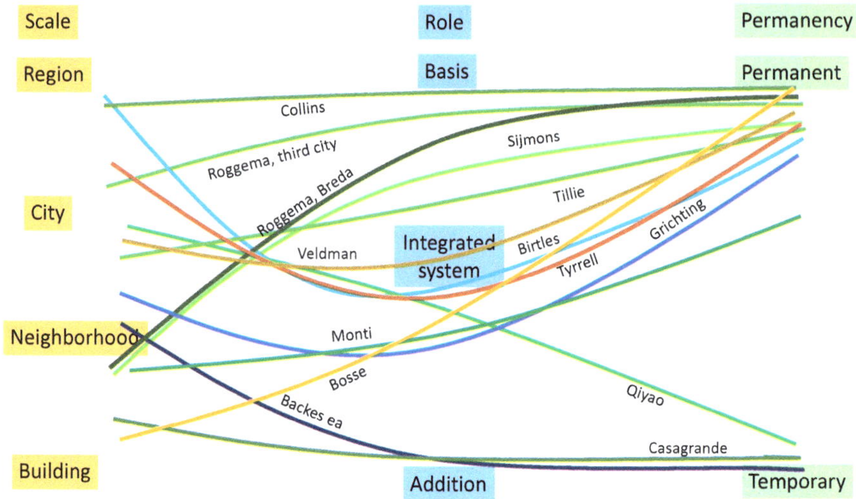

Fig. 1.4 Scales, roles and permanency of nature in urbanism

Looking at the role nature plays in the different chapters three different ones can be identified (Fig. 1.4). The first and most fundamental role is to use landscape, nature and natural systems as the driving force for spatial design. They determine the lay-out of the urban layer and urban functions are responsible for the quality of nature and landscape. At any scale high quality nature and green is designed first before embedding and adding other urban and non-urban uses. Good examples of this philosophy are found in Chaps. 2, 5, 8 and 14). The majority of thinking however aims to integrate ecological and landscape systems in the urban fabric. Grids and networks are connected and as such made an integral part of the city (Chaps. 4, 6, 10, 11, 12 and 13). A third way of looking is to see green, nature or landscape as an addition to an urban structure or system. This is reflected in Chaps. 3, 7, 9 and 15).

Finally, there are different ways to look at temporality of nature, Backes et al., Casagrande, and Qiyao/Keeffe all define nature as a temporary function at a certain space, while others hold the opinion that nature should take a permanent position in the city.

In this book the different chapters altogether show (Fig. 1.4) that a nature-driven approach to urbanism can be or should be applied at every urban scale; in architecture at the building scale, to the urban design level of neighborhoods, and city planning and landscape architecture at the city and regional scale. At every scale, nature-driven approaches add to the quality of the built structure and the quality of life of the people living there. To add nature/green to these built structures is a good starting point and can add value, however most of the authors seem to have more fiducia in giving nature a more fundamental role as an integrated network in city design or to make nature the entrance point of the design process and base the design on the needs and qualities of nature itself. The highest existence of nature is a permanent ecosystem that is keeping itself alive under all circumstances for a prolonged period. However, especially in an urban context this is not always possible, and temporality could be an interesting concept to explore when nature cannot stay forever. The ecological contribution to the environment, and indirect dispersion of species, from a temporary location will, overall add biodiversity to the entire system. Maybe second-best nature, but still essential to support overall ecological qualities.

References

Ahuja A (2016) Integration of nature and technology for smart cities. Springer, Cham
Backes C, Krefeld A, Schoukens H (this volume) Chapter 3: Temporary Nature – a win-win for nature and developers: tinkering with the law in order to combat biodiversity loss. In: Roggema R (ed) Nature-driven urbanism, Contemporary urban design thinking, vol 2. Springer, Dordrecht
Birtles P (this volume) Chapter 10: South Creek in Far Western Sydney: Opportunities for a new waterway focused city. In: Roggema R (ed) Nature-driven urbanism, Contemporary urban design thinking, vol 2. Springer, Dordrecht
Bosse C (this volume) Chapter 15: Bio-inspiration: merging nature and technology. In: Roggema R (ed) Nature-driven urbanism, Contemporary urban design thinking, vol 2. Springer, Dordrecht
Brook, B. (2019) The Australian suburbs and council areas most vulnerable to extreme heat
Buijs A, Nieuwenhuizen W, Langers F, en Kramer H (2019) Resultaten Nationale Landschapsenquête; Onderzoek naar visies en waardering van de Nederlandse bevolking over het landelijk gebied in Nederland, Wageningen Environmental Research, Rapport 2937. Wageningen University, Wageningen
Casagrande M (this volume) Chapter 7: From urban acupuncture to the third generation city. In: Roggema R (ed) Nature-driven urbanism, Contemporary urban design thinking, vol 2. Springer, Dordrecht
Collins G (this volume) Chapter 14: A contemporary approach to the design of road transport infrastructure in balance with the landscape. In: Roggema R (ed) Nature-driven urbanism, Contemporary urban design thinking, vol 2. Springer, Dordrecht
Garrard C, Williand N, Bekessy S (2017) Here's how to design cities where people and nature can both flourish. The Conversation. Published] 4 May 2017. https://theconversation.com/higher-density-cities-need-greening-to-stay-healthy-and-liveable-75840
Grichting A (this volume) Chapter 9: Blue design for urban resilience in drylands: the case of Qatar. In: Roggema R (ed) Nature-driven urbanism, Contemporary urban design thinking, vol 2, Dordrecht, Springer
Han Q, Keeffe G (this volume) Chapter 4: Stepping-stone city: process-oriented infrastructures to aid forest migration in a changing climate. In: Roggema R (ed) Nature-driven urbanism, Contemporary urban design thinking, vol 2, Dordrecht, Springer
Holling CS, Gunderson LH (2002) Resilience and adaptive cycles. In: Holling CS (ed) Panarchy. Island Press, Washington, DC
LOLA (2016) Stadsnatuurkaart Rotterdam. Vereniging Deltametropool, Rotterdam
Monti S (this volume) Chapter 11: Nature-inclusive cities: concepts and considerations. In: Roggema R (ed) Nature-driven urbanism, Contemporary urban design thinking, vol 2, Dordrecht, Springer
News.com.au published 5 January 2019. https://www.news.com.au/technology/environment/climate-change/the-australian-suburbs-and-council-areas-most-vulnerable-to-extreme-heat/news-story/0df25ff17daedbb5793b20da70968671
Reumer J (2014) Wildpark Rotterdam. De stad als natuurgebied. Historische Uitgeverij, Groningen
Roggema R (this volume-a) Chapter 5: Landscape first! Nature-based design for Sydney's third city. In: Roggema R (ed) Nature-driven urbanism, Contemporary urban design thinking, vol 2, Dordrecht, Springer

Roggema R (this volume-b) Chapter 8: Urbanism on water and ecology: the early example of Westerpark, Breda. In: Roggema R (ed) Nature-driven urbanism, Contemporary urban design thinking, vol 2, Dordrecht, Springer

Roijackers V (2017) Nature based solutions – Inspiratieboek. gemeente Eindhoven, Eindhoven

Sijmons D (this volume) Chapter 2: Contrast, contact & contract, pathways to pacify urbanization and nature. In: Roggema R (ed) Nature-driven urbanism, Contemporary urban design thinking, vol 2, Dordrecht, Springer

Swinbourne R, Rozenwax J (2018) Green infrastructure. A vital step to brilliant Australian cities. AECOM, Sydney

Tillie N (this volume) Chapter 6: From urban green structure to tidal river in Rotterdam: testing grounds for urban ecology. In: Roggema R (ed) Nature-driven urbanism, Contemporary urban design thinking, vol 2, Dordrecht, Springer

Tyrrell M (this volume) Chapter 12: Exploring new urban futures through Sydney's hidden grids. In: Roggema R (ed) Nature-driven urbanism, Contemporary urban design thinking, vol 2, Dordrecht, Springer

Veldman D (this volume) Chapter 13: A bold Vision for Sydney's future. In: Roggema R (ed) Nature-driven urbanism, Contemporary urban design thinking, vol 2, Dordrecht, Springer

Chapter 2
Contrast, Contact, Contract; Pathways to Pacify Urbanization and Natural Processes

Dirk Sijmons

Abstract This chapter develops design speculations how to mitigate the effects of urbanization on biodiversity. Some pathways are explored.

First, we show that nature can not be switched on and off at will and there is always a natural expression in – and of – our urban landscapes. Looking at urbanization at a global scale the world acknowledges biodiversity hotspots to coincide with the expected fastest growing cities of the Earth. The best proven way in these situations might still be to traditionally plan for strict nature reserves.

Infrastructures of different kinds in the urban landscape are the main agents in fragmenting habitats, hence remediation strategies such as ecological infrastructure and ecological connecting are briefly explored as the second pathway. This green infrastructure can act as an additional strategy for the conservation pathway and as an auxiliary for all other situations.

The third pathway is the landscape architecture' best bet: making parks. Next to being favourite public spaces and the lungs of a city, parks can act as stepping stones in the green infrastructure.

The fourth pathway, and the core of this chapter, aims to (re)shape the configuration of our urban landscapes. Here we muse the thought that the classic urban utopian models, the garden city, the lobe city and broad-acre city, all three stand for a specific way of interaction between urbanization and landscape. Making contrast, seeking contact and entering into a new contract between city and nature, we will interrogate the significance for the present day situation in helping to reweave the urban landscapes into a more nature inclusive way. International examples are touched upon, while three metropolitan Amsterdam cases show that the strong

The author likes to thank Harma Horlings from the Dutch State Forestry Service and Geert Timmermans, Urban ecologist of the municipality of Amsterdam for commenting on an earlier version of this chapter.

D. Sijmons (✉)
TU-Delft, Delft, The Netherlands

H+N+S Landscape Architects, Amersfoort, The Netherlands

R. Roggema (ed.), *Nature Driven Urbanism*, Contemporary Urban Design Thinking, https://doi.org/10.1007/978-3-030-26717-9_2

points of all three models can be brought into action in a single urban landscape, thus making a vital contribution to nature-driven urbanism.

Keywords Biodiversity hotspots · Landscape architecture · Urban nature · Urban density · Occupation strategy · Rewilding · Configuration of Urban Landscapes · Amsterdam metropolitan area · Paris · Mumbai · Houston

2.1 Introduction

To our profession the pictures of our urbanized world, seen from satellites at night had almost the same impact as the famous 'Earthrise' photo taken by Frank Borman aboard Apollo 8 in 1968. For the first time many professionals saw the real size, shape and character of urbanization (Fig. 2.1) [ill 1]. Not only the dense and dynamic

Fig. 2.1 Urban landscapes by night, Western Europe bathing in electric light (NASA)

pits of urbanity but the shape of enormous and seemingly endless urban landscapes literally bathing in electric light. An endless filigree of light lines spins a web with large and small nodes, embodying the pervasiveness of human activity in the urban fabric. It is more than just built up area. These urban carpets blend in with agricultural areas, 'contain' and enclose natural areas, industrial zones, agro-production clusters, airports, water extraction areas, mining areas, and recreational areas connected and bisected by a range of infrastructural routes ranging from pipelines and radio links to roads and railways. Especially along coasts and in deltas, these huge artefacts are true urban landscapes. If we look at the earth from this far, it immediately becomes obvious that many global environmental problems are associated with urbanization, or at least have urban roots. This means that the solutions to these problems will also have to be found in the city. This is also partly true for the relation between city and nature, or more precisely phrased the complex and layered relationship between natural processes and the processes of urbanization. I use this somewhat laborious formulation because nature and city are too easily positioned as antagonists. *City* and *Nature* may be distinguished but cannot be separated.

The PENN project 'Atlas for the End of the World' opens our eyes for a head on collision in the making (Weller et al. 2017). Public data from the IUCN, UNESCO, Critical Ecosystem Partnership Fund and other sources form the basis of a series of world maps with which combinations can be made. First and foremost, there is the map of the biodiversity hotspots. These are the regions of the world recognized by the global scientific and conservation community as containing an exceptional and irreplaceable diversity of life – and a high percentage of endemic species – that is threatened with extinction. Taken together, the hotspots are the sum of the world's genetic inheritance. One qualification is important here: one must not be deceived by biodiversity as an absolute measure stick. It can be used as a criterium of natural value but it may not be synonym with the health of an ecosystem. An ecosystem is something rather different then a rare stamp collection. As an example from my moderate climate zone of the world: the red list species Ophrys apifera evolved to attract bees for its pollination, but these same bees will need a mass of e.g. very common willow trees to stay alive during the long period that the orchid isn't flowering (Fig. 2.2) [ill. 2].

Secondly, there is a world map on expected urban growth. One can see it is unevenly distributed and concurs with the growth rate of the population of countries. In order of their contribution to growth: India, Nigeria, the Democratic Republic of the Congo, Pakistan, Ethiopia, Tanzania, the United States of America, Uganda and Indonesia. Combining these two maps shows that the world biodiversity hotspots and the epicentres of the urban growth of the twenty-first century almost coincide or are riddled with the worlds fastest growing urban regions (Fig. 2.3) [Ill 3]. This is a very alarming observation given the fact the world population will grow to 9,8 billion in 2050 and possibly 11,2 billion in 2100 and currently 54% lives in cities, a percentage that will rise to 66% urban in 2050 and 85% urban in 2100 (UN Department of Economic and Social Affairs 2017). The atlas allows zooming in on the different hotspots, their most important eco-regions and the

Fig. 2.2 Pollination of Ophrys apifera and willows

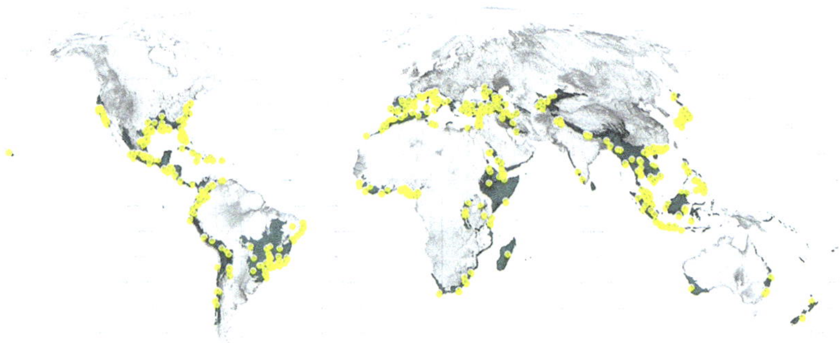

Fig. 2.3 The worlds biodiversity hotspots coincide with urbanization zones. (Source: Weller et al. 2017)

embedded cities in question with their growth trajectory. It is thus possible to identify the natural areas most threatened by urbanization. In these situations it is urgent to put protective strategies in place and nature reserves are designated. Although conservationists engaged in, sometimes heated, debates on the effectiveness, on form, size, connectedness (Diamond 1975), the role of indigenous people, including or excluding tourist income, the nature reserve still is the proofed model for in situ conservation. The Convention on Biodiversity (UNEP 1992), now agreed upon by 196 countries, formulated targets of a minimum percentage of protected areas in the ecological hotspots of 17% in 2020. The atlas makes it clear that in most regions these targets will not be met before 2020 despite the agreements made. So, designating nature reserves has not always proven to be successful.

I want to make three additional remarks to the research undertaken in the Atlas. The coincidence of the hotspots and the fastest growing cities tells only one part of the story. It shows the spectacular, but relatively recent, growth of cities that (partly) border natural areas. A geography that leapfrogged the reclamation and agricultural stages, one could say. In other geographies there might be a reclamation history that sometimes goes back millennia. Ages of agricultural use formed the substratum for urbanization. Think Eastern China, think Euphrates and Tigris, think the European Delta's (Fig. 2.4). [ill. 4] These urban areas are embedded in agricultural landscapes. Sometimes these patchwork mosaics of different forms of land use have considerable natural value and biodiversity too. These situations also have to be given attention and possibly need other strategies when it comes to pacification of city and nature.

Even if the population is not growing at a spectacular pace it does not mean that the built up area of a city won't grow. Urban growth might be fuelled by a decreasing mean occupancy rate of houses and/or an increase of surface area per capita of the mean houses when wealth increases. Suburbanisation driven by living preferences and market forces might also be a factor here.

The third remark is about a thin silver lining around the urbanization cloud. The massive migration from the countryside to the cities is having an unexpected side effect. For the third time in western history agriculture seems to be retreating regionally,because of rural depopulation. The two earlier periods being the confusing times following the collapse of the Roman Empire round 400 AC and the period of the medieval plague epidemics fourteenth to fifteenth century (Scott 2017). Generally, places where very traditional agriculture can't keep up with modern (industrial) stan-

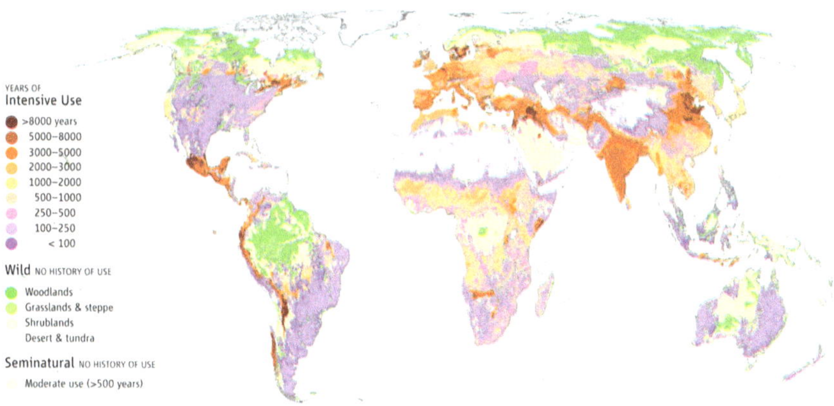

Fig. 2.4 Antropogenic Transformation of the Terrestrial Biosphere. (Source: Ellis 2011)

Fig. 2.5 Europe's living proofs of the rewilding trend: Grey Wolf, Northern Chamois, Eurasian Lynx, Brown bear

dards because of natural handicaps like mountains, regular flooding, boulder clay but also suboptimal opening up, can make regions lack behind and force people to leave. Think regions in middle-Europe, North of Scandinavia, parts of the Apennines and the Iberian peninsula but also parts of North America. In these regions the ousted wild animals seem to be coming back by making use of this (temporary?) void of human presence. Some progressive nature conservationists even developed programs – Rewilding America (Forman 2004), Rewilding Europe (Allen et al. 2017) – to speed up these processes by planning migration zones for animals and by offering alternatives for the bleak local economy in the form of wildlife tourism (Fig. 2.5).

2.2 Leadership Needed

The ideals of urbanism have long been pivoting around the relation between city and landscape. Landscape architecture is a discipline that is said to be able to mediate between man and nature and could play a more prominent role in the urbanization process. When confronting the urgency to mitigate the tensions between urbanization and biodiversity, shouldn't we critically take a new look at our role, and re-assess it? A defeatist look at urbanization would be that the original sin of architecture (Betsky 2006) by taking land and building upon it, can never be compatible with nature and biodiversity. Every time we build a new building, we do so at the expense of land. Buildings replace land that could otherwise be used for nature and communal enjoyment; architecture's – and urbanism surpassing that – original sin. But even if this conundrum cannot be solved, can we at least pacify the tensions and set out strategies for mitigation of the bio-diversity effects of urbanization? This chapter is a reconnaissance into the possibilities and potential instruments we have at our disposition and has no other objective than to produce a typology of possible strategies. Non of the presented instruments are quite new, some even have a track record of more than 100 years. It is clear that we have to do more than our utmost in the projects we are working on to ease the tensions between urbanization and biodiversity. If we want to produce something like 'Nature-Driven Urbanism' we have to look into the fundamentals of our disciplines and find out, through the development of theory and research by design, how such a seemingly contradiction in terms could be conceptualized. Possibly a new mix of instruments already at our disposal, dusting off old ideals and strategies could shed new light on this everlasting antithesis. This mix of instruments must be tuned to adapt to the local situations and the biomes involved.

The timing for this introspection seems right. The international nature conservation community is building up momentum to their own 'Paris-moment'. The Wilderness conference hosted by Beijing in 2019 is the stepping stone to Marseille that is the venue of the next IUCN World Conference on Biodiversity in 2020. In 2020,in Beijing, China, the Convention on Biological Diversity will hold the 15th meeting of the Conference of the Parties to the Convention on Biological Diversity.

What contribution have our disciplines to offer? Which levers can we use? Can we show some hard-needed leadership?

2.3 Baseline: Urban Nature

A very consoling thought is that we share the city with an impressive variety of species. It proves again and again that 'nature' cannot be switched off and at will. In other words: there will always be a natural expression of the artificial abiotic habitat the city has to offer and the enormous amount of food and waste available there for all kinds of organisms (Fig. 2.6). [ill 6]. It is true to say that we have very long been blind for this natural aspect of our built environment. We always thought of the urban landscape as our exclusive habitat. Nature and landscape were to be found outside the city-boundaries. Nature was something 'out there'. The city is just an unhealthy stone desert. But it proves to be that even our urban ecology as a species is linked to a web of other organisms. The biome of the urban habitat is a relatively recent 'discovery' and urban ecology is a just as new a specialisation that studies its characteristics and, sometimes surprising, biodiversity that differs from city to city. The interest in urban ecology rose in the late 1970s of the twentieth century. Some

Fig. 2.6 Urban nature: Swan's nest on building site, ferns between canal bricks, Nest of a seagull at a refinery made of steel and isolation material, mosquito

pioneers in this field (Pelt 1977; Sukopp et al. 1979) followed some botanists who worked on adventives in cities. The nineteenth century Finnish Lichenologist Wilhem Nylander was the first to notice the specific character of lichen flora in the urban environment (Nylander 1866). The differences are partly caused by differences in the local quality of the water or the air. European and American cities in the 1980's and the Chinese cities of the last decade and now India are almost derelict from lichens caused by extreme air pollution of sulphur-dioxide from coal-fired plants and traffic (Li et al. 2017).

But most important to boost the biodiversity of cities is their specific (bio) geographical position (such as for instance Istanbul), their variety in microstructures, the specific building materials and to put a long story short, their a-biotical diversity that generates an even diverse biotical livestock. Close inspection and long year monitoring of the biodiversity of metropolitan areas by urban ecologists and social science projects of urban areas during the last twenty years reveal that the Dutch cities are really diverse over a broad front (Amsterdams Bureau Stadsecologie 1996; Denters 1994; Melchers and Timmermans 1991; Melchers and Timmermans 1998). Apart from plants, birds and insects, the bryologist Marinus van der Sande Lacoste already made inventories in the nineteenth century of the mosses and lichens in and around

Amsterdam, his work is being picked up by Henk Timmerman and others in recent days. A long timeline of distribution of these inconspicuous species has become available and more recently mammals were covered in the documentary *The Wild City* (2018), shot in Amsterdam. Due to the intensification of agriculture and the erosion of the biodiversity of the Dutch landscape – the modern meadow is mockingly said to be down to 1: tetraploid perennial English Rye grass – the biodiversity might counterintuitively be higher within the city limits than outside (de Jong 2012).

As designers we can greatly enhance urban nature by choosing the right materials, apply nature driven solutions, use ecosystem services, fine-tuning the urban metabolism, making green stepping stones in the form of parks, defragmenting measures hence connecting isolated populations, making the blue infrastructure of the city into living water networks. All these elements will be addressed in other paragraphs and in more depth in other chapters of this book.

Let us limit ourselves here to the observation that enhancing urban nature seems a promising path for pacifying urbanization with natural processes. There is a growing body of knowledge on urban nature, as well as the ways to elicit desired 'natural expressions' from human interventions to inform and inspire designers. Some of you might consider applying this knowledge in your design too passive, or find these angles to anthropocentric. Another angle on urban nature is virtually changing yourself in a migrating fish and consider all the barriers in the urban water system you encounter. Or change yourself in a Peregrine falcon and look around what opportunities you see that, with a little addition, could offer a breeding possibility. To radicalize this approach let yourself be inspired by a writer who tried living like a badger, an otter, a fox, stag and swift, in order to better understand their lives (Foster 2016) or the artists who allowed himself a vacation from being human and tried living as a goat (Thwaites 2016), (Fig. 2.7) [ill7] Urban and landscape design

Fig. 2.7 Goat-man Thomas Thwaites interacts with an alpine goat. (Photo: Tim Bowditch)

from the standpoint of different organisms could give a boost to the quality of the design and thus to the biodiversity in our urban landscapes.

2.4 Densities

In the professional discourse the dominant line of reasoning seems to be that density in or densification of existing urban areas is the right answer to most sustainability dilemmas. The denser the city the less oil, gas, electricity and water people individually consume and the higher the propensity to use public transport, because in denser areas this is a public service with a business case, and the less time they spend in (their own) automobiles. In these dense cities cycling or walking are important means of daily transportation (Owen 2009). These are of course all 'grey' environmental criteria of sustainability. There is hard evidence that the energy use for transportation in low-density American and Australian cities is considerably higher per capita that the denser European cities (Jones 2014). In terms of the pacification between urbanization and natural processes the reasoning is that the denser the city the less space is lost because of sprawl. The dense city thus generously grants nature it's own space by packing people within small areas (Farr 2007). This is the theory of the calculus. Some indicators of sustainability may point in different directions. We may have to look at different scales as density is allways scale related. The quality of life and or important future indicators like urban heat islands and – indeed – urban biodiversity may score differently depending on the amounts of green and blue space on block, neighbourhood, district and regional level. Moreover, where densification might be a sound general recipe, the global reality is that – counterintuitively – urban densities are decreasing. A combination of these factors might explain why globally the mean urban densities are decreasing. Of course there are high density peaks locally (think Mumbai, think Manhattan, (Fig. 2.8) but these are always surrounded by urban landscapes with considerable lower densities. Mass mobility with private cars even made its appearance in countries like India. Mean densities are decreasing worldwide with an annual 1,5% since their peak in the late nineteenth century (UN-Habitat 2012) Doubling the population will result in tripling the urban surface (Angel et al. 2012).

The influence of urbanists on increasing densities must not be overrated, real estate developers or land prices might be more dominant factors. Still one could say that architects, urbanists and planners are the core of the density fan club. They are mainly drawn by the dense and dynamic aspects of the city. These are positive and fascinating, and it is true that dense cities are the generators of the economy and the hotspots of creativity and the breeding ground for young talent because of their intensive networked interrelatedness (Florida 2002). But in the light of biodiversity we must not look away from the fact that most of the built environment consists of urban landscapes, Zwichenstädte, in-between-worlds, tapijtmetropolen or whatever this sub-urban condition is called in different parts of the world. (Sieverts 1997;

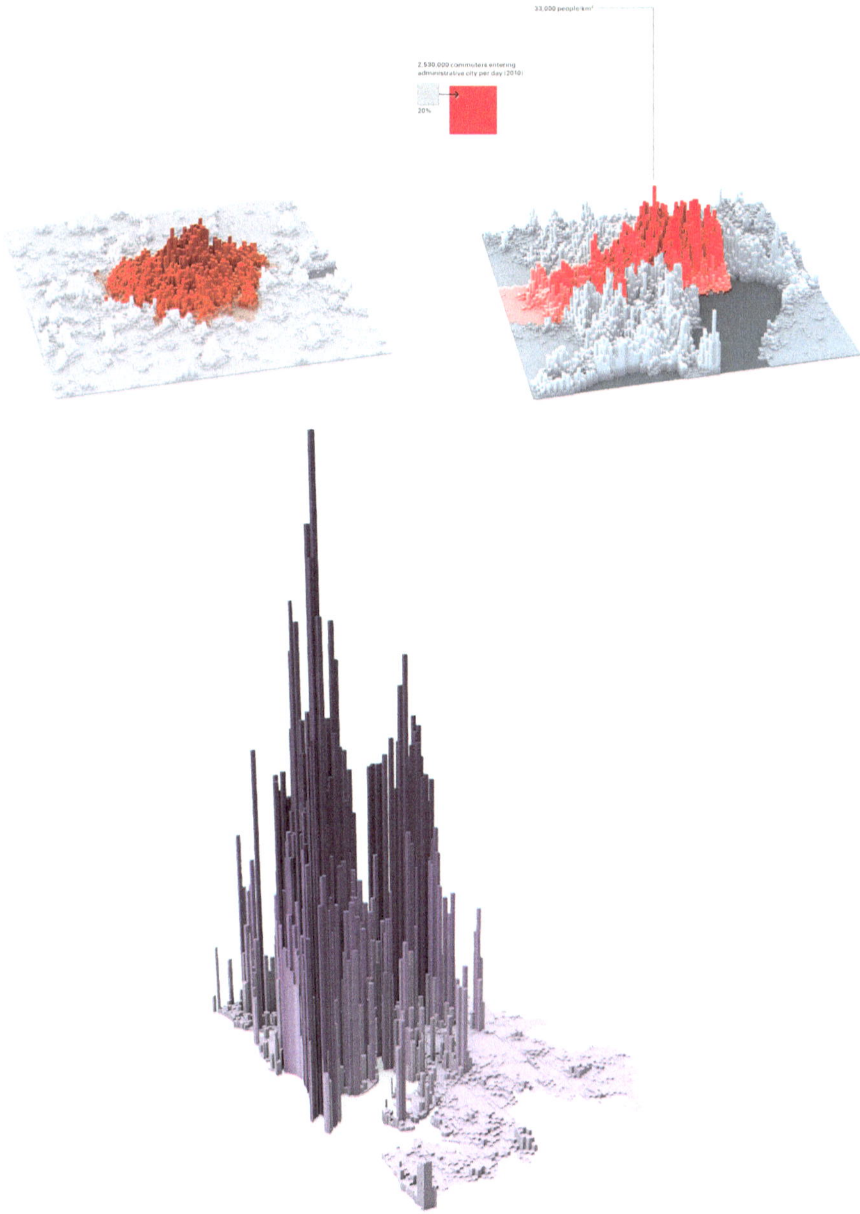

Fig. 2.8 (**a**, **b**, **c**) The densities of the centers of the megacities are only part of the story: London, Istanbul, Mumbai. (Source: LSE 2011)

Forman 2008) Density and densification are usually a good idea but not the panacea that can be applied as a miracle medicine. We also have to look at the specific problems and specific potentials of the urban landscapes around the world when it comes to pacification of urbanization and natural processes.

2.5 Urban Metabolism and Nature Based Solutions?

Many global environmental problems are associated with urbanization, or have at least urban roots (Sijmons 2014a), this means that the solutions to these problems will have to be found in the city. Urban planning itself would need an overhaul to meet these challenges (Hajer 2014). An inspiration is the work of system ecologists of the late seventies and early eighties of the twentieth century, where cities, whole countries and even the whole planet were analysed as open systems and where the human economy was embedded in a larger ecological framework (Meadows et al. 1972; Odum 1983) (Fig. 2.9).

One of the ways to do so is reframing the city from the primarily physical entity of the built up environment to a more liquid view where the city is (also) the playing ground as well as the complex result of numerous flows and processes. This metabolism of the city is made up of very different flows. Building material, surface water,

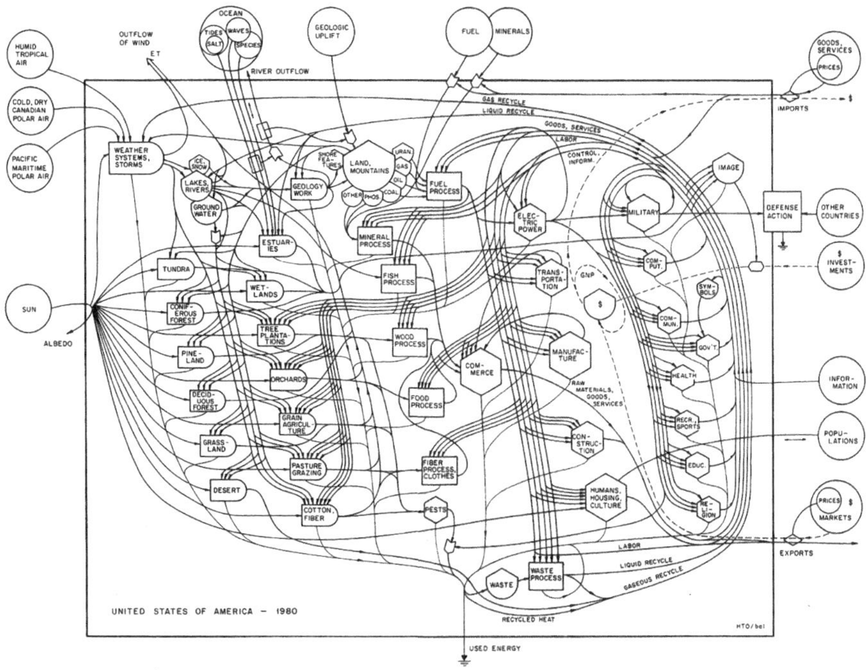

Fig. 2.9 A systems view on the ecology and economy of the USA in 1980. (Source: Odum 1982)

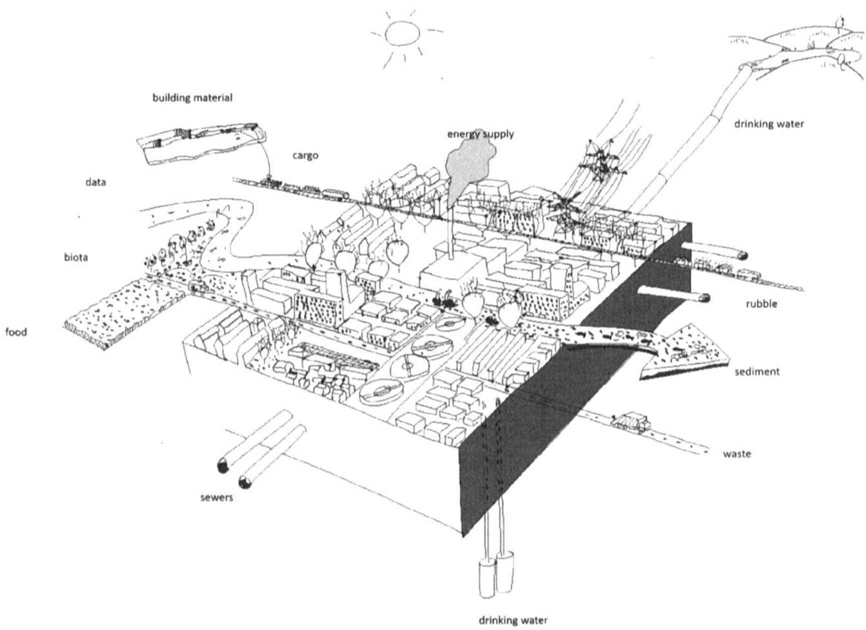

Fig. 2.10 Urban metabolism, nine formative flows, Source: Sijmons (2014b)

sediment, drinking water, heat and air, waste, food, cargo, persons, data, and finally the most invisible for us, biota (Fig. 2.10). Finetuning of these flows (reducing them if needed or stop losses), regionalizing the source and the sink and in some cases make the flow territorial again can mitigate the environmental effects.

The International Architecture Biennal Rotterdam Urban-by-Nature in 2014 showed the 38 best practises from all over the world how the different flows can be optimized. Moreover, the projects convincingly showed that the metabolic angle on planning was able to produce an interface between spatial and environmental planning (Sijmons 2014b). Stock was taken of the quantities and qualities of the flows on the level of the individual (Dutch) household, on the national level and the global level. Maps were added and all were represented in diagrams and other infographics. The metabolism angle produces ample criteria to construct something like a LEED sustainability standard on a city level.

But most of the flows are objects of 'grey' environmental policy. Regarding biodiversity there are two lines that might be of interest to build out:

1. Finetuning the flows could also be optimized by minimalizing and mitigating effects of urban metabolism on natural values and biodiversity. Looking at surface water management for instance minimizing the barriers that are created for migrating fish could be minimized, or fish ladders could be provided.
2. A more pro-active attitude could bring a rich harvest in added biodiversity by using nature based solutions to provide the services metabolic flows deliver to the functioning of the city. One can think of using constructed wetlands, artificial

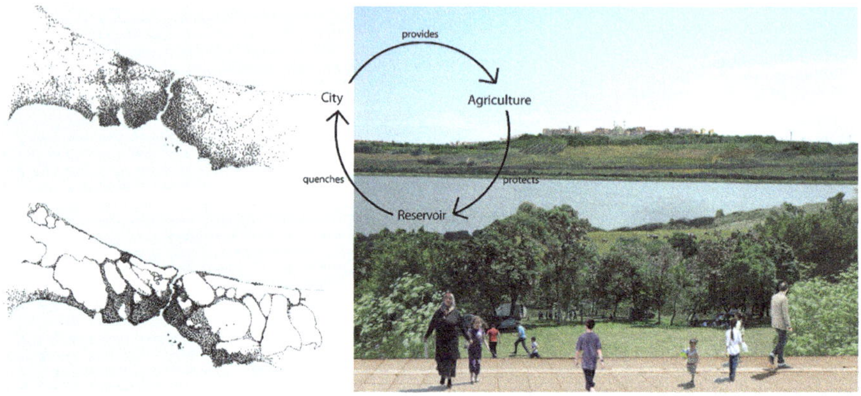

Fig. 2.11 Istanbul Arnavutköy*: Ridge city.* (Source: Kornaropoulou et al. 2017)

wetland to treat municipal or industrial wastewater, greywater or stormwater runoff. Vegetations of carefully selected plants and indeed also natural vegetations are able to remove superfluous nitrogen, phosphorous, metals, pathogens et cetera. These constructed wetlands are miniature ecosystems in their own right and attract all kind of other organisms.

A research by design project for Istanbul showed that the flow of drinking water production – on the critical path for growth of this megacity – could even be formative for the city shape of this fast growing urban landscape. A new way of protecting the water basins, in combination with precision agriculture, irrigated with purified grey water from new housing projects, could produce a sustainable occupation strategy resulting in high quality public space, interspersed with robust green/blue elements (Kornaropoulou et al. 2017) (Fig. 2.11).

Blue and green infrastructure on a larger scale in the city is a very good antidote to the urban heat island affect. Green roofs, with their specific succulent vegetations, by their water retention capacity, slow down water run off thus relieving sewer systems. Making more natural river shores could offer the same safety as concrete infrastructure and offer the bonus of biodiversity. Regionalizing (part of) the food production, one of the most important metabolic flows, reduces foodmiles and moreover, offers new unexpected opportunities for field weeds that become ever more rare in the intensive agricultural landscapes.

Finally, one of the 'flows' is the constant migration, settling, extinction of biota. Biota in the form of seed, sperms, spores, living animals, plants and other migrating animals. The success percentage of settlement is dependant on the habitats offered and, what I would like to call, the permeability of the urban landscape. We will separately focus on this aspect in the next paragraph.

2.6 Size and Configuration

The role of size and scale can hardly be overestimated. Functional surface is needed if we want natural processes to play their role in the web of life. Size matters: e.g. sedimentation and erosion produce their abiotic structures on a landscape scale. Animals have different demands to the sizes of their habitats but a rule of thumb is that the larger the animal the larger their territory. Animals that are able to be formative to ecosystems like herds of large herbivores living and even more so predators, demand extensive terrains to 'do their thing'. Not only is space often a scarce commodity in and around urban landscapes, the existing areas are also often being fragmented by bundles of infrastructure, e.g. roads, pipelines, railroads, electricity lines, et cetera. In the context of the urban landscape it is important to talk in terms of scale, configuration and connectivity of the remaining 'functional surface' for natural processes. Insights from biogeography offers both the theoretical background and the right terminology.

One of the breakthroughs in the science of ecology was the seminal 'equilibrium island theory' (MacArthur and Wilson 1963). It was based on the work of Wilson researching the distribution of ant-species on Melanesian islands. The theory was mathematically formulated by MacArthur and translated the colonization and extinction processes and equilibrium into a simple graphical representation of immigration and extinction curves, from which one can determine the equilibrial species number on an island. Determining factors are, the size of the island (strongly influencing the success rate of migration), and the distance of the island to a continent or another island (determining the chance of migrating species arriving on the island). The theory promised to promote biogeography and ecology to the elite class of predicting sciences. It also led to heated debates on the precision of the theory. In the congress celebrating the 50th anniversary of the theory, Wilson himself stated that the theory had it's flaws. We have to credit Jared Diamond for the observation that the situation of island nature could be a working analogy for isolated nature reserves floating in a sea of agriculture (Diamond 1975). It was translated in diagrammed recommendations for the design or bounding of nature reserves in the World Conservation Strategy of the IUCN in 1980. 'The bigger the better', 'the closer together the better', 'the stronger the interconnection the better' (Fig. 2.12). The effectiveness in terms of extinction risk reduction of these recommendations were intensively discussed in the conservation science literature, also known as the SLOSS-debate (Some-Large-Or-Several-Small). What should be learned from the theory, the accompanying scientific debate and the applications in practice in both nature conservation and nature development projects is that thinking in terms of isolation and connection, in terms of ecological infrastructure to connect (threatened) isolated populations proved to be a strong instrument. And, most important for our subject, the analogy of the islands might also work for the position of scattered green-blue areas in the urban landscape.

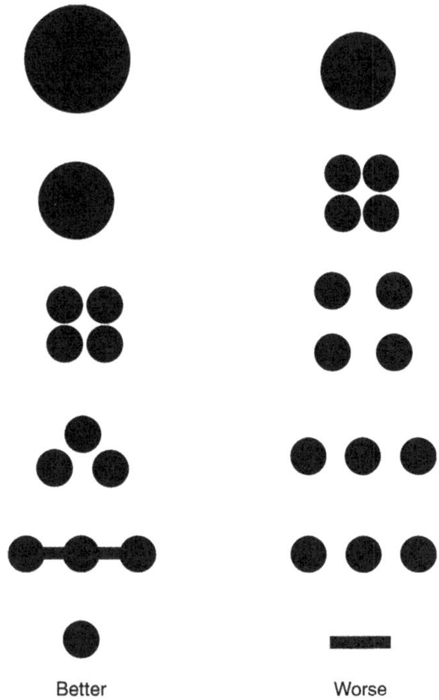

Fig. 2.12 Island biogeography theory generated six guidelines for the design of conservation reserves. (Source: Diamond 1975)

2.7 Permeability

Organisms of any kind 'scan' the urban landscape on opportunities and suitability to forage, occupy, leap-frog, or cross. For all these possible courses of action permeability of the fabric of the urban landscape plays a crucial role. Permeability of the urban landscape will be dependent of the size and configuration of green-blue patches (the islands), their interconnectedness, the extent to which infrastructure in the urban landscape is bundled determines whether the urban environment is passable for different kinds of organisms as well as the perviousness of the (abiotic) built up elements for the distribution and settlement of species.

In many cities parks and park systems comprise the largest green 'islands' in the urban fabric. They could play an important role as stepping stones and refugia for species. But we should not forget that in many aspects parks are more cultural than the urban fabric. Some traditional parks look like the spitting image of nature because they are crafted as a representation of nature. They contain leisure, enhance wellbeing and play a pivotal role in calming down citizens in a hyperventilating metropolitan surrounding. That being said, parks play an indispensable role in the urban green/blue infrastructure.

In landscape design a lot of experience and know how has been amassed on overcoming fragmentation and barriers by with aqueducts, cerviducts, be it wildlife ways or tunnels, and overcoming barriers for migrating fish (Fig. 2.13). This body

Fig. 2.13 Defragmentation projects: A2 Best Netherlands, Banff National Park in Canada, Landbridge Vancouver, Fish passage, Afsluitdijk Netherlands, salt-sweet gradient restored.

of knowledge and experience can be applied in the urban context. The urban landscape is crisscrossed by infrastructure (roads, pipelines, railroads, canals, airfields, etcetera) that hamper or block spontaneous migration in one direction but which verges, roadsides and banks could function as a migration zone in the other direction enhancing the green/blue structure of the urban fabric. Fragmentation can be remediated by adding small green infrastructural elements on strategic places.

Being hospitable, permeable, passable and pervious to organisms can also be considered on a higher level of scale: that of the urban landscape itself.

2.8 Urban Configuration

The perhaps the most promising but also most demanding path to pacification, is that of influencing, planning/designing the urban configuration in a way that important natural areas might be spared and integrated in the tissue of the urban landscape. It could be demanding because the prerequisite is some basic kind of planning of the urbanization. If we look back at the recent history of ideals in urbanism there are basically three types that, after close inspection, construct a specific relationship with landscape and nature, or stronger, seem to be pivoting around the dichotomy between city and nature. The oldest one is the ideal of the Garden City (Howard

1902). The city was considered an unhealthy making organism, quite rightly so in that time. The healthy answer was to bring satellites of the city, the garden cities, to nature. Not very much younger is the archetype of the 'lobe city'. It stems from the Grossstadt competition (Eberstadt et al. 1910). They share the diagnosis of the city with the garden city movement, but offer a reverse medicine, bring nature to the city. The third utopian model for urbanity and nature might be 'the Broadacre City' (Wright 1932). It was both a planning statement and a socio-political scheme by which each U.S. family would be given a one-acre plot of land to live and also produce part of their own food.

All three ideals materialized in some form around the world. The garden cities as garden quarters but also as satellites and new towns. The most successful application of the lobe city model are possibly Copenhagen and Amsterdam where it guided (green) planning and city extensions for more than half a century. The Broadacre city has been the Printed Circuit Board of, and sometimes legitimation for, suburban sprawl after Pandora' box of mass mobility opened. Stripped from it's original intension of self-sufficiency it can be found all over the world but, admittedly, very beautiful suburbs with high natural value that come close to the spatial ideal of Wright also exist. With a little sublimation one could say that these three archetypical urban concepts stand for contrast, contact and contract between city and nature (Fig. 2.14).

What can we do with this 'old school stuff'? My proposition would be to interrogate these three archetypes on their possible significance for the present day, in order to mitigate the tense relationship between red and green, between urbanization and natural processes. As a basis for this analysis I constructed a table (Luiten and Sijmons 1998) (Fig. 2.15). Let me try to run you through it without elaborating on each cell of the table. The top and the bottom dark rows contain the Image of Nature and The Vision of the City in the Contrast, Contact and Contract model. So

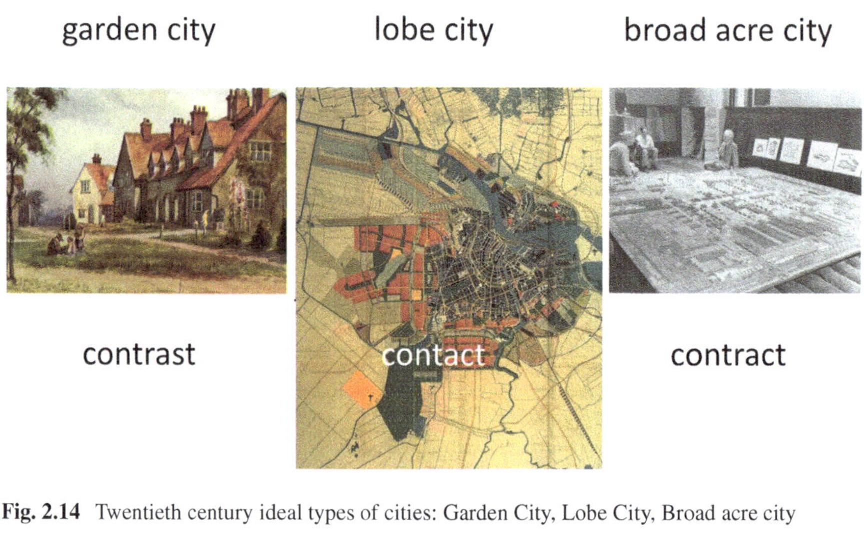

Fig. 2.14 Twentieth century ideal types of cities: Garden City, Lobe City, Broad acre city

	CONTRAST	CONTACT	CONTRACT
IMAGE OF NATURE	wilderness	accessible nature	ecosystem services
Formal Interaction	city and nature have sharp boundaries, protected areas	city and nature intertwine	city and nature take each others form
commentary	*bring the city to nature 'satellites' and 'garden cities'*	*insert nature into the city 'green wedges' and 'parks'*	*go for a complete mix, 'reweaving the urban tapestry' and 'broadacre city'*
Functional Interaction	city and nature are each others jungle	city and nature come to each others rescue	city and nature take on each others form
commentary	*'places to get lost'*	*regulated leisure in nature*	*produce food on your own gardenlot*
Physical Interaction	city and nature keep their distance	city and nature exchange information	city and nature take on each others construction
commentary	*natural expression of the city 'non human' outside*	*natural expression of the city 'well tempered' environment outside*	*expression of city and agriculture 'new hybrids' in- and outside the city*
VISION OF THE CITY	from 'Cabanes' to 'Metropolis'	'Green-Blue infrastructure' to 'Lobe city'	from 'Subtopia' to 'Metabolic City'

Fig. 2.15 Three ways of conceptualizing the city-landscape-nature relationships

Contrast would be 'Wilderness', the Contact model would refer to 'Accessible Nature' and the Contract is represented best by something like 'Ecosystem services'. The other rows contain the keywords on the Formal, the Functional and the Physical interaction between city and nature of the Contrast, the Contact and the Contract-model. To all rows a commentary line is added dedicated to the linked planning action.

From a perspective of nature conservation the 'Contrast' model would presumably have the best qualifications. Paris and it's Forêt de Fontainebleau, one of the oldest designated Nature Reserve (1853, twenty years before Yellowstone) spring to mind. Forêt de Fontainebleau also makes us aware that Nature Conservation is an urban idea from the nineteenth century that took urban innovations to materialize. The tramline from Paris to Fontainebleau to connect the natural area with a public. Secondly, the invention of oil paint in tubes that made it possible the Barbizon school masters could paint in the open air that eventually turned Fontainebleau into a Reserve Artistique that protected it from the ax of foresters (Fig. 2.16).

A second example of the Contrast model showing an even more direct relationship between a natural reserve and the urban metabolism is the Sanjay Gandhi National Park. With its almost 90 km^2 encircled in less then 40 years by the urban landscape in less then 40 years of megacity Mumbay, it is boasting a healthy population of Leopards. The area it is also being protected by the fact that most of Mumbai's water come from the lakes in that nature reserve. Temporary overpopulation of leopards created some animal-man incidents in 2000–2003. Research showed that capturing 'problem' leopards from farmlands and releasing them in the reserve was probably the primary cause for the attacks. Leopards are now also being credited as useful animals because they predate on dogs gone wild decreasing rabies (Nair 2016) (Fig. 2.17).

The second model, that of 'Contact' could offer real perspectives for pacification between city and half natural landscapes. Through its planned configuration every

Fig. 2.16 Forêt de Fontainebleau, the World's first designated nature reserve (1853) or to be more precise: Reserve Artistique. Nature conservation has urban and cultural roots.

inhabitant of Kopenhagen only has to bike half an hour to find him/herself in a medieval landscape. And if well designed, even in an unexpected way for the problems of the head-on collision between rapid urbanization and biodiversity hot-spots in situations where strict preservation in reserves is impossible (Fig. 2.18).

Even the 'Contract' model with its sub-urban inclination could have an interesting natural expression as one can see in one of the most extreme sprawled urban landscapes, Houston that locally even developed a canopy. The low density offers possibilities for all kind of ecosystem services, a different way of giving substance to Wright's ideal. When designed and maintained well the contract model could offer chances to specific biomes. (Forman 2008) Changing maintenance is the possible trump card here, to extend and detox the gardening – however difficult to achieve – could work miracles in terms of enhancing biodiversity and permeability of the urban fabric (Fig. 2.19).

Fig. 2.17 Leopard (Panthera pardus fusca) walking in alley between houses. Aarey Milk Colony in unofficial buffer zone of Sanjay Gandhi National Park, Mumbai, India. January 2016. (Photo: Nayan Khanolkar).

2.9 The Best of Three Worlds?

On the scale of present day metropolitan urban landscapes reality is that the strong points of all three models, contrast, contact and contract, can be used in a strategic mix to pacify. I will elaborate on the example of the metropolitan region of Amsterdam where, over the course of its recent urban history all three models did play a role (Fig. 2.20).

2.9.1 Contrast

The garden city movement, whose influence was restricted to some garden quarters in the North-part of Amsterdam (Garden villages like Oostzaan and Nieuwendam), got a new youth in the multi nodal new town of Almere in the brand new polder of Flevoland. Almere is a configuration of separate new towns in a green setting linked together by high quality public transport and offers a garden city environment for some 200.000 people. Their daily urban system is varied because it is only half an hour travel distance to/from Amsterdam and 45 min to/from Utrecht. The green structure of Almere is designed for leisure and sports and not primarily focussed on biodiversity. The urban forests that were planted with mostly poplars half a century ago formed a wood humus and now second generation is planted with more interesting

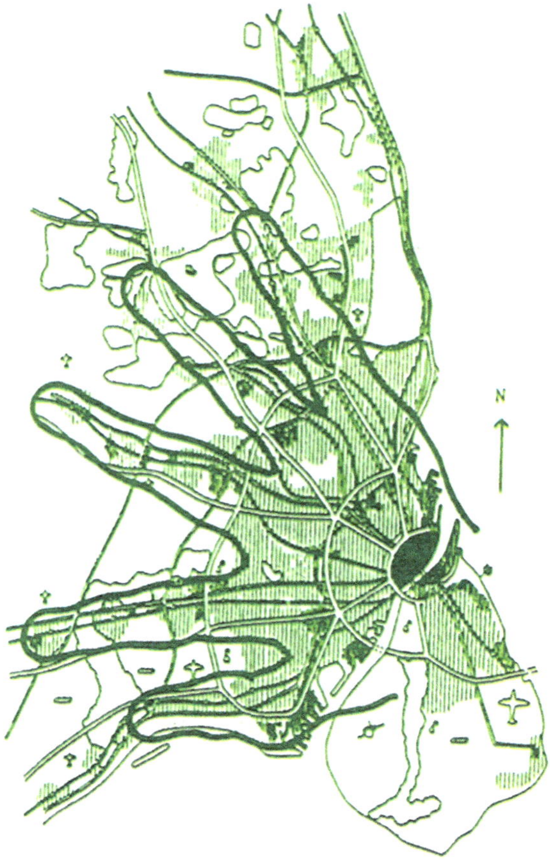

Fig. 2.18 Kopenhagen, the "five finger plan" of Kopenhagen (1947) to control urban and suburban growth in the city was devised under the direction of Danish architect and urban planner Steen Eiler Rasmussen.

species as oak. Moreover, because of their loose setting of the different nodes of Almere the interconnectedness and the size of the green islands is optimal and has a serious long-term potential. The most spectacular 'green island' is the Oostvaardersplassen between Almere and Lelystad, the other new town in Flevoland. Stretching along the Markermeer dike for 10 km the area was meant to be an industrial area with the name Lage Vaart 2. It was not drained completely because it was thought that was a waste of money because it was needed after 2020 when Almere and Lelystad would have developed to their expected size. What happened is that 'nature' stepped in this rare void in Dutch planning frenzy like a squatter movement. The marshland was occupied by pioneers and first years vegetation successions swept over the 6.250 hectare of wetland. Everybody expected that the marshland would soon develop into a forest. But a massive reappearance of the Grey Goose, a bird that wasn't seen breeding in Holland since the end of the nineteenth century, effected in a relative stable state. The geese being herbivores stopped short the succession and opened the eyes of many ecologists for the pivotal role of herbivory in eco systems (Vera 2009). In line with this observation larger grazing animals, Heck-cattle, Konicks

Fig. 2.19 Houston, possibly the most sprawled city in the USA has an almost closed canopy.

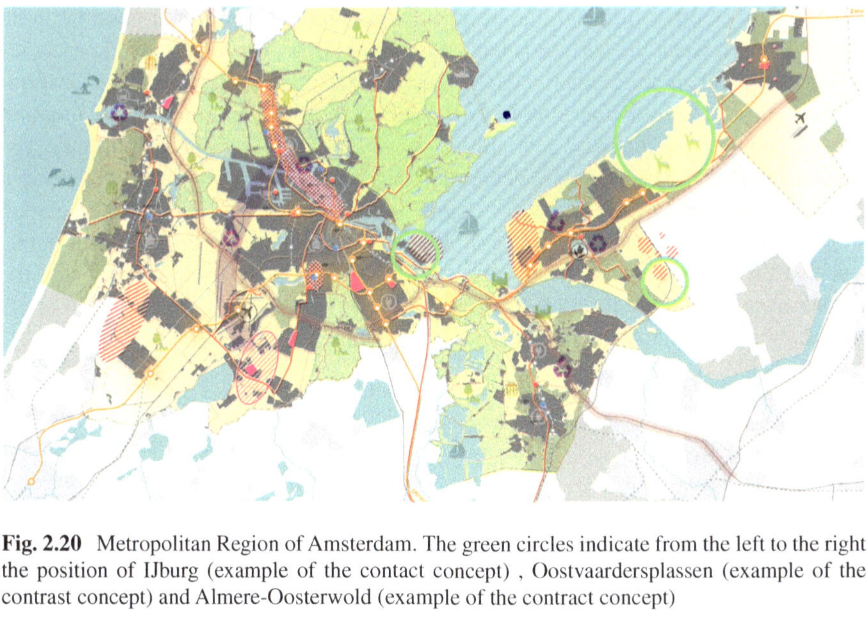

Fig. 2.20 Metropolitan Region of Amsterdam. The green circles indicate from the left to the right the position of IJburg (example of the contact concept) , Oostvaardersplassen (example of the contrast concept) and Almere-Oosterwold (example of the contract concept)

Fig. 2.21 Herds of Konicks horses, Heck cows and Red deer grazing the Oostvaardersplassen between Almere and Lelystad.

horses and Red deer were introduced. These animals are adapted for year-round grazing (Fig. 2.21). The Oostvaardersplassen flourished as an experiment of the effects of large herbivores that lived in social bands and were allowed to die. Biodiversity on all trophic levels was greatly enhanced with as a spectacular height the comeback of the White-tailed eagle – aka 'the flying door' – as a breeding bird. This example shows – among a lot of other things – that spectacular nature development is possible in an urban context, when the surface is large enough to allow formative natural processes to play their role. In the Netherlands the Oostvaardersplassen was a source of inspiration. It showed how nature is able to bounce back after the almost dead situation brought about by intensive agriculture when given the right conditions. The Oostvaardersplassen works as a generator for further distribution of bird species in the wide surroundings but is also feeding in on the green infrastructure of the garden city of Almere with beavers and otters. The urban ecologist Ton Eggenhuizen points out that the connection zone between the Kromslootpark in the southwest and the Lepelaarplassen in the north of the city is actively being used by beavers and otters (https://almerenatuur.wordpress.com/tag/stadsnatuur/).

2.9.2 Contact

The outline of the urban configuration of Amsterdam is mostly determined by extension plan (AUP) made by Cornelius van Eesteren in the 1930s and the way it has been leading the planning of the city's extensions and the green planning ever since (Van Eesteren and Van Lohuizen 1934). Inspired by the work of Martin

Wagner (Wagner 1915) Amsterdam is one of the examples of a lobe city, designed to penetrate landscape, nature and leisure into the city fabric. The city form breathes the ideals of the modern movement where work, living and leisure are separated and large green spaces are the necessary complement for urban life. How the lobe philosophy is still alive and how it could still be topical when thinking about the relation between urbanization and nature became clear in the planning of IJburg, the new eastward city extension of the Dutch capital. IJburg is built on an archipelago of artificial islands. The land making started in 2001 and is still going on. When executed it will host 18.000 dwellings and some 45.000 inhabitants.

To understand the natural context I have to elaborate on the character of the IJmeer lake where this extension is planned. In 1932 the former Zuyderzee was cut off from the North sea by a large compartmentalizing dike to enhance water safety after the disastrous flood of 1916. It also made us able to make large polders on the former seabed for modern agriculture. Fed by the river IJssel a fresh water lake gradually formed, in fact, the largest fresh water lake in NW-Europe. IJburg is located in one of the pockets of this large lake system (Fig. 2.22).

The Dutch Delta has always been a crossroad in the bird migration routes of the western pale-arctic flyway. The addition of a large freshwater lake enhanced this position as the most important waterfowl area in Europe (Saeijs and Baptist 1980) It is only logic that in the late 1990's the whole Ijsselmeer/Markermeer lake was designated a protected waterfowl area, by the European Natura 2000 legislation. A good question is of course, why build in an area like that in the first place. Why not rely on protection only. The point in Amsterdam was that already in the early sixties

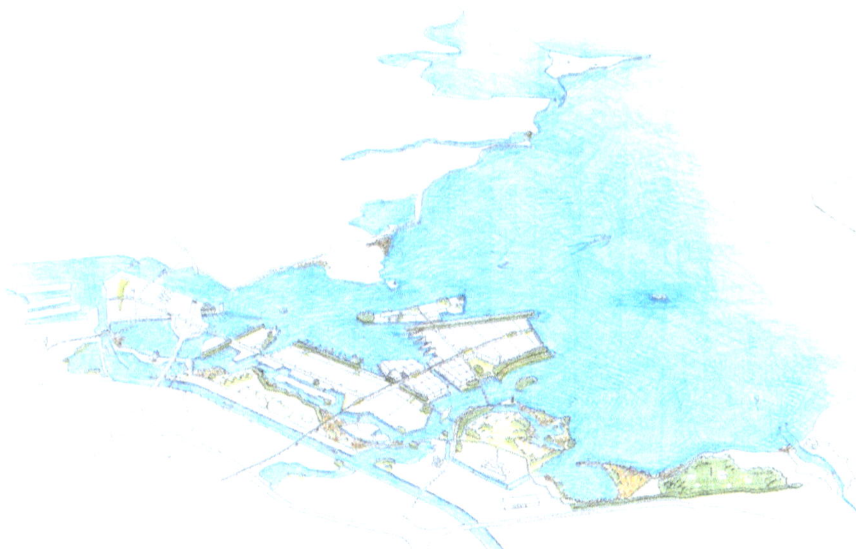

Fig. 2.22 IJburg Masterplan 1998 (Drawing Frits Palmboom). (Source: Municipality of Amsterdam)

there had been plans for an eastern extension. A polemic plan, Plan Pampus,1963 by the Dutch architect and urbanist Jaap Bakema made an unsuccessful plea to build in the IJmeer instead of the south-eastern Bijlmermeer polder. Although his advice was not followed it rose the interest of Amsterdam policymakers that claimed the IJmeer in a later Structure plan. Amsterdam had earlier rights on the area and legally claimed successfully that they had preference over the Nature legislation. A long discussion and a referendum in the city followed. Given this contested situation the designers claimed that they could strike a compromise by building in dialogue with the aquatic ecosystem and by adding specific, missing, biotopes to the lake system. No hiding, no contrite beforehand but self-consciously try to bring both worlds together and bring the city dwellers in as close as possible contact with the richness of the water(fowl)-world. It was pivotal to make an in depth analysis of the aquatic ecosystem and see which levers could be influenced to boost the system (Fig. 2.23). The most important potentials were manipulating with water depth and water regime and a judo-like use of the wind and wave energy to introduce lee elements where reed can sprout to gradually create a healthy shoreline vegetation. It also enabled us to create undeep areas where new underwater vegetation makes a chance of expansion. This creates coverage for predator fishes controlling the population of bream. Hard substrate was also created where Dreissena polymorpha – the staple food for diving ducks and other water fowl – can find suitable abiotic conditions.

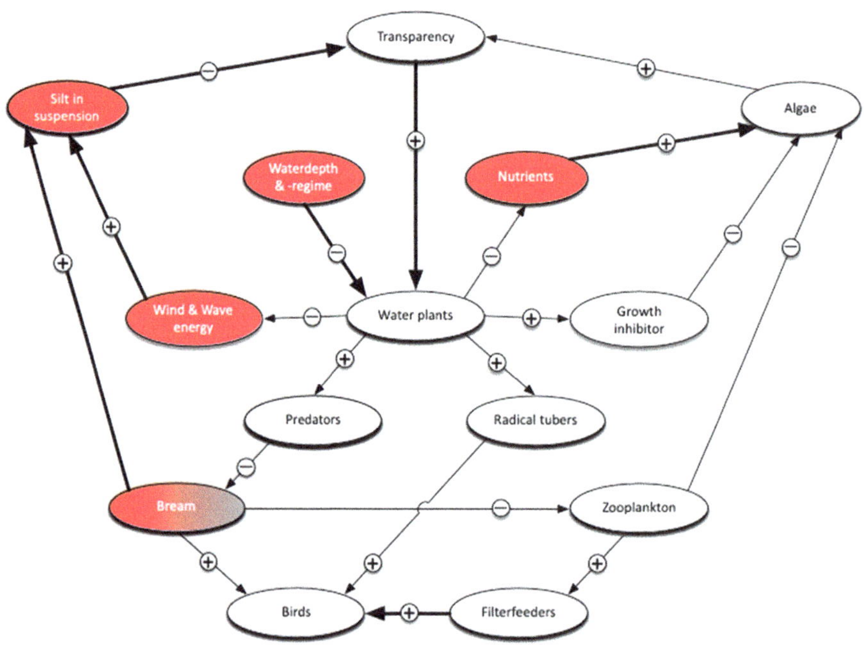

Fig. 2.23 Schematic structure of the aquatic ecosystem of the Markermeer. In red the most important levers to influence the performance of the ecosystem.

Fig. 2.24 (**a**) The Hoeckelinge dam, a compensation project in the vicinity of IJburg, produces a lee zone, the water depth was reduced to stimulate growth of specific under water vegetation (**b**) The canals inside IJburg do the same on a larger scale

For IJburg, the water depth and regime were of paramount importance because the IJsselmeer as a whole has an artificial and quite unnatural water regime. In the winter the water is low, to allow large run offs from the river IJssel and in the summer it is high to have a fresh water retention lake that can irrigate almost all of the Dutch agriculture. This can't be changed. But this unnatural regime comes at a price: reed and other littoral plants don't shoot there spontaneous, only in the most wind lee places one can see a fully developed lake bank vegetation. Nature development projects in the vicinity of IJburg were executed but even more so, IJburg with it's inner water system, produced hundreds of hectares of lee and shallow water thus contributing a missing type of biotope to the ecosystem (Fig. 2.24a, b).

Wind- and water energy were made instrumental in concurrency with the configuration of the islands of the archipelago in the making. Different configurations where modelled to optimize the waterflows inside IJburg in the hope to produce the right – hard substrate – biotope for Dreissena, the fresh water mussel benthos that forms the feed stock for most of the winter water fowl ducks.

The Master Plan was approved in 1997, the first houses were surrendered in 2002 building in the archipelago will go on to 2024. Not only nature inclusive design is a criterium, but other sustainable development goals are realized as well. IJburg is an all-electric city quarter, collects its rain water, has a good modal split and high quality public transport. Twenty years after the polarized discussion over IJburg it can be concluded that this nature-inclusive urban design strategy is fruitful and even got compliments from former nature conservationist opponents (Melchers 2016).

2.9.3 Contract

The Netherlands is a densely populated country but has also been characterized as a low density urban landscape because real metropolitan centers are missing. Because of the tradition of strict planning it's nevertheless hard to find real sprawl and examples that might illustrate the potential of the contract model between nature and city.

There is one exception though. Fuelled by the unflagging zeal of alderman Adri Duivesteijn to promote the freedom of private commissioning of housing, as a counterbalance of Dutch state and large building firms dominated housing market, a large 'free' urban development project has been designed by MVRDV in the east of Almere: Oosterwold (https://www.mvrdv.nl/projects/32/almere-oosterwold). It is an experiment that has to grow to an area of 4.300 hectare giving space to 15.000 new dwellings – that brings it in quantities on an even foot with IJburg – on rather large lots that will not only maintain the green and agrarian character, but requires its new inhabitants to be active in the field of (urban) agriculture or at least have a plan for their lot. Everyone is asked to generate their own energy, to store surpluses of rainwater and to clean their waste water. The role of the municipality is restricted to the outline and only formulates the ambitions, the framework, the conditions and the (few) rules of the game. People have to collaborate with different other lots to plan the roads and even the sewer-system. In the hands of MVRDV – ironically enough an outspoken advocate for high densities – this form of bottom-up development comes very close to the ideals once stated in Broad-acre city by Frank Lloyd-Wright. Or more recently to the ideas of Italian architect Andrea Branzi who in his 1995 project Agronica—Weak Urbanisation develops a logic of conflating agricultural and energy infrastructures: "(…) An urbanism in which the component of time returns as a variable in an imperfect and incomplete equation that adapts itself to change" (Branzi 2006). It is way too early to tell what the natural expression of this new way of gardening/urban agriculture will look like. But it could offer unexpected chances for the return of the arable weeds that have been banished by intensive professional agriculture (Fig. 2.25a, b).

Fig. 2.25 (**a**) Oosterwold plan source MVRDV (**b**) Oosterwold arial view of the situation in 2018. (Source: Municipality of Almere)

2.10 Planning Context

I tried to make a coarse inventory of potential instruments at the disposal of urban planners and landscape architects. Whether they will or can be applied successfully is very much dependent on the planning context. Can we build on the humus-layer of a planning tradition or do you have to start from scratch? What exactly is the span of control of local governance and planning? Is the political will present to work on a nature and landscape informed urbanism? How dominant is the position of real estate developers? If the interest in the many advantages of 'nature-driven urbanism' will sink in on the decision makers, many of the strategies sketched in this book will be used, separately or in a combination depending on the local situation and planning context.

It might seem more problematic to pacify nature and rapid urbanization in the global south. Even more so, if we take a realistic and sobering view on what urbanism might be able to do in planning mega-cities at large, let alone picking up nature friendly strategies. Koolhaas already diagnosed the position of urbanism in 1995 with the sharp observation: "How to explain the paradox that urbanism, as a profession, has disappeared at the moment when urbanization everywhere – after decades of constant acceleration – is on its way to establishing a definitive, global "triumph" of the urban condition? (…) For urbanists, the belated rediscovery of the virtues of the classical city at the moment of their definitive impossibility may have been the point of no return, fatal moment of disconnection, disqualification. They are now specialists in phantom pain: doctors discussing the medical intricacies of an amputated limb." (Koolhaas and Mau 1995).

Twenty years later, his partner in OMA, Reinier de Graaf, observes that 'thought production' by the architectural profession has come to a grinding halt, that since the 1990's large urban visions or manifestos are no longer being produced. By linking Koolhaas' observations to the dominant economic drivers of urbanization he adds an even darker shade of black. He observes that talking about governance the megacity is nonexistent as an administrative entity and defies political guidance and therefore in the final instance even the public sector's sharply reduced repertoire (infrastructure, services) cease to apply: "The supposed rehabilitation of urban planning within the architecture profession has done little to help. It's focus on the city serves only to mask an acute lack of mandate; it knows deep down that, in the market economy like a Faustian bargain, urban growth has come to imply the relinquishing of urban planning –- that the principle from which the megacity emerges is the same as the one that denies it guidance." (De Graaf 2017).

Of course it is true that any hope of mitigating the effects on nature of rapid urbanization can only be founded on the possibility to take this form of land use change off the automatic pilot. The mechanism of economic/geographical succession that land prices and real estate profits drive the process where urban land use can pay more for the land than agriculture and that agriculture has to move and reclaims nature areas ad infinitum. And that takes political will and courage (Fig. 2.26). This might prove very difficult as the elites and the ruling classes of

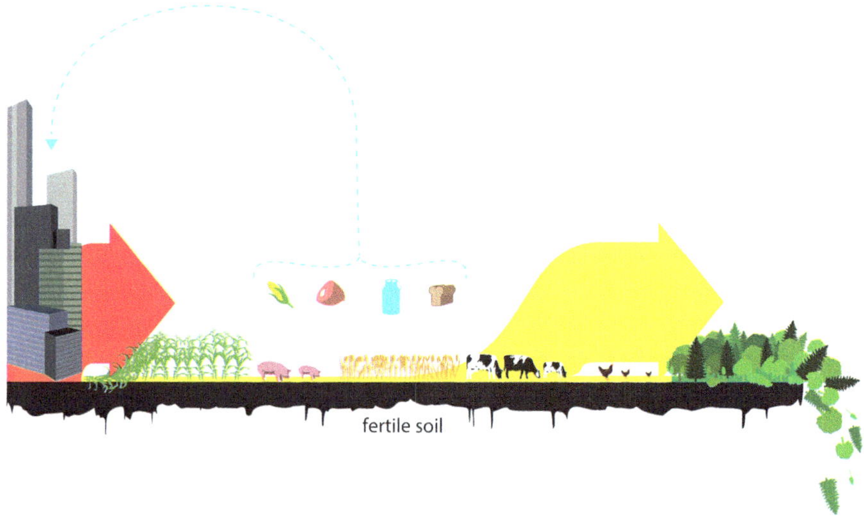

Fig. 2.26 The automatic pilot of Urbanization: the city can make a higher return on the land than agriculture can possibly produce, is being bought out and reclaims natural area: a succession of land use.

many countries are closely connected with the real estate or building industry (Marcinkoski 2015). This is not only true for the viral urban growth in the global south, but almost everywhere in the world. As nature based solutions require always some or considerable functional space it is a prerequisite for nature-driven urbanism to break this cycle. Thisnis the only way a stable place in the urban fabric can be established for low-economic performing forms of land use. There might be some ways, positive or negative, that could force nature based principles into the neoliberal development logic.

A positive incentive might be that the proximity of green areas, forests, parks or water bodies has a measurable effect on real estate prices and tax income. This mechanism might spur the insertion of (more) green elements in plans. In a more indirect way the effects on wellbeing, health and leisure is seen more and more as a crucial asset for cities. This criterium makes it's way in the yearly rankings of 'Most Liveable Cities' (Monocle, 2009–2018)that are a yearly phenomena and seem to have some influence on foreign investments and establishment of branches or headquarters of businesses. In this very indirect way, using nature based solutions in the city, is a good investment. But only a municipality would invest out of enlightened self-interest, rather than an individual company or an individual project.

A strong incentive comes from a completely different side: the necessary adaptation of (mega)cities to the effects of climate change. Here we see a completely different kind of ranking: which cities are most vulnerable for rising sea levels, ever more forceful tropical storms, wildfires, heatwaves, periods of drought and unpredictable rivers, heat island phenomena that seem to constitute the 'New Normal' (Ovink and Boeijenga 2018). Adaptation to these effects of climate disruption could force metropolitan regions into a (new) planning mode. Some sort of planning, however basic, will be highly necessary to effectuate preventive measures or, more dramatic, disaster control after the worst happened. The investments needed to face a different kind of world than we were used to, are almost un-imaginable and are counting in trillions rather than billions of dollars. But we know that even these figures fade in the face of the possible damage done by disasters. Prevention is cost-effective, but nearly 87% of disaster-related spending on aid goes into emergency response, reconstruction and rehabilitation, and only 13% toward reducing and managing the risks before they became disasters (Kellet and Caravani 2013). It is my conviction that nature based solution will play an important role in prevention strategies. We have been trusting hard civil engineering for too long and thought we could ignore natural forces and boundaries. Natural systems that enhanced water safety but were removed or reclaimed – like mangrove forests – will make their come back. A €2,3 billion river flood risk prevention plan in the Netherlands shows that giving more room to the river not only improves water safety but also offers chances for nature development and landscape quality (Sijmons et al. 2017). It is interesting to see that in the last two decades comparable projects have been planned and executed in Switzerland, Germany and France all with a different framing to fit them in their specific national discourses and all with considerable gains for natural quality (Rossano 2017). In urban areas, using river corridors as the backbone of ecological and social resilience can be extremely successful when the history film of malinformed civil intervention can be strategically rewound (Kuzniecow et al. 2014, Forgaci 2018). The idea to create living breakwaters for the South Shore of Staten Island to prevent damage like Hurricane Sandy to repeat itself is another example of the specter of nature based solutions available in coastal protection (SCAPE 2014). Disaster prevention could work as a powerful leverage for nature development in and around urban landscapes.

2.11 Conclusions

It might seem almost impossible the pacify biodiversity with the seemingly all-overpowering process of rapid urbanization. I might seem naïve, utopian or overly theoretic to even raise this subject when expanding cities and their responsible administrators have other worries and priorities. But I think that given the seriousness of the global problem of the erosion of biodiversity as a discipline we have to research and conceptualize possible answers. It is a problem that causes irreversible damage and can be ranked at least at the same level as climate change. Urbanization is not a natural disaster. It is manmade and it can be mitigated by man. I am convinced that with the tool box at our disposition and with political will, we will be able to construct a multidisciplinary endeavour capable of navigating the motivations and consequences of these pursuits.

Urbanists, landscape architects and planners can weave nature based solutions into the urban carpet in their daily practise. Designers can also conceptualize possible acting perspectives in research-by-design projects. In Biennales, competitions and at universities the design community can take on problems that will never have a real client. Because the scale is to large or the problems involve to many interests or seem impossible to take on because they are too 'wicked' (Sijmons 2016).

Ecologists and other life science practitioners have to be involved in the administrations of the municipalities. In research terms they face important questions to raise urban ecology to a higher level. If it helps to make urban nature more biodiverse, to what degree does it help to stop the erosion of biodiversity?

To what extent can we call urban nature an ecosystem, or much more positively restated what natural processes, species and species interconnections are crucial to improve on the performance of 'urban nature'? What role does scale play in the size of the (biggest and smallest) green islands? Functional space for natural processes could be of paramount importance and how to assess to role of spontaneous succession? What must be the role of herbivory? If e.g. soil formation is on the critical path what strategies do we have against the compression of the soil? What is our attitude on invasive species in urban nature?

For all disciplines involved, the matrix presented here is purely theoretical but to my mind interesting because in actual urbanization processes we can learn from all three concepts, Contrast, Contact and Contract. Should all fail that we can console ourselves that nature can't be switched off and on at will and that there will always be an urban nature.

References

Allen D, Schepers F et al (2017) Rewilding Europe, making Europe a wilder place, annual review 2017, Nijmegen

Amsterdams Bureau Stadsecologie (1996) Sijsjes en Drijfsijsjes. Fontaine Publishers, Amsterdam

Angel S, Parent J, Civco DL, Blei AM (2012) Atlas of urban expansion. Lincoln Institute of Land Policy, Cambridge, MA

Betsky A (2006) Landscapers: building with the land. Thames and Hudson, London

Branzi A (2006) Weak and diffuse modernity: the world of projects at the beginning of the twenty-first century. Skira editori, Milano

De Graaf R (2017) Four walls and a roof – the complex nature of a simple profession. Harvard University Press, Cambridge, MA

de Jong, T.M. (2012) Stedelijke verscheidenheid in Zoetermeer (Urban Biodiversity in the town of Zoetermeer) http://www.taekemdejong.nl/Publications/2012/Stedelijke%20verscheidenheid2.pdf

Denters T (1994) Van muurbloem tot straatmadelief, wilde planten in en rond. KNNV Publishers, Amsterdam

Diamond JM (1975) The island dilemma: lessons of modern biogeographic studies for the design of natural reserves. Biol Conserv 7:129–146

Eberstadt R, Möhring B, Petersen R (1910) Gross-Berlin: ein Programm für die Planung der neuzeitlichen Grossstadt. Ernst Wasmuth a.-g, Berlin

Ellis E (2011) Antropogenic transformation of the terrestrial biosphere. Proc R Soc Math Phys Eng Sci 369(1938):1010–1035

Farr D (2007) Sustainable urbanism: urban design with nature. Wiley
Florida R (2002) The rise of the creative class. Basic books, New York
Forgaci C (2018) Integrated urban river corridors. Spatial design for social-ecological resilience in Bucharest and beyond. PhD-thesis, TU-Delft
Forman D (2004) Rewilding North America, a vision for the conservation in the 21st century. Island Press, Washington, DC
Forman RTT (2008) Urban regions: ecology and planning beyond the City. Cambridge University Press, Cambridge
Foster C (2016) Being a beast, an intimate and radical look at nature. Profile Books Ltd., London
Hajer M (2014) The challenge of the century; the need for a complete overhaul of urban planning. In: Brugmans G, Strien J (eds) IABR-2014 urban-by-nature. IABR, Rotterdam, pp 160–163
Howard E (ed) (1902) Garden cities of tomorrow (2nd edition of To-morrow: a peaceful path to real reform 1898). S. Sonnenschein & Co, London
Jones C (2014) Spatial distribution of U.S. household carbon footprints reveals suburbanization undermines greenhouse gas benefits of urban population density. Environ Sci Technol 48(2):895–902
Kellet J, Caravani A (2013) Financing disaster risk reduction; A 20 year study of international aid, GFDRR Global Facility for Disaster Reduction and Recovery. World Bank, Washington, DC/Brussels,Tokyo
Koolhaas R, Mau B (1995) S,M,L,XL, OMA. The Monicelli Press, New York, pp 959–971
Kornaropoulou S, Van der Salm J, Sijmons D (2017) Making City. Aravutköy, Istanbul. In: Ranzato M (ed) Water vs urban scape. Exploring integrated water-urban arrangements. Jovis Publishers, Berlin, pp 97–114
Kuzniecow Bacchin T, Ashley R, Sijmons D, van Timmeren A, Zevenbergen C (2014) Green-blue multifunctional infrastructure: an urban landscape system design new approach. https://doi.org/10.13140/2.1.2061.504. Conference paper: 13th international conference on urban drainage at: Sarawak, Malaysia
Li C, et al (2017) India is overtaken China as the World's Largest Emitter of Antropogenic Sulphur Dioxide. Sci Rep 7. Article number: 14304
LSE Cities (2011) Where people live. London School of Economics, London
Luiten E, Sijmons D (1998) Contrast, contact en contract. In: Feddes F (ed) Oorden van Onthouding, new nature in urbanising Netherlands. NAi Publishers, Rotterdam, pp 169–174
MacArthur RH, Wilson EO (1963) An equilibrium theory of insular zoogeography. Evolution 17(4):373–387
Marcinkoski C (2015) The city that never was. Princeton Architectural Press, New York
Meadows DH, Meadows DL, Randers J, Behrens WW III (1972) Limits to growth, a report for the Club of Rome's project on the predicament of mankind. Universe Books, New York
Melchers M (2016) Van Elburg tot IJburg. Uitgeverij IJburg, Amsterdam
Melchers M, Timmermans G (1991) Haring in het IJ. Stadsuitgeverij, Amsterdam
Melchers M, Timmermans G (1998) Paardenbijters en Mensentreiters. Fontaine Publishers, Amsterdam
Monocle, Top 25 liveable cities (2018, June 22) The liveable City index 2010 17 June 2010 and top 25 most liveable Cities, 17 June 2009
Nair A (2016) Mumbai's urban leopards, http://nayankhanolkar.com/mumbais-urban-leopards/
Nylander W (1866) Les lichens du Jardin du Luxembourg. Bull Soc Fr 13(S):364–372
Odum HT (1982) Pulsing, power and hierarchy. In: Mitsch WJ, Ragade RK, Bosserman RW, Dillon JA (eds) Energetics and systems. Ann Arbor Science Publishers, Ann Arbor, p 33e59
Odum HT (1983) Systems ecology: an introduction. Wiley, Hoboken
Ovink H, Boeijenga J (2018) Too big: rebuild by design: a transformative approach to Climate Change. NAi010 Publishers, Rotterdam
Owen D (2009) The Green Metropolis: what the city can teach the country about true sustainability. Riverhead Publishers, New York City

Pelt J-M (1977) L'Homme re-naturé, vers la societé ecologique. Editions Seuil, Paris
Rossano F (2017) Flood-scapes. Contemporary landscape strategies in times of climate change. PhD thesis, ETH Zürich
Saeijs HLF, Baptist HJM (1980) Coastal engineering and European wintering wetland birds. Biol Conserv 17(1):63–83
SCAPE (2014) Living breakwaters, rebuild by design competition. In: Ovink, H. and Boeijenga, J. (2018) Too Big, Rebuild by Design: a Transformative Approach to Climate Change. Rotterdam: NAi-010 Publishers
Scott JC (2017) Against the grain. A deep history of the earliest states. Yale University Press, New Haven/London
Sieverts T (1997) Zwischenstadt. Zwischen Ort und Welt, Raum und Zeit, Stadt und Land. Vieweg, Braunschweig, 1997
Sijmons D (2014a) Waking up in the anthropocene. In: Brugmans G, Strien J (eds) IABR-2014 urban-by-nature. IABR, Rotterdam, pp 13–20
Sijmons D (2014b) Urban metabolism. In: IABR—2014 urban-by-nature. IABR, Rotterdam
Sijmons D (2016) When research by design takes politics on a sabbatical detour. In: Brugmans G, Strien J (eds) The next economy. Catalog IABR, Rotterdam, pp 132–137
Sijmons D, Feddes Y, Luiten E (2017) Room for the river; safe and attractive landscapes. Blauwdruk Publishers, Wageningen
Sukopp H, Blume HP, Kunick W (1979) The soil, flora and vegetation of Berlin's wastelands. In: Laurie IC (ed) Nature in cities. Wiley, Chichester, pp 115–132
Thwaites T (2016) GoatMan, how I took a holiday from being human. Princeton Architectural Press, Hudson
UN Department of Economic and Social Affairs (2017) World population prospects: the 2017 revision. United Nations, New York
UNEP (1992) UN Conference on the Environment, Rio "earth summit", June 1992
UN-Habitat (2012) United Nations Human Settlement Programme (UN-HABITAT), Nairobi
Van Eesteren C, en Van Lohuizen TK (1934) Algemeen Uitbreidingsplan (AUP) van Amsterdam. Municipality of Amsterdam
Vera FWM (2009) Large-scale nature development – the Oostvaardersplassen. British Wildlife 20(5):28–36
Wagner M (1915) Das sanitäre Grün der Städte, ein Beitrag zur Freiflächentheorie. Dissertation. Berlin
Weller RJ, Hoch C, Huang C (2017) Atlas for the end of the world, University of Pennsilvania. http://atlas-for-the-end-of-the-world.com/world_maps/world_maps_hotspots17.html
Wright FL (1932) The disappearing city. W.F. Payson, New York

Chapter 3
Temporary Nature - A Win-Win for Nature and Developers: Tinkering with the Law in Order to Combat Biodiversity Loss

Chris Backes, Arnold van Kreveld, and Hendrik Schoukens

Abstract Temporary Nature has been pitched as a recent illustration of a more collaborative, reconciliatory approach to nature management in human-dominated landscapes. In essence, the novel concept is focused on providing more opportunities for nature development on temporarily available lands, which will subsequently be turned into a housing zone or an industrial site. By opening up these sites for nature development on a temporary basis, without hampering future developments, the concept might lead to net gains for endangered pioneer species. In doing so, Temporary Nature stands out as a remarkable win-win approach, which might help to enhance nature on lands which would, in lieu of such an instrument, remain out of reach for nature. The recent Dutch experiences with Temporary Nature have already revealed that such long-term beneficial effects effectively materialize on the ground. Even so, additional research will have to reveal the ideal circumstances under which this concept can yield an optimal outcome in terms of biodiversity gains and local acceptance.

Keywords Temporary Nature · Safe harbor agreement · Reconciliation ecology · Pioneer species · EU Nature Directives

C. Backes (✉)
Professor of Environmental and Planning Law, Utrecht Centre of Water, Oceans and Sustainability Law (UCWOSL), Utrecht, The Netherlands
e-mail: c.w.backes@uu.nl

A. van Kreveld
Temporary Nature Foundation, Houten, The Netherlands

H. Schoukens
Ghent University, Ghent, Belgium

R. Roggema (ed.), *Nature Driven Urbanism*, Contemporary Urban Design Thinking, https://doi.org/10.1007/978-3-030-26717-9_3

3.1 Developers Are People Too

People love threatened species, especially the charismatic threatened ones. Developers are people too, that goes without saying. And many developers, when on holiday, really enjoy seeing threatened, charismatic species. Typically, however, their enthusiasm is much less when these species show up at their building sites. In such a context, endangered nature is often exclusively approached as a 'liability', which could give rise to a potential obstacle course when seeking to obtain planning permits for new developments. This more reluctant view is understandable; many threatened, charismatic species are strictly protected, and their presence could indeed spell trouble for a building project. This will manifest itself in longer procedures to obtain a permit to remove the protected species or extra costs for compensation for example. In some instances, the future economic development plans will have to be placed on the back burner to execute the substantive protection duties attached to certain endangered species.

To prevent this from happening, developers increasingly decide to implement avoidance actions to keep their vacant lots 'nature-free'. Such measures are often very costly and involve actions such as intensive mowing, the use of pesticides and the placing of fences on areas suitable for future development. In turn, this gives rise to a certain paradox: one should expect nature conservation legislation to spur measures that are beneficial for biodiversity. However, the effect of conservation legislation in these specific circumstances is counterproductive: it leads to actions that are damaging for nature. Instead of promoting win-win scenarios, which seems to be in line with the recovery objectives upon which most nature conservation laws are predicated (Cliquet et al. 2015), a stringent application of protection schemes seems to give rise to perverse incentives, which favour unsustainable management practices on lands over interesting win-win scenarios. Nevertheless, the survival of many pioneer species, such as the Natterjack Toad and the Fen Orchid (Fig. 3.1),

Fig. 3.1 The strictly protected Natterjack Toad and Fen Orchid regularly benefit from dynamics circumstances at building sites. (Photo credit: Rudmer Zwerver)

has become increasingly dependent on the seizure of this unhidden potential on lands that are often not located in designated protected sites. Doing nothing is not an option.

In the Netherlands an innovative legal solution - beneficial to developers, nature and local people – has been proposed to turn this lose-lose situation around. The solution is dubbed 'Temporary Nature' and aims to reconcile the possibility to create new opportunities for endangered species with the quest for additional legal certainty at the developers' side. To do this, it was suggested to grant project developers the possibility to obtain a derogation to remove endangered species before they actually decide to open up their lands for them. In other words, they receive additional legal certainty about future development actions prior to their decision to allow nature to settle on these lands for a provisional time. Key to this approach is that no additional mitigation and/or offset measures will arise when developers opt for Temporary Nature on their lands. In 2006, a team of Dutch legal experts (Chris Backes and Hans Woldendorp), a conservation organisation (ARK Nature) and an innovative consultancy firm (Stroming) jointly worked out the relevant legal, ecological and social issues of this novel approach. From early on, these experts were supported by the Dutch government, resulting in a guidance document, which further spells out the details of this innovative nature conservation concept.

From Idea to Reality

Innovations often take time to land, and Temporary Nature is no exception. However, as of today, it is widely applied in the Netherlands and has also been included in guidelines in other countries such as the Flemish Region of Belgium (Agentschap Natuur en Bos 2018) and Germany (Becker et al. 2018). There is a foundation promoting the concept with a board of companies (Port of Amsterdam), a foundation of conservation organisations: 'LandschappenNL' (a network representing 20 provincial nature and landscape conservation organizations) and the Dutch Butterfly Conservation, with many others supporting their work.

To date, almost 50 derogations have been granted for a total of over 3500 ha of Temporary Nature (tijdelijkenatuur.nl, undated). A few sites have already been cleared and developed, but most are still Temporary Nature. While these areas are found all over the country, in a large variety of settings, the majority are located on former agricultural lands destined for houses or business parks. Some, mostly the smaller ones, are found in cities or towns. The harbours of Rotterdam, Amsterdam (Fig. 3.2) and Groningen province contain the largest areas.

In the Netherlands alone, there are more than 40,000 ha of empty land which could potentially become Temporary Nature. And although each Temporary Nature area will obviously be temporary, the gain for nature in general is permanent. Seeds and young animals will spread out into the surrounding environment, helping to preserve and strengthen populations in the wider landscape.

Fig. 3.2 Official opening of the 1st Temporary Nature site at Port of Amsterdam in 2010. (Photo credit: Arnold van Kreveld)

3.2 The Legal Framework

3.2.1 The Legal Dilemma

Within all EU member states, wild birds and many other species are strictly protected on the basis of the European Birds Directive (European Parliament 2010) and Habitats Directive (Council of the European Communities 1992). These two directives constitute the bedrock of the EU nature conservation policy (Schoukens and Bastmeijer 2015). Killing and intentionally disturbing protected animals or gathering fruit or seeds from protected plants is forbidden. For many activities, a derogation from these prohibitions is not easy to obtain, if at all. If such a derogation is applied for at the moment the developer wants to clear a site to realise his plans, the criteria for granting exceptions are very restrictive, especially in the case of birds. As explained earlier, developers will try to prevent protected species to occur on their sites. In essence, the application of the European Birds Directive and Habitats Directive depends on the actual presence of protected species. And accordingly, one can freely implement mowing and ploughing practices aimed at preventing such species to settle in the first place. As a result, in such instances, nature protection law does not protect nature, but on the contrary it prevents that nature can develop (Fig. 3.3).

Fig. 3.3 Mowing a vacant lot to keep it free from protected species. (Photo credit: Ingrid Roerhorst)

3.2.2 The Legal Solution: Antedating Requesting a Derogation

The legal solution for this dilemma is antedating requesting and granting of the derogation to the moment before the nature develops (Schoukens 2017). Hence, the derogation is immediately applied for when the developer purchases a plot of land or when a previous use, like for example agricultural use, is stopped and nature is given room to further develop. Granting a derogation before opening up lands for nature enhancement is obviously not thought of when the legal provisions in the Birds and Habitats Directives and their national equivalents were drafted. Usually, someone who applies for a derogation, for example a developer, does exactly know which species are present at his site and which nests, birds or plants he wants to remove. The developer will then apply a derogation for a precise list of species. However, the respective EU and national legal provisions do not force such a reading and application. As such, the Directives do not stand in the way of antedating the granting of a derogation, as depicted above. Antedating the derogation and allowing to remove all kinds of species which are likely or might occur on the site is not explicitly forbidden by the law (Woldendorp and Backes 2011).

Antedating the application and granting of the derogation solves three problems at a single blow.

> First, the abovementioned approach can be aligned with a specific derogation ground, mentioned in the Birds and Habitats Directives. As is obvious from the above, removing nature – even if only intended to be temporary – is prohibited by the strict protection duties set out

in both directives. However, according to the Birds Directive, a derogation can be justified for 'the protection of flora and fauna' (Article 9(1), sub a Birds Directive). Similarly, the Habitats Directive allows the issuance of derogations 'in the interest of protecting wild fauna and flora and conserving natural habitats' (Article 16(1) sub a Habitats Directive). Both justification grounds can serve as legal basis for derogations which will be necessary to remove the nature which has developed on sites which temporarily were not used for their primary (and definite) purpose. Allowing nature to develop on sites which temporarily are not used for other purposes creates, even if this nature is 'removed' in the end, a long-term surplus for flora and fauna and therefore contributes to 'the protection of flora and fauna', respectively is 'in the interest of protecting wild fauna and flora'. These justification grounds cannot be used if the derogation is only applied for, as usually was done, just before the Temporary Nature is to be removed. In such a scenario, the removal itself has only negative effects on the protection of wild flora and fauna. Only if the application for the derogation is antedated and the removal in the future and the occurrence of temporarily nature is seen altogether, as a "single act", the sum of the effects on nature is positive and the mentioned reasons for derogating from the prohibitions of species protection law can be used. This approach also rests upon the assumption that without this win-win solution, no additional opportunities for the enhancement of nature will arise. In such an instance, developers will give precedence to the continuation of the shoot, shovel and shut-up approaches.

Second, the antedating of the derogation creates additional legal certainty for the developer. Prior to opening up lands for nature development he has obtained his derogation to remove the nature in a subsequent stage (see Fig. 3.4). This entails that the developer knows that he will not face any additional legal problems due to the presence of nature, when the nature is removed, and the site is economically developed.

The third advantage has also to do with legal certainty. Even if a developer tries to keep a site "nature free", he or she cannot be sure this will be successful. If, despite using pesticides, deep ploughing and other measures, a certain protected species appears nevertheless, the developer will have to ask for a derogation to remove it, which will not be easy or even impossible to obtain. These concerns will disappear if the developer chooses to obtain a derogation for a case of Temporary Nature, as he already has the derogation in his hands and does not have to worry about protected species that may appear before the site is used for its definite purpose.

3.2.3 Ensuring a Net-Positive Effect for Nature

To ensure that allowing Temporary Nature to develop and to be removed has a net-positive result for nature and effectively favors 'the interest of protecting wild fauna and flora and conserving natural habitats', the derogations come with a number of conditions. The most important conditions that are attached to the derogations are the following:

- Do not disturb breeding birds;
- The derogation is limited to the species and activities mentioned. The derogation is applied for all species that may, given the local circumstances, occur. When assessing applications, a distinction will be made between the different biogeographical regions of the Netherlands. For each of the different biogeographical

Fig. 3.4 Port of Rotterdam opened up lands for nature development. (Photo credit: Niels de Zwarte)

regions a species list covers all of the species that may be found in that region. The application for the derogation can simply include a reference to the biogeographical map and the associated species list;
- If other (newly established) strictly protected species are found, or other activities planned than those mentioned in the derogation, the authorities need to be notified immediately;
- Monitoring must take place before clearing an area;
- A so-called 'Ecological Working Protocol' must be drawn up by a trained ecologist, before clearing the site, and then adhered to. Often the ecologist must be present during (a part of) the clearing of the site.

These conditions are common and in general added to all derogations granted under the Dutch nature protection law. There are two additional conditions, specific to Temporary Nature:

- Measures that could restrict colonization of the site by the species for which the derogation was granted must be omitted as much as possible;
- Measures that could restrict dispersal from the site of the species for which the derogation was granted must be omitted as much as possible.

These two extra conditions are aimed at increasing the effectiveness for nature of using Temporary Nature on a site. In the meantime, the Dutch government developed a guideline on Temporary Nature, which explains the concept of Temporary Nature and lists the conditions for its use (Ministerie van Economische Zaken 2015).

3.3 Collaborative Policies

In recent years it has become clear that an increasingly strict application of nature conservation laws, such as the Birds and Habitats Directives, gives rise to counterproductive results when it comes to biodiversity protection in private lands. While a strict enforcement of nature protection statutes is an evident key to environmental success-stories, the case of Temporary Nature reveals that 'out of the box'-thinking is required in order to avoid unintended consequences to arise, especially in the context of human-dominated landscapes where the room for additional nature development is scarce. Temporary Nature must be singled out as a rare, effective example of and inspiring template for more collaboration-based approaches to nature conservation.

3.3.1 Deadlocks and Obstacle Courses: Command and Control Leading to Perverse Incentives

Before describing how the case of Temporary Nature fits with other trends towards more collaborative approaches to nature conservation, one needs to understand the mounting criticism to which the latter fell victim over the past decades. In essence, many of the environmental regulatory statutes dating back to the 1970s to 1990s – such as the Habitats and Birds Directives – are grounded upon a so-called 'command and control' approach (Schoukens 2015). The rules concerning strictly protected species under EU law provide a poignant illustration thereof. In essence, this set of rules bans a certain number of inherently harmful practices and activities. They principally reflect a preventative approach to nature conservation. As can be derived from the analysis above, these rules appear to be more focused on the protection of individual specimens than on the preservation of the wider population of a species; they also apply both inside and outside protected sites. Only under a very limited set of circumstances can a derogation be granted for economic developments that run counter to the protection duties under the Birds and Habitats directive. This leaves very limited room for bargaining, even when economic considerations are at play.

Given the increasingly tight application of the rules on species protection in countries such as the Netherlands, Germany, the United Kingdom and Belgium, the Habitats and Birds Directive were increasingly framed as an obstacle course than an instrument to achieve clear wins from the vantage of biodiversity (Schoukens 2015). This was especially the case in the Netherlands, where the relatively strong and prominent role of environmental NGOs led to an increasing number of legal actions against planning permits based upon EU nature conservation law. Hundreds of legal challenges based upon species protection law created additional fear amongst project developers. For instance, the presence of several highly endangered hamsters led to long delays when developing a cross-boundary industrial estate due to lengthy legal procedures (Dutch Council of State 2000). When a colony of spoonbills settled in the Vlissingen Port Area, a Dutch NGO unsuccessfully tried to force the Dutch government to designate the area as protected site (Dutch Council of State 2010). Even when the bulk of these challenges did not lead to a definitive permit refusal for project developers, the impression was created that nature conservation law seemed to punish private landowners who had species habitats on their lands by restricting future development. This finding fuelled the resistance amongst business people and project developers against nature conservation laws. A much-shared criticism was that modern nature conservation laws did not put forward sufficient incentives to compel or encourage private landowners to restore lost habitats. In the 1990s, several private landowners in the United States openly opted for a so-called 'shoot, shovel and shut up'-approach, which resulted in the clear-cutting of areas in order to prevent protected species to settle there in the first place (Paulich 2010). Also in the EU, especially in the Netherlands and Belgium, several cases of pre-emptive habitat destruction have emerged over the past 10 years. For instance, in Belgium courts reasserted the legality of the actions of a harbour company aimed at preventing sea gulls to roost on plots of land intended for the enlargement of an industrial estate (Court of First Instance (Bruges) 2014), while in the United Kingdom the technique of 'newt fencing' (see Fig. 3.5), poised to preventing the arrival of Great crested news on sites destined to become industrial estates and housing zones.

3.3.2 *The Shift towards More Collaborative Approaches*

Against the backdrop of these increasingly antagonistic stories regarding nature conservation on private lands, a new, more collaborative and reconciliatory environmental paradigm emerged. In literature, the concept of 'reconciliation ecology' was

Fig. 3.5 Fence aimed at keeping protected amphibians out of an area to be developed. (Photo credit: Arnold van Kreveld)

pitched, which specifically aims at fostering nature conservation in human-dominated landscapes (Lundholm and Richardson 2010). Instead of focusing on what is bad for nature, the new approach tried to tackle the underlying incentives when it comes to nature conservation on private lands. While not all of the currently unused or undeveloped lands might offer additional opportunities for nature conservation, it became clear that merely focusing efforts on protected sites will not lead to a more sustainable solution for the ever-increasing biodiversity loss. Whereas many of these private lands will continue to lay fallow for several years – often awaiting a future economic development – such areas might still serve as important safe harbors, especially for pioneer species, such as Natterjack toads and Little plover (see Fig. 3.6). Seeing that the wider landscape is increasingly built up, such species lack sufficient pioneer habitats to thrive. By encouraging private landowners and project developers to open up their lands – even on a temporary basis – for these species, nature conservation laws could effectively make the difference between imminent extinction and much-needed recovery.

Given that a large share of the actual and potential habitats of endangered species are located on private lands, the question arises whether modern nature conservation law can be interpreted so as to foster more facilitative approaches to nature conservation. This quest started in the United States, back in the 1990s. The fierce opposition to the Endangered Species Act – which dated back from 1973 – prompted the legislator to include additional derogation clauses. Yet, in addition, other guidelines were promulgated which allowed for additional 'bargaining in the shadow of the law' (Wheeler and Rowberry 2009).

With the arrival of the so-called 'safe harbor agreements' in the mid-1990s, an instrument was finally available to encourage habitat restoration and conservation

Fig. 3.6 Little plover depends on pioneer habitats not often found in Dutch nature. (Photo credit: Arnold van Kreveld)

amongst landowners, who do not necessarily want to develop their land in the short run, but want to reserve the right to do so at a later point in time. According to the U.S. Federal Fish & Wildlife Service's Policy document (FWS) 'A safe harbor agreement is a voluntary agreement involving private or other non-Federal property owners whose actions contribute to the recovery of species listed as threatened or endangered under the Endangered Species Act' (US Fish and Wildlife Service 1999). Under a safe harbor agreement landowners who voluntary use their property for the benefit of species will, in return, be provided with a 'safe harbor guarantee', implying that no additional conservation measures will be imposed on their lands, even if the number of threatened or endangered species grows as a result of the actions of the landowner. The first safe harbor agreements were concluded in the U.S. back in 1995, in the absence of further administrative guidelines on how to reconcile the actions with the Endangered Species Act. The Policy itself only became officially effective according to the Federal Register of June 1999. In exchange for additional recovery actions, the participating landowners receive formal assurances that no additional restrictions will be imposed if the number of species increases through the landowner's actions. The landowner or farmer may, at the end of the agreement period, return the enrolled property to the baseline conditions that existed at the start of the safe harbor agreement. In the past decades, several safe harbor agreements have been concluded between the FWS and private landowners (Schoukens 2015). At least some of these agreements have reached remarkable successes. According to recent data, 4 million acres of private lands are now covered by these agreements, which harbor approximately 63 rare species. While many of the safe harbor agreements have a relatively short running time, the agreement for the Aplomado falcon has yielded the most impressive results, which is partly the result of robust reintroduction measures. However, other agreements have given rise to mixed results (Kishida 2001).

Either way, it is hard not to notice the parallels between the safe harbor agreements concluded under the Endangered Species Act and the concept of 'Temporary Nature', as put forward by the Dutch government in recent years. Admittedly, the territorial scope of the safe harbor agreements is notoriously larger than Temporary Nature, which basically focuses on vacant lots that have been accorded an economic destination of the applicable land-use plans. The former also apply to woodlands and prairies. Safe harbor agreements are also less preoccupied with going back to a baseline scenario. Whereas a return to the baseline is permissible, it is expected that in many instances private landowners are already content with the theoretical possibility to remove the additional nature at their own discretion. With Temporary Nature the focus is more on industrial plots of land which will inevitably be returned to nature-free zones on the short term. This also explains the differences in term of duration between safe harbor agreements and Temporary Nature. Yet, by and large, both instruments aim at providing more legal guarantees for private landowners when opening up their lands for additional nature enhancement actions. Moreover, the legal foundations of both approaches are quite similar. Both approaches are framed within one specific derogation ground, granting additional leeway for actions which, when approached in a wider perspective, might ultimately enhance the survival of endangered species.

3.4 Ecological Effectiveness

The ecological effects linked to the usage of Temporary Nature-instruments have been the topic of many investigations. In 2006 a first report was prepared looking at the potential ecological effects of Temporary Nature in general (Linnartz 2006). The bottom-line of the findings was the following: Temporary Nature has many winners and no losers. On a plot of land, opened up for the development of Temporary Nature, plants and animals settle and their numbers evidently increase when no actions are contemplated aimed at the removal of these species. However, logically, these beneficial effects disappear whenever the nature is removed, and the site is being built. But in the wider scheme of things the positive effect linked to the usage of Temporary Nature is permanent (Reker 2006). When approached at population level, opening up potential industrial sites and housing zones for Temporary Nature makes sense because young animals and plant seeds spread out from the temporary habitat into the surrounding environment. Since areas used for Temporary Nature can function as centre of colonisation and stepping stone, they have a permanent beneficial effect on the populations of plants and animals in the wider environment. The risk that some species may ultimately decline due to the development of Temporary Nature areas is negligible when assessed at population level (Linnartz 2006). In other words, the overall populations of the targeted species will never be smaller compared to a zero-scenario of doing nothing.

For many species, including terns, Natterjack toads and various orchids, the impermanence of the sites is not a major issue in terms of survival conditions (Linnartz 2006). On the contrary, these so-called pioneer species thrive in areas where the conditions experience significant variation. For these 'dynamic environment specialists' in particular, Temporary Nature constitutes a welcome addition to permanent natural areas, where maintenance usually focuses on stability. At the end of the day, the mentioned species would also disappear as a result of natural succession.

The 2006 report concluded that Temporary Nature offers a place to settle, breed, forage, spend the night or pass the winter for pioneer species, species from early and later succession stages, migratory birds and winter visitors. The area can also function as a stepping stone or ecological connection, making it easier to reach other temporary and permanent nature areas. This research was subsequently backed up by more recent Flemish findings, which additionally stressed that Temporary Nature could be framed within the so-called metapopulation theory (Vriens et al. 2013). It is underlined that when framed within a meta-population approach Temporary Nature will lead to an increase of local populations of pioneer and early species. It was noted that the risk of creating additional 'ecological traps' – by opening sites for species which will be economically developed at the end of the day – is not higher when compared with biodiversity in permanently protected sites.

3.4.1 *Reality Check*

Even though in theory the ecological risks tied to the usage of Temporary Nature appeared limited, a reality check is never a bad idea. In the past few years a number of field studies were performed looking at the effects of Temporary Nature in the real world. The two cases that stand out in terms of ecological knowledge are the Temporary Nature that has been developed at the Port of Amsterdam and at the Eeserwold (near the city of Steenwijk). These two sites are discussed in the following paragraphs.

Still a disclaimer has to be made. It is important to note that Port of Amsterdam and Eeserwold might not be representative for other sites. In both cases, (some) active measures were taken to improve ecological conditions, and these have had a notable positive impact on the biodiversity. Implementing such measures is not mandatory nor a prerequisite on other sites. Also, many sites have a much poorer starting point (e.g. very rich agricultural soil, which is much less interesting from an ecological point of view) and/or will be developed within 1–2 years, leaving less time for species to locate and colonize these sites.

3.4.1.1 Port of Amsterdam (Fig. 3.7)

This was the first Temporary Nature area to be established, with its derogation granted on 15 July 2009 (FF/75C/2009/0068.toek.mo). Ecological development and the clearing of this site have been well-documented (Vliegenthart 2012; Smit and Melchers 2016), making this an interesting case. It is a small (9 ha) site. Though being a small site, it is ecologically interesting for diverse reasons. The poor, sandy soils are ideal for a diverse vegetation with many flowers and warmer micro habitats. This is attractive for many insects, including a number of relatively rare species. The harbour also hosts a few strictly protected species (under EU or only under the then applicable Dutch law).

Although the granted derogation did not include a duty to actively restore biodiversity, the Port of Amsterdam nevertheless decided to dig a pool for Natterjack toads (Fig. 3.8) and to erect a wall, in which Sand martins could breed. The site was not actively managed since this does not constitute a general obligation when working with Temporary Nature. In this case, the soil (sand) was so poor that the vegetation remained open. Natterjack toads, while rare in the Netherlands, are common in this area. Fen orchid and the then in the Netherlands strictly protected Western or Broad-leaved marsh orchid are also found frequently at the Port of Amsterdam. A number of strictly protected bats forage in the harbor, but they are not dependent on the Temporary Nature site.

Fig. 3.7 The first Temporary Nature area at Port of Amsterdam. (Photo credit: Arnold van Kreveld)

Fig. 3.8 Natterjack toads propagate in pioneer ponds like these at the Port of Amsterdam. (Photo credit: Arnold van Kreveld)

Ecological Effects

A first evaluation of the ecological results at this site (and two adjacent Temporary Nature sites in the immediate vicinity) was published in 2012 (Vliegenthart 2012). The report by Dutch Butterfly Conservation focuses on insects. It concludes:

> 'Temporary Nature' develops fast and in the right direction. There are already flowery meadows with high grassland butterfly diversity, which is positive since they are under high pressure. The investigated area of 'Temporary Nature' functions as important stepping stone in the region for this group and other species associated with open sand and pioneer habitat. These are usually dynamic systems from where species can disperse.

And:

> The pilot project of Amsterdam Harbor creating the artificial relief and ponds in the Temporary Nature area, achieved a very good positive development for the biodiversity in the area. At this moment the area is a very important habitat and stepping stone for species of pioneer habitat and grasslands, which are currently threatened in the Netherlands.

The appearance of the rare (albeit not protected) blue-winged grasshopper (*Oedipoda caerulescens*) was tagged as a major surprise.

The Amsterdam site was cleared in 2016 and 2017, and this process has been well-documented by Bureau Waardenburg (Smit and Melchers 2016). The report concludes:

> The development of Temporary Nature since 2009 has been successful. A total of four more strictly protected species are found nearby. Of these, three have turned up in the Temporary Nature site. The site has become the most important area for Natterjack toad. Western or Broad-leaved marsh orchid and Bee orchid are well-established too. A number of bird species have also used the site, such as sand martin and kingfisher.

The clearing of the site was carried out by catching high numbers of Natterjack toads, small rodents and Smooth or Common newt and moving them to suitable areas in the vicinity. Orchids were replanted. Monitoring will take place in upcoming years to see if the species have successfully established themselves in their new habitats.

Over the years, tens of species, including a small number of strictly protected ones, have successfully reached the original site and subsequently propagated. Undoubtedly some of their offspring and seeds have dispersed from here, thereby increasing the chance for these species of reaching new suitable areas.

3.4.1.2 Eeserwold

The case of Eeserworld also constitutes another useful case, with the derogation granted on 1 July 2010. The site covers 172 ha. A derogation for working with Temporary Nature was granted for 113 ha, whereas the rest of the area is still under intensive agricultural use (corn) and a lake. The future development plans include areas for living (60 ha), a business park (32 ha), offices (8 ha) and public green areas used for recreation and water storage. Eeserwold is located directly northeast of the A32 highway, with the city of Steenwijk on the other side. On its southern border flows a small river (the Steenwijker Aa). To date, a few houses and some office building have been built, but most of the area is still Temporary Nature (Fig. 3.9).

Fig. 3.9 Temporary Nature at Eeserwold. (Photo credit: Arnold van Kreveld)

As mentioned earlier, the area has been subject to agricultural activities during the past years and, as a result, is relatively poor in species. Yet in the vicinity of and within the lake a few interesting species have been noted. Most notable was the occurrence of the pool frog (*Pelophylax lessonae*), which is protected under the Bern convention and the Habitats Directive. After the implementation of Temporary Nature, the site is much richer, with high numbers of Red List breeding birds. The project generated much enthusiasm amongst provincial and local nature organization, who decided to proactively collaborate with the other stakeholders of the project. Their activities include management advice, organizing excursions, conducting inventories and documenting ecological results.

Under the terms of the derogation for Temporary Nature, the area is managed naturally, partly through extensive grazing (Hereford cows), partly through extensive mowing and some areas are not managed at all. Management aims at creating and maintaining diverse, favorable circumstances for pioneer nature and for other natural values. Pool frog and Weatherfish (*Misgurnus fossilis*) are present at the site, mostly in parts where no developments will take place. Pool frog could (possibly only temporarily) benefit from the developments.

Ecological Effects

In 2011 an interesting study was published which focused on the breeding birds present in the said area (Zekhuis and Van der Weele 2011). It compares the Temporary Nature site to a reference area. This nearby reference area shares many of the characteristics of the Eeserwold before it was opened up for

Temporary Nature development. 36 bird species were found to breed on the Eeserwold, compared to only 7 species in the reference area. The main reason for this enormous difference is the fact that Eeserwold is much more diverse. The growth of higher vegetation in areas that were not managed (herbs, shrubs, young trees and reeds) attracts a high number of songbirds, some of which occur here in high densities.

The difference between the numbers of Red List species was much smaller; 6 at the Eeserwold (including good numbers of yellow wagtails, Fig. 3.10) and 4 at the reference area. Interestingly, the densities of meadow birds appearing in both areas were generally higher at the reference site than at the Eeserwold. However, as a result of early mowing in the (agricultural) reference area no young birds were thought to have survived. This in fact makes this agricultural site, like many other agricultural lands in the Netherlands, effectively an ecological trap.

Other groups are represented at Eeserwold in good numbers as well, including 147 species of plants (with 7 from the Red List), dragonflies and other insects and a number of mammals (hare, rabbit, roe deer and mice). No comparative studies were performed in the reference area, but most agricultural areas in the Netherlands have very low numbers of plant and animal species.

The authors of this study conclude that certain types of management could further increase numbers of species and their densities. Two years later, after 4 years Temporary Nature 'development' at Eeserwold, a follow-up study was published (Zekhuis and De Gelder 2013). The surveys covered breeding birds, butterflies, dragonflies, grasshoppers, mammals, amphibians and plants. The results were positive, with significant growth in the number of species and specimen. The rare Siberian winter damselfly (see Fig. 3.10) a species from the Red List, and otter appeared in the area.

Fig. 3.10 Yellow wagtail breeds in good numbers at Eeserwold and the rare Siberian winter damselfly is found here as well. (Photo credits: Arnold van Kreveld)

3.4.2 *Conclusion on Ecology*

The authors conclude that scale, location and accessibility have undoubtedly generated a major effect on the number of species and specimen in Temporary Nature. They advise to keep some parts of the area closed for people (e.g. for birds breeding on the ground). Not mowing some areas has provided good habitat for quite a few songbirds. Grazing also has positive effects on a number of species, but this is only the case when done extensively.

As with the Amsterdam site, undoubtedly offspring and seeds of tens of species, including protected ones, will have dispersed from here, thus increasing the chance for these species of reaching new suitable areas. However, a subsequent study might provide additional insights of the net-effect of the area when economically developed.

The above-treated research has aptly demonstrated benefits of Temporary Nature can be substantial. How much so depends on size, location, diversity of habitats (e.g. availability of open water), the soil, duration of the derogation, management, etc. As was predicted by theoretical studies, Whether or not this results in a stronger regional population in the long term is unknown. It may, as a higher number of offspring increases the chance of a species finding new suitable areas.

3.5 Conclusions and Discussion

In recent years, the application of the collaborative policies such as Temporary Nature has gained considerable traction, with at present over 3000 ha of lands covered in the Netherlands and other Member States implementing similar policies (Agentschap Natuur en Bos; Becker et al. 2018). This is not surprising, seeing that recent Dutch practices have shown that Temporary Nature has a positive effect on biodiversity, reduces the legal risk for landowners and also has created additional recreational opportunities for neighbors. Some of the above-mentioned benefits should be addressed more into detail.

> First and foremost, by taking away the fear among landowners of facing additional restrictions when opting for more favourable nature management techniques on their lands, novel policies and concepts, such as Temporary Nature, could open new doors for the recovery and reintroduction of endangered species on large acreages of land which traditionally remained off the chart for traditional nature management actions. In recent literature the importance of having put into place strategies to foster nature conservation on urban and industrial sites is highlighted (Lundholm and Richardson 2010).

> Second, collaborative instruments such as Temporary Nature also allow governmental bodies to strike sensible deals with private landowners in order to enhance biodiversity within urban or industrial zones, where nature often only plays a secondary role. In times of increasing resistance against environmental protection, especially whenever it touches upon ownership rights, shifted approaches such as Temporary Nature can help to further enhance the legitimacy of nature conservation laws without undermining its core principles, such as the preventative approach.

Third, while comprehensive research on the effectives of these novel policy approaches is lacking due to their relatively young age, recent studies indeed reveal that recently created Temporary Nature sites in the Netherlands appear capable of attracting many endangered species.

Fourth, it is widely known that funding shortages are seriously compromising the effectiveness of nature conservation law. In this regard, another important benefit to be mentioned precisely relates to the funding of Temporary Nature. In comparison with traditional conservation instruments, such as the concept of 'protected sites', the habitat creation for pioneer species is entirely supported by private landowners. In some cases, Temporary Nature might even be framed as a simple positive externality of an inherently damaging activity, such as mining or harbour development. Thus, in times of budgetary constraints, Temporary Nature steps in as a relatively cheap and attractive policy instrument to achieve quick wins for endangered species.

On a concluding note, it can be maintained that novel, more collaborative instruments – such as Temporary Nature – are not to be approached as a panacea for all ills. Of course, there might also be drawbacks and pitfalls. For instance, it needs to be ensured that Temporary Nature is not prioritized over more lasting efforts to preserve existing nature. In addition, Temporary Nature will only manage to create net effects when used in combination with a well-functioning environmental and conservation policy, which is based upon robust and well-protected ecological networks (Schoukens 2017). However, in times of persistent biodiversity crisis, more collaborative approaches definitely stand out as striking illustration of the recently emerged branch of reconciliation ecology. So great is the threat of imminent extinction, that out of the box-thinking is required to stave off new cases of extinction. Innovative tools, such as Temporary Nature, might serve as a useful counterpoint to command and control policies which, while much-needed, might in some instances lead to perverse incentives. It is crystal-clear that such novel approaches can serve as additional extras for species faced with imminent extinction. With their focus on nature enhancement in human-dominated landscapes, the above-discussed concepts open up new avenues for many endangered species. As of today, ordinary nature is often poorly protected beyond the ambit of protected sites and therefore any additional instrument capable of fostering additional protection is to be cherished. It must be stressed that caution is in order to avoid that concepts, such as Temporary Nature, are abused for the destruction of permanent nature. However, when sufficient oversight is put in place, this risk is negligible and Temporary Nature might continue to function as one of the most promising win-win approaches for nature in the context of human-dominated landscapes. Additional research and monitoring needs to provide additional background on the exact conditions under which such concepts might yield the most optimal results.

References

Agentschap Natuur en Bos. www.natuurenbos.be/beleid-wetgeving/vergunningen/tijdelijke-natuur. Accessed 27 Dec 2018

Becker N, Handke J, Muchow T, Wellens C (2018) "Natur auf Zeit: Rechtliche und fachliche Rahmenbedingungen" Abschlussbericht des F+E-Vorhabens "Natur auf Zeit: Rechtliche und fachliche Rahmenbedingungen" (FKZ 3516 81 0800). (Hrsg.) Bundesamt für Naturschutz. Bonn

Cliquet A, Decleer K, Schoukens H (2015) Restoring nature in the EU: the only way is up? In: Born CH et al (eds) The habitats directive in its EU environmental law context: European nature's best hope? Routledge, Oxford, pp 265–285

Council of the European Communities (1992) Council Directive 92/43/EEC of 21 May 1992 on the conservation of natural habitats and of wild fauna and flora OJ L 206, 22.7.1992, p. 7–50. URL: https://eur-lex.europa.eu/legal-content/EN/TXT/?uri=CELEX%3A31992L0043. Accessed 30 Dec 2018

Court of First Instance (Bruges) (2014), case no. 14/119/C

Dutch Council of State (2000), case no. 199901039/1

Dutch Council of State (2010), case no. 20090772/1/R2

European Parliament (2010) Directive 2009/147/EC of the European Parliament and of the Council of 30 November 2009 on the conservation of wild birds OJ L 20, 26.1.2010, p. 7–25. URL: https://eur-lex.europa.eu/legal-content/EN/TXT/?uri=CELEX%3A32009L0147. Accessed 30 Dec 2018

Kishida D (2001) Safe Harbor agreements under the endangered species act: are they right for Hawaii? Univ Hawaii Law Rev 23:507–539

Linnartz L (2006) Tijdelijke natuur en beschermde soorten: permanente winst Een ecologische onderbouwing Bureau Stroming en ARK Nature InnovatieNetwerk rapportnr 06.2.134, ISBN 90-5059-306-2

Lundholm JT, Richardson PJ (2010) Habitat analogues for reconciliation ecology in urban and industrial environments. J Appl Ecol 47:966–975

Ministerie van Economische Zaken (2015) Temporary nature policy, 10 September 2015, Staatcourant 2015, 29016

Paulich N (2010) Increasing private conservation through incentive mechanisms. Stanford J Animal Law Policy 3:106–158

Reker J (2006) Tijdelijke Natuur, permanente winst. Bureau Stroming en ARK

Schoukens H (2015) Habitat restoration on private lands in the United States and the EU: moving from contestation to collaboration. Utrecht Law Rev 11(1):33–60

Schoukens H (2017) Reconciliation ecology in practice: legal and policy considerations when implementing temporary nature on undeveloped lands in the European Union. Land Use Policy 67:178–199

Schoukens H, Bastmeijer K (2015) Species protection in the European Union: how strict is strict? In: Born CH et al (eds) The habitats directive in its EU environmental law context: European nature's best hope? Routledge, Oxford, pp 121–146

Smit GFJ, Melchers M (2016) Tijdelijke natuur Westpoort. Ontwikkeling 2009–2016. Bureau Waardenburg. Rapportnr, Culemborg, pp 16–259

Tijdelijkenatuur.nl (undated) Overzicht tijdelijke natuur per Provincie. URL: http://www.tijdelijkenatuur.nl/projecten/. Accessed 27 Dec 2018

US Fish and Wildlife Service (1999) Safe Harbor agreements and candidate conservation agreements with assurances; Announcement of final Safe Harbor policy; announcement of final policy for candidate conservation agreements with assurances; Final Rule and Notices

Vliegenthart A (2012) Kwaliteit van tijdelijke natuur in de haven van Amsterdam. Wageningen, De Vlinderstichting. Rapport VS2012.022

Vriens L, Gyselings R, Mergeay J, Van den Bergh E (2013) Advies betreffende de ecologische effecten van tijdelijke natuur. INBO, Brussel. Rapportnr: A.2012.168

Wheeler, D.P. and Rowberry, R, M. (2009) Habitat conservation plans and the endangered species act. In: Baur, D.C. and Irvin, W.R. (Eds.) Endangered species act: law, policy and perspectives. Chicago: ABA Publishing. pp. 223–224

Woldendorp H., Backes Ch W (2011) Tijdelijke Natuur - Advies over de juridische aspecten, Utrecht 2006, ISBN 90-5059-308-9. (The concept finally was accepted by the District Court of Amsterdam in 2011: Rechtbank Amsterdam 27 May 2011, ECLI:NL:RBAMS:2011:BU4250.

This judgment was upheld in higher appeal at the Dutch Council of State in 2012: ABRvS 25 July 2012, ECLI:NL:RVS:2012:BX2544.)

Zekhuis M, De Gelder A (2013) Monitoring vier jaar Tijdelijke Natuur Eeserwold. Landschap Overijssel. Gefinancierd door Ministerie van Economische Zaken & Team Natuurlijk! Ondernemen, Dalfsen

Zekhuis M, Van der Weele J (2011) Broedvogels van braakliggende bedrijventerreinen. In: Schelhaas H, Dekkers G (eds) Vogels in Overijssel. Zwolle, WBooks, pp 118–125

Chapter 4
Stepping-Stone City: Process-Oriented Infrastructures to Aid Forest Migration in a Changing Climate

Qiyao Han and Greg Keeffe

Abstract Large-scale urbanisation has posed extreme challenges to the biota of the planet by creating non-permeable barriers to movement, especially in the context of global climate change. From a multi-scale perspective, this chapter discusses the importance of landscape connectivity in facilitating ecological processes and develops a conceptual framework of process-oriented green infrastructures. A study in the Greater Manchester area, UK is used to demonstrate the application of this framework to improve urban landscapes for climate-driven forest migration. The result reveals that the migration process at the metropolitan scale can be facilitated by a large number of stepping stones formed by small landscape interventions at site scales.

Keywords Stepping stone · Forest migration · Functional connectivity · Green infrastructure · Ecological process · Urban permeability

4.1 Introduction

Cities are usually considered as barriers to ecological processes (Dullinger et al. 2015; Lazarus and McGill 2014; Tomiolo and Ward 2018). Modern urbanisation causes land exploitation that replaces natural habitats by artificial surfaces, resulting in landscape fragmentation and habitat loss. In the context of global warming, particularly, such fragmented landscapes have greatly impeded the range shift of species for more suitable climate conditions (Wessely et al. 2017; Robillard et al. 2015). Moreover, according to predictions by the United Nations, global urban populations are expected to increase by 72% by 2050, reaching 6.3 billion (United Nations 2012). To support the demands of urban population growth, cities will continue to sprawl and fragment landscapes, further exacerbating the problem.

Within this context, urban green infrastructure is gaining increasing acceptance as a nature-driven solution for reducing landscape fragmentation and facilitating

Q. Han (✉) · G. Keeffe
School of Natural and Built Environment, Queen's University Belfast, Belfast, UK

R. Roggema (ed.), *Nature Driven Urbanism*, Contemporary Urban Design Thinking, https://doi.org/10.1007/978-3-030-26717-9_4

ecological processes. Urban areas are basically covered by two components, built-up spaces and green spaces. Examples of urban green spaces include gardens, golf courses, green roofs, and public parks. These human-modified landscapes provide important ecosystem services for cities, often approached with the concept of green infrastructure (Lovell and Taylor 2013; Maes et al. 2019). However, it should be noted that not all green spaces are able to qualify as green infrastructure. As indicated by the term "infrastructure", green infrastructure should be regarded as necessary for the city as traditional "grey" infrastructure, such as highways, bridges, and sewage systems, whereas green spaces are often viewed as something that is nice to have (Pauleit et al. 2011; Benedict and McMahon 2006; Wright 2011).

This chapter discusses the main feature of urban landscapes as green infrastructure in terms of ecological processes. Particularly, a study in the Greater Manchester area, UK is used to demonstrate how to design process-oriented green infrastructure to facilitate forest migration under climate change.

4.2 Process-Based Connectivity

Green infrastructure works as a landscape network rather than a set of free-standing green spaces in the city. Interconnected landscapes are able to mitigate the negative effects of habitat fragmentation and mediate multiple ecological processes that individual ones cannot support on their own (Douglas and Ravetz 2011; Staddon et al. 2010). The ecological processes could then, in turn, improve landscape diversity and connectivity (Colding 2007; Lafortezza et al. 2013). For example, well-connected urban forests promote the movement of animals, which in turn aid the range shifts of trees by dispersing seeds and pollen. Accordingly, the feature of connectivity links urban landscapes and ecological processes in a feedback loop (where one thing affects a second, which affects the first), and therefore can be seen as the overriding feature of green infrastructure.

Connectivity is often viewed from two perspectives: structural and functional. Structural connectivity describes the physical connections (e.g., corridors) or distance between landscape fragments (Fig. 4.1a), whereas functional connectivity refers to the degree to which a landscape facilitates or impedes ecological processes (e.g., species migration, seed dispersal, and hydrological process) (Fig. 4.1b) (Vimal et al. 2012; Baguette and Van Dyck 2007).

While structural connectivity is often used as a proxy for functional connectivity, in cities and other human-dominated environments, functional connectivity is the predominant perspective. This is because extremely heterogeneous land cover (small, narrow, and scattered spatial units) makes it difficult to create continuous landscapes, especially across high-density residential areas or city centres. In addition, many ecological processes in cities do not necessarily depend on physical connections, although a continuous green corridor may facilitate them (Hejkal et al. 2016; Forman 2014).

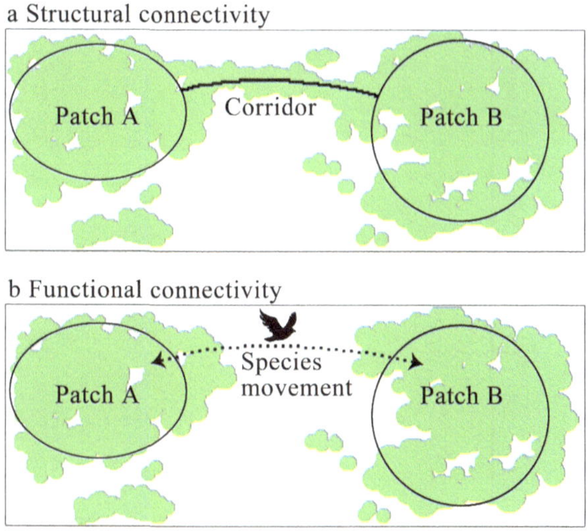

Fig. 4.1 An illustration of structural and functional connectivity. (Source: Author)

4.3 Process-Oriented Infrastructures

While ecological processes in urban areas can be maintained and facilitated by functionally connected landscapes, incorporating them into the management and design of green infrastructure is still challenging. This is due to the fact that ecological processes operate at nested scales, thereby requiring an integrative approach from a multi-scale perspective. Here, scales refer to physical dimensions, in terms of space or time, of landscapes or processes (Fig. 4.2).

Urban landscapes are nested hierarchical systems in both structure and function: the landscape at a given scale is composed of interacting components at a smaller scale and is itself a component of the landscape network at a larger scale (Forman 2014). This is in line with our nested urban systems, which ranges from individual dwellings, blocks and neighbourhoods, to cities and finally bioregions (Newman and Jennings 2012; Marcotullio and Boyle 2003; Keeffe 2014).

The hierarchical organisation of urban landscapes leads to landscape connectivity at multiple scales, which affects different ecological processes operating at each scale (Rayfield et al. 2016). For example, individual trees provide fruit or shade at a local scale, but when connected as lines of street trees, they can serve as corridors connecting neighbourhood parks, and when further aggregated to the level of urban forests, they may provide habitat refuges that support long-distance dispersal of populations in regions with intensive human activity. In general, small-scale connectivity provides species access to spatially distributed resources in the short term, while large-scale connectivity facilitates species movement and gene flow across entire species ranges, often related to climate change adaptation in the long term (Dilt et al. 2016).

Due to the nested structure of landscape scales, a given ecological process of interest (at a specific scale) is actually a synergistic result of the processes at its

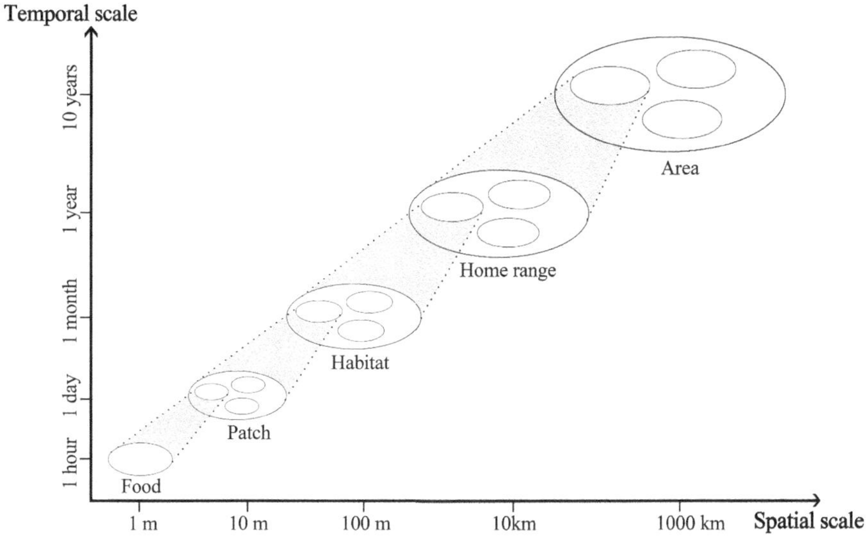

Fig. 4.2 Example of temporal and spatial scales: scales in the hierarchy of decisions made by large wading birds. Source: author, based on the work by Holling (1992)

adjacent scales (O'Neill 1986; Wu and Li 2006; Scholes et al. 2013). To be more specific, processes at a smaller scale provides initiating conditions to the focal scale, whilst processes at a larger scale exert constraints (e.g., boundary, direction) to the focal. As a result, incorporating ecological processes into the design of green infrastructure requires a consideration of landscape connectivity at multiple scales.

Here, we present a conceptual framework of process-oriented green infrastructures (Fig. 4.3). To account for the nested organisation of urban landscapes and the cross-scale dynamics of ecological processes, two adjacent scales (smaller and larger than the focal one) are included in the framework. At the same time, the closed loop of urban landscape, functional connectivity, and ecological process is highlighted at each scale. A main practical advantage of the framework is that it indicates a process-oriented design approach, in which small interventions in the landscape can reshape connections and processes within green infrastructure, which might further affect the processes at a larger scale. In the following sections, we demonstrate an application of this framework for improving urban landscapes for forest migration, using Greater Manchester, UK as a case study area.

4.4 Forest Migration Under Climate Change

Redistribution of tree species is one of the most serious challenges related to climate change (Dyderski et al. 2018). By the end of this century, more than half of plant species in Europe are projected to lose climatic suitability in existing conservation areas (Araujo et al. 2011). Similarly, in the western United States, 55% of the

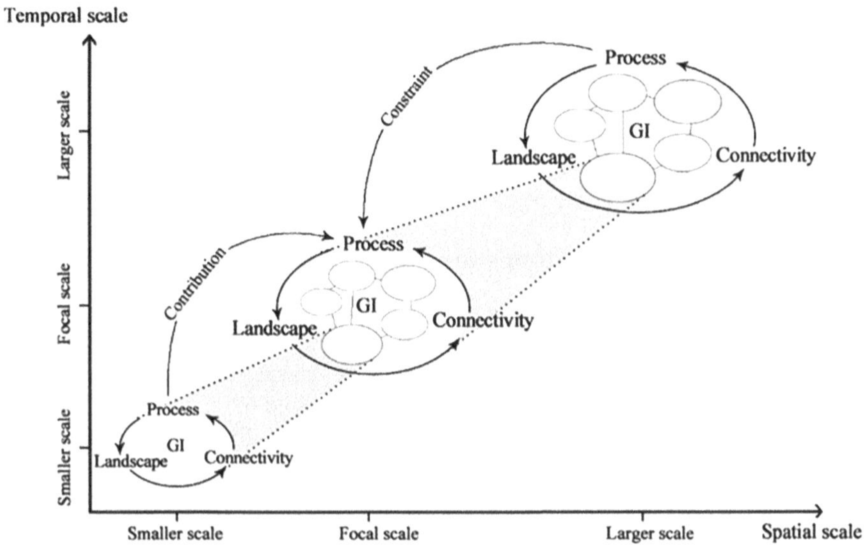

Fig. 4.3 The proposed conceptual framework of process-oriented green infrastructures. One-way arrows represent direct causal influences of one scale on another. (Source: Author)

landscape will exhibit climates that are incompatible with the same vegetation that it has today (Rehfeldt et al. 2006). Within this context, many tree species have started to shift their geographic distributions towards higher latitudes or altitudes (Woodall et al. 2009; Parmesan and Yohe 2003; Walther et al. 2002). Nevertheless, such movement cannot guarantee their survival if they are not able to move fast enough to keep pace with rapid climate change.

Large-scale urbanisation is expected to be a major obstacle to forest migration. Successful forest migration depends on effective seed dispersal between forest fragments, which is affected by the ways in which seed dispersers move and interact with the landscape (Clobert et al. 2012). In regions that have already been substantially transformed by human activities, such as metropolitan areas, fragmented landscapes and substantial anthropogenic barriers (e.g., agricultural land, buildings, and highways) dramatically reduce the population and activity of animals responsible for long-distance seed dispersal, restricting the ability of trees to migrate through the landscape. The pressure caused by landscape fragmentation will be exacerbated by rapid climate change. The rate of observed climate warming suggests that environmental conditions are changing faster than tree populations can adapt, acclimate or migrate (Montwe et al. 2018), indicating a high risk of species extinction (Petit et al. 2008).

The mismatch between climate shifts and species migrations will have radical commercial, biological, and climatological consequences. Forests are one of the most important ecosystems on the planet, which cover about 30% of the land surface (approximately 42 million square kilometres) (Bonan 2008) and remove about one-third of the anthropogenic CO_2 emissions (Grassi et al. 2017). Therefore, the failure of trees to track climate change will not only lead to the loss of wood resources, but also further contribute to climate warming through reduced carbon sequestration (Montwe et al. 2018). Furthermore, due to the importance of trees for sustaining biodiversity, delays in forest migration could also slow the movement of animals that depend on them for habitat or food (Hampe 2011).

4.5 Urban Stepping Stones for Migration

To help tree species traverse existing human-created barriers, increasing the functional connectivity of urban landscapes is expected to be a necessary action (De La Pena-Domene et al. 2016). Interconnected urban landscapes could enhance the probability of seed dispersal by facilitating the movement of seed dispersers.

A common strategy to increase landscape connectivity is constructing continuous corridors between habitat areas (Hansson et al. 2014). However, for tree species which depends on long-distance seed dispersal, corridors may not be effective as connectivity providers (Pearson and Dawson 2005; Robillard et al. 2015). Recent studies suggest that seed dispersal events would be unlikely to occur when dispersers use corridors connecting distant habitats, because corridors might induce dispersers to move large distances along them while increasing the likelihood that seeds are deposited in the corridor before reaching the habitats that are suitable for establishing new populations (Pouzols and Moilanen 2014). In addition, since different species move as individuals along their own trajectories rather than as communities, increasing connectivity by creating corridors is unlikely to provide a universal solution for all species affected by climate change. Furthermore, numerous constraints in urban environments also make the creation of large linear green spaces difficult. As such, an alternative strategy of stepping stones is recommended.

Compared with corridors, stepping stones might be more effective in terms of promoting seed dispersal. Stepping stones refer to small, scattered patches of vegetation that create potential dispersal paths for the movement of species across open spaces (De La Pena-Domene et al. 2016). In regions with intensive human activity, stepping stones, such as street trees and forest remnants, can play a pivotal role in the persistence and movement of seed dispersers by mitigating the negative effects of landscape fragmentation (Saura et al. 2014; Pena et al. 2017; James Barth et al. 2015). Indeed, empirical evidence suggests that birds and mammals prefer stepping stones to corridors when moving across human-modified landscapes (Doerr et al. 2010). Moreover, according to the Cities and Biodiversity Outlook (Secretariat of the Convention on Biological Diversity 2012), there is substantial room for increasing tree cover in most cities around the world, indicating a great potential for creating stepping-stone patches. Therefore, improving connectivity by adding stepping stones is suggested to be an effective and practical solution for facilitating seed dispersal in the short term and species range shifts in the long term.

Based on the framework of process-oriented green infrastructures (Fig. 4.3), we develop a framework for the addition of stepping stones (Fig. 4.4). Since the aim of

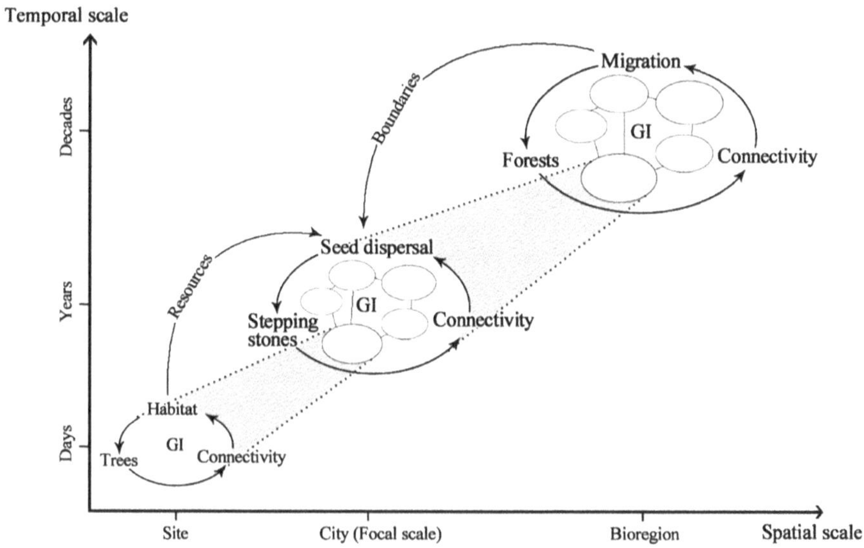

Fig. 4.4 The proposed framework of green infrastructure (GI) for forest migration. (Source: Author)

this study is to facilitate seed dispersal through the city (focal scale), bioregion (larger scale) and site (smaller scale) are identified as two adjacent scales. Here, the bioregion is defined as the reproductive region within which similar ecological and climatic characteristics are found. Effective seed dispersal is influenced by the landscapes and ecological processes at both bioregion and site scales: large forest fragments at the bioregion scale determines the directions and pathways (boundaries) of migration imposed by climate change, while trees at the site scale offer food and shelters (resources) that are attractive to seed dispersers (Fig. 4.5). This process-oriented framework is then applied to a case study in Greater Manchester.

4.6 Process-Based Design

Greater Manchester is a metropolitan region of approximately 127,600 ha in North West England, UK (Fig. 4.6). The region comprises ten districts: Bolton, Bury, Manchester, Oldham, Rochdale, Stockport, Tameside, Trafford, Wigan, and Salford. The total forest area in Greater Manchester is about 4695 ha, only representing 3.7% of the land area. Broadleaved forests are the dominant woodland type, representing 74.6% of all woodland, followed by mixed forest 8.0% and conifer forest 7.8%. In such highly fragmented landscapes, it is important to not only preserve existing forest fragments but also to improve landscape connectivity for the range shift of trees as a response to climate change. According to the Forestry Commission (see https://www.forestry.gov.uk/fr/infd-837f9j), a number of tree species need to

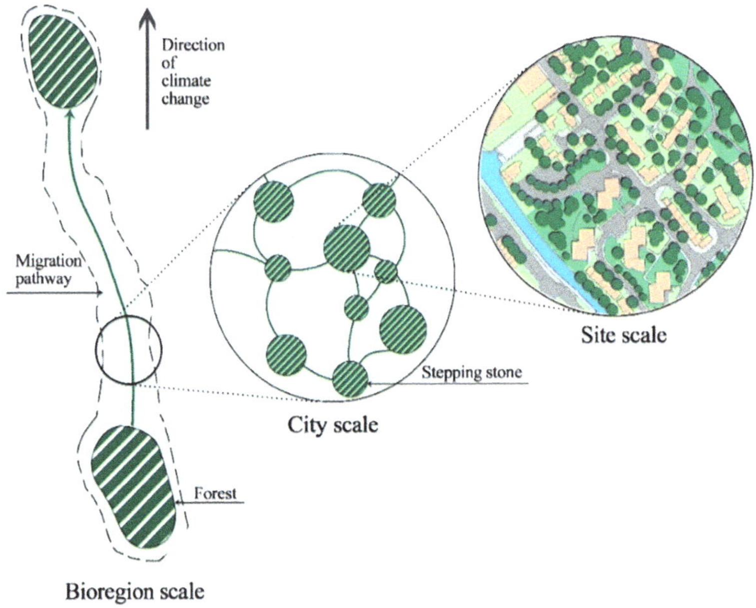

Fig. 4.5 An illustration of the landscapes and processes at three spatial scales in relation to forest migration. (Source: Author)

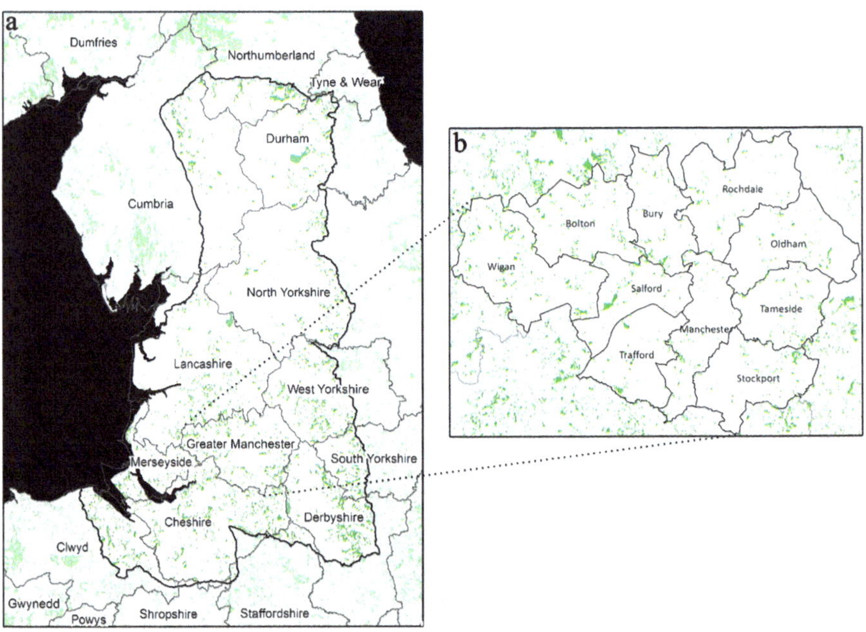

Fig. 4.6 Greater Manchester and its bioregion. (**a**) The bioregion of Greater Manchester. (Source: Author); (**b**) Greater Manchester. (Source: Author)

migrate through Greater Manchester in this century, including European larch (*Larix decidua*), Sitka spruce (*Picea sitchensis*), sweet chestnut (*Castanea sativa*), lodgepole pine (*Pinus contorta*), Scots pine (*Pinus sylvestris*), sessile oak (*Quercus petraea*), and beech (*Fagus*). Most of them are dispersed by frugivorous birds. Here, Eurasian jay (*Garrulus glandarius*) is considered as the main seed disperser for the demonstration of our approach, although there are a number of seed dispersers available in the case study area, such as Eurasian siskin (*Spinus spinus*), coal tit (*Periparus ater*) and grey squirrel (*Sciurus carolinensis*).

4.6.1 Forest Migration at the Bioregion Scale

Studies at the scale of bioregion aim at identifying directions of forest migration and areas where efforts to improve landscape connectivity should be concentrated. The range shift of species in the long term depends on the connectivity between habitats that are suitable at present and those that will be suitable in the future. Since climate change is considered to be the main factor shaping the distribution of tree species, improving landscape connectivity along the expected direction of climate change is suggested to be important for forest migration.

Given the considerable uncertainty and variability in projections of future climate change, the direction of climate change is identified based on spatial temperature gradients rather than present/future climate conditions. Temperature gradients over extensive geographic areas (from several kilometres to several hundred kilometres) are driven largely by topography and are expected to maintain their geographic directions as climate changes (Daly 2006). Therefore, if forest fragments of different temperatures are connected along temperature gradients, forest trees can move to the nearest fragments with relatively cooler present-day climates as the climate warms and continue occupying their climatic niches (Nunez et al. 2013). In this respect, landscape connectivity along temperature gradients could also offer flexibility to the migration process if the climate warms more or less than projected.

Multiple bioclimatic variables are taken into account in the analyses of temperature gradients, considering that the distribution of different tree species might be driven by different variables. For example, the range of Aspens is most influenced by the maximum temperature of the warmest month (Rehfeldt et al. 2015), while Palms are sensitive to the minimum temperature of the coldest month (Walther et al. 2007). For the sake of simplicity, we take four extreme weather variables from the WorldClim Version2 (Fick and Hijmans 2017) as the key climatic determinants of species distributions, including the maximum temperature of the warmest month (BIO5), the minimum temperature of the coldest month (BIO6), mean temperature in the warmest month (BIO10), and mean temperature in the coldest month (BIO11) (Walther et al. 2005). These extreme weathers are more important than mean annual temperature, which only have a minimal impact on tree species (Park and Talbot 2018).

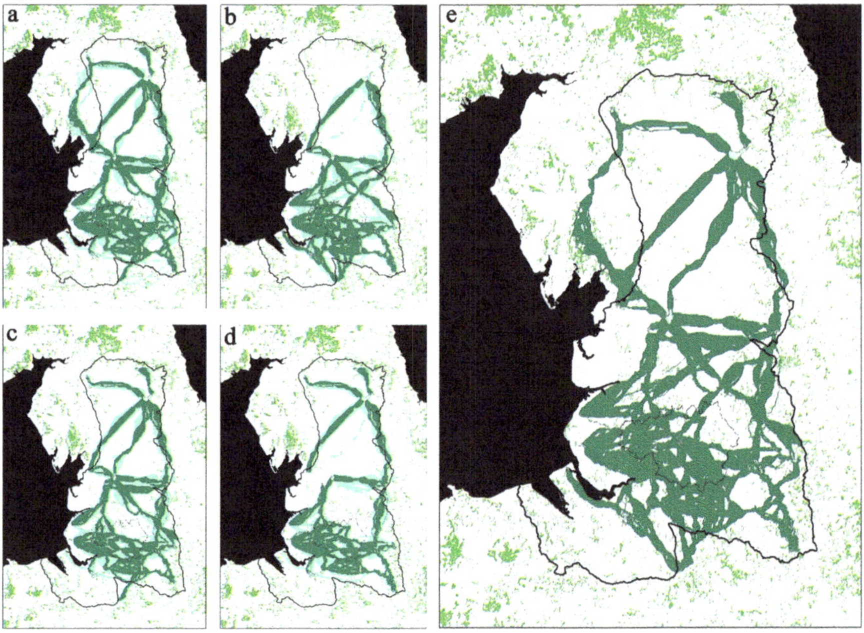

Fig. 4.7 Forest migration pathways along climatic gradients at the bioregion scale. (**a**) Pathways along BIO5, (**b**) Pathways along BIO6, (**c**) Pathways along BIO10, (**d**) Pathways along BIO11, and (**e**) Combined pathways of a–d. (Source: Author)

Accordingly, migration pathways between forest fragments are modelled along the temperature gradients of the four bioclimatic variables (Fig. 4.7), using an ArcGIS-based tool, climate linkage mapper (Nunez et al. 2013). The extent of the pathways is also mapped to identify the potential areas within which the foraging activities of seed dispersers are expected to contribute most to forest migrations.

4.6.2 Foraging Activity at the Site Scale

Although small-scale landscapes might not be able to support viable populations of seed dispersers in the long term given the scarcity of resources, they can provide critical resources for nesting and foraging on a daily basis (Holling 1992). Hence, studies at the site scale focus on identification of existing habitat patches for seed dispersers and potential reforestation sites that could serve as habitat for them.

In highly fragmented landscapes, birds that cannot find habitats large enough to support their resource requirements may be able to utilise nearly continuous fragments of suitable habitat as a single patch of habitat (Albert et al. 2017). Accordingly, we aggregate small, scattered urban woodlands in Greater Manchester into 869 large, contiguous habitat patches, based on the minimum habitat requirement (4 ha) of Eurasian jays. Details of this method can be found in Han and Keeffe (2019).

Besides, to identify potential habitat patches for improving connectivity, we simulate a reforestation campaign in the study area. It is suggested that tree-planting campaigns without explicit spatial targets but enacted at a large enough scale are able to accelerate the migration rate of a forest biome (Lazarus and McGill 2014). Two famous examples are the Green Belt Movement in Kenya founded by the late Nobel Laureate Prof. Wangari Maathai and the Billion Tree Campaign launched by the United Nations Environment Programme. In Greater Manchester, particularly, private gardens and streets are considered to be available for tree planting (Fig. 4.8). Two reforestation strategies are proposed:

- Strategy A: planting one tree (with crown radius of 4 m) in each garden bigger than 99 m^2 (the median size of all the gardens) and every 10 meters along each street;
- Strategy B: planting one tree (with crown radius of 4 m) in each garden and every 10 meters along each street.

Strategy A results in 1,350,616 trees planted in gardens and streets, only 23,455 (1.7%) of which contribute to 233 additional habitat patches for Eurasian jays (Fig. 4.9a). Comparatively, 800 additional habitat patches are yielded in strategy B, resulted from 288,590 (12.5%) of 2,303,357 trees planted in gardens and streets

Fig. 4.8 An example of the potential reforestation sites in Greater Manchester. (Source: Author)

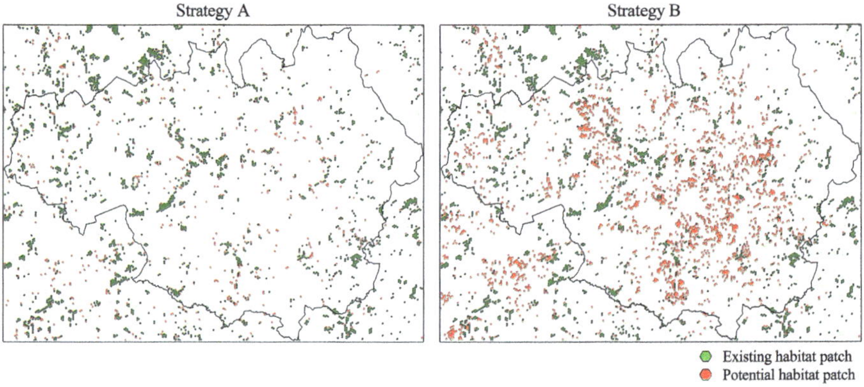

Fig. 4.9 Existing and potential habitat patches for Eurasian jays. (Source: Author)

(Fig. 4.9b). These patches might act as stepping stones that help integrate the dispersal networks of Eurasian jays, promoting seed dispersal at the city scale.

4.6.3 Seed Dispersal at the City Scale

Although habitat patches at the site scale could improve landscape connectivity and thus facilitate the movement of seed dispersers, not all of them are able to act as stepping stones for effective seed dispersal under climate change. According to the framework (Fig. 4.4) in Sect. 4.5, efforts to improve connectivity should be concentrated within projected migration pathways (Fig. 4.7e) where the dispersal of seeds can contribute most to forest migration. As a result, habitat patches within regional migration pathways are identified as stepping stones (Fig. 4.10). This results in 151 and 492 new stepping stones in strategy A and B, respectively.

To evaluate the potential effect of these new stepping stones, we calculate the accessibility of seed-dispersal networks and the corresponding permeability of the urban matrix. Seed-dispersal networks are composed of the movement paths of dispersers across non-habitat areas, functionally connecting scattered habitat patches. Methods for mapping seed-dispersal networks are described in Han and Keeffe (2019). We use a graph-based approach, Space Syntax, to assess the accessibility of each path, which in turn is used as a weight to calculate the kernel density of the dispersal networks, with a search distance of the daily foraging limit (1 km) of Eurasian jays (Fig. 4.11). Areas with high density indicate a relatively high probability of movement, and thus can be regarded as highly permeable to seed dispersal.

As is shown in the figure, both reforestation strategies in Greater Manchester lead to a significant improvement in the ability of urban landscapes to aid forest migration. Strategy A contributes to a 30.8% increase in the average permeability of

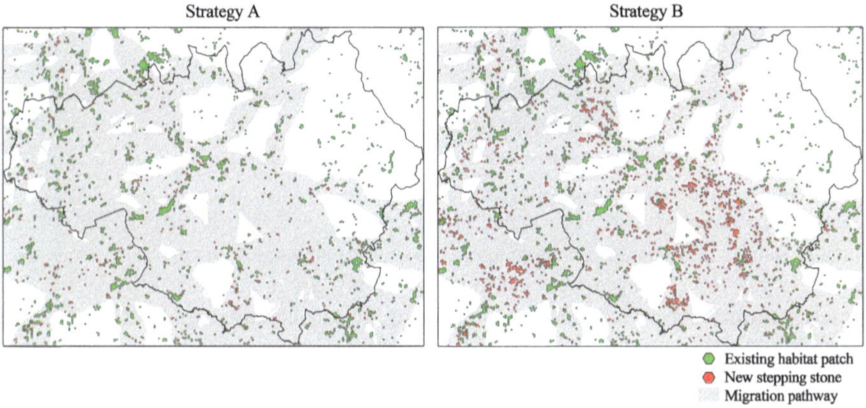

Fig. 4.10 Additional stepping stones for seed dispersal. (Source: Author)

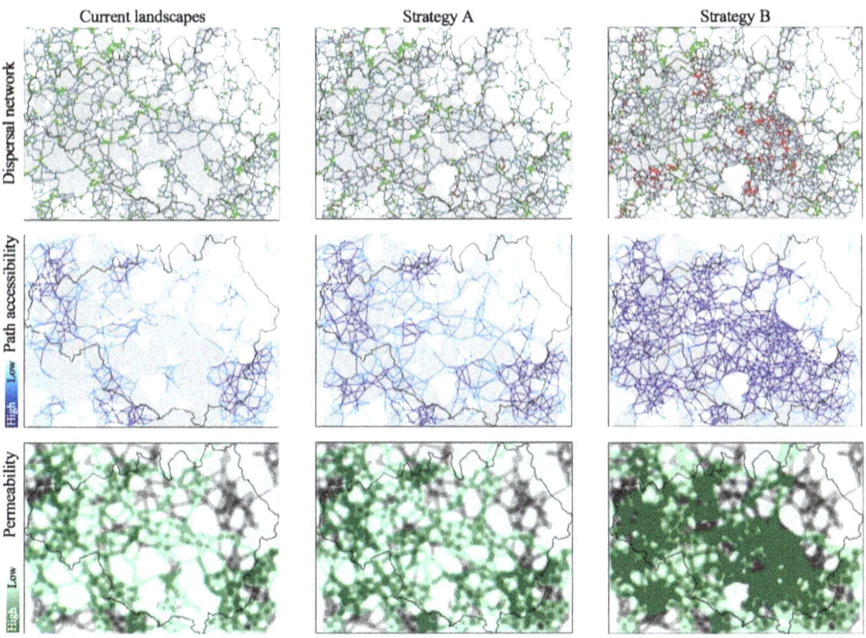

Fig. 4.11 Seed-dispersal networks, path accessibility, and landscape permeability of Greater Manchester. (Source: Author)

the urban matrix within regional migration pathways, in contrast to strategy B which improves permeability dramatically by 200% on average. This is because when all the gardens are considered for tree planting regardless of their sizes, the small ones can generate more aggregated tree canopies and thus more stepping stones for seed dispersers.

4.7 Conclusion

This paper proposes a process-based design approach that allows landscape designers and architects to re-visualise urban landscapes as a series of interconnected stepping stones, which in turn allow for a more piecemeal form of design to improve urban landscapes for climate adaptation. Graph analysis is used in the approach to map the connectivity between urban landscapes as well as the permeability of the urban matrix. Moreover, the process-based design approach incorporates both top-down and bottom-up approaches into the design of green infrastructure. While targeted planting at specific locations (within migration pathways) assumes a top-down approach to climate adaptation, such adaptation is conducted through a bottom up approach in which people act locally (e.g., planting trees in their own gardens). From this perspective, a process-oriented project might be more practical and manageable than traditional goal-oriented ones.

Our exploratory experiments in Greater Manchester suggest that ecological processes at a metropolitan scale can be facilitated by a large number of small landscape interventions at site scales. While this chapter focuses on Eurasian jays, our further experiments considering other seed dispersers (including Eurasian siskins, coal tits, and grey squirrels) show that different dispersers offer different ways of seeing the city as landscape networks, as a result of their different dispersal abilities.

References

Albert CH, Rayfield B, Dumitru M, Gonzalez A (2017) Applying network theory to prioritize multispecies habitat networks that are robust to climate and land-use change. Conserv Biol 31:1383–1396

Araujo MB, Alagador D, Cabeza M, Nogues-Bravo D, Thuiller W (2011) Climate change threatens European conservation areas. Ecol Lett 14:484–492

Baguette M, Van Dyck H (2007) Landscape connectivity and animal behavior: functional grain as a key determinant for dispersal. Landsc Ecol 22:1117–1129

Benedict MA, McMahon ET (2006) Green infrastructure: linking landscapes and communities. Island Press, Washington

Bonan GB (2008) Forests and climate change: forcings, feedbacks, and the climate benefits of forests. Science 320:1444–1449

Clobert J, Baguette M, Benton TG, Bullock JM (2012) Dispersal ecology and evolution. Oxford University Press, Oxford

Colding J (2007) ‘Ecological land-use complementation’ for building resilience in urban ecosystems. Landsc Urban Plan 81:46–55

Daly C (2006) Guidelines for assessing the suitability of spatial climate data sets. Int J Climatol 26:707–721

De La Pena-Domene M, Minor ES, Howe HF (2016) Restored connectivity facilitates recruitment by an endemic large-seeded tree in a fragmented tropical landscape. Ecology 97:2511–2517

Dilt TE, Weisberg PJ, Leitner P, Matocq MD, Inman RD, Nussear KE, Esque TC (2016) Multiscale connectivity and graph theory highlight critical areas for conservation under climate change. Ecol Appl 26:1223–1237

Doerr VAJ, Doerr ED, Davies MJ (2010) Does structural connectivity facilitate dispersal of native species in Australia's fragmented terrestrial landscapes? CEE review. Collaboration for Environmental Evidence
Douglas I, Ravetz J (2011) Urban ecology—the bigger picture. In: Niemelä J, Breuste JH, Elmqvist T, Guntenspergen G, James P, McIntyre NE (eds) Urban ecology: patterns, processes, and applications. Oxford University Press, New York
Dullinger S, Dendoncker N, Gattringer A, Leitner M, Mang T, Moser D, Mücher CA, Plutzar C, Rounsevell M, Willner W, Zimmermann NE, Hülber K (2015) Modelling the effect of habitat fragmentation on climate-driven migration of European forest understorey plants. Divers Distrib 21:1375–1387
Dyderski MK, Paź S, Frelich LE, Jagodziński AM (2018) How much does climate change threaten European forest tree species distributions? Glob Chang Biol 24:1150–1163
Fick SE, Hijmans RJ (2017) Worldclim 2: New 1-km spatial resolution climate surfaces for global land areas. Int J Climatol *37:4302–4315*
Forman RT (2014) Urban ecology: science of cities. Cambridge University Press, Cambridge
Grassi G, House J, Dentener F, Federici S, Den Elzen M, Penman J (2017) The key role of forests in meeting climate targets requires science for credible mitigation. Nat Clim Chang 7:220–226
Hampe A (2011) Plants on the move: the role of seed dispersal and initial population establishment for climate-driven range expansions. Acta Oecol 37:666–673
Han Q, Keeffe G (2019) Mapping the flow of forest migration through the city under climate change. Urban Plan 4:139–151
Hansson L-A, Åkesson S, Brönmark C, Chapman BB, Hedenström A, Hollander J, Johansson C, Liedvogel M, Lindström Å, Lundberg M (2014) A synthesis of animal movement across scales. In: Hansson L-A, Åkesson S (eds) Animal movement across scales. Oxford University Press, Oxford
Hejkal J, Buttschardt TK, Klaus VH (2016) Connectivity of public urban grasslands: implications for grassland conservation and restoration in cities. Urban Ecosyst 20:511–519
Holling CS (1992) Cross-scale morphology, geometry, and dynamics of ecosystems. Ecol Monogr 62:447–502
James Barth B, Ian Fitzgibbon S, Stuart Wilson R (2015) New urban developments that retain more remnant trees have greater bird diversity. Landsc Urban Plan 136:122–129
Keeffe GP (2014) Super Suburbia: developing resilience by closing resource cycles in North Belfast. 2nd international conference on urban sustainability and resilience, London
Lafortezza R, Davies C, Sanesi G, Konijnendijk CC (2013) Green Infrastructure as a tool to support spatial planning in European urban regions. iForest Biogeosci For 6:102–108
Lazarus ED, McGill BJ (2014) Pushing the pace of tree species migration. PLoS One 9:e105380
Lovell ST, Taylor JR (2013) Supplying urban ecosystem services through multifunctional green infrastructure in the United States. Landsc Ecol 28:1447–1463
Maes J, Thijssen M, Raynal J, Zulian G, Günther S (2019) Enhancing Resilience Of Urban Ecosystems through Green Infrastructure (EnRoute). European Commission
Marcotullio PJ, Boyle G (2003) Defining an ecosystem approach to urban management and policy development. United Nations University Institute of Advanced Studies (UNU/IAS, Tokyo
Montwe D, Isaac-Renton M, Hamann A, Spiecker H (2018) Cold adaptation recorded in tree rings highlights risks associated with climate change and assisted migration. Nat Commun:9, 1574
Newman P, Jennings I (2012) Cities as sustainable ecosystems: principles and practices. Island Press, Washington
Nunez TA, Lawler JJ, McRae BH, Pierce DJ, Krosby MB, Kavanagh DM, Singleton PH, Tewksbury JJ (2013) Connectivity planning to address climate change. Conserv Biol 27:407–416
O'Neill RV (1986) A hierarchical concept of ecosystems. Princeton University Press, Princeton
Park A, Talbot C (2018) Information underload: ecological complexity, incomplete knowledge, and data deficits create challenges for the assisted migration of forest trees. Bio Science 68:251–263
Parmesan C, Yohe G (2003) A globally coherent fingerprint of climate change impacts across natural systems. Nature 421:37–42

Pauleit S, Liu L, Ahern J, Kazmierczak A (2011) Multifunctional green infrastructure planning to promote ecological services in the city. In: Niemelä J, Breuste JH, Elmqvist T, Guntenspergen G, James P, McIntyre NE (eds) Urban ecology. Oxford University Press, Oxford

Pearson RG, Dawson TP (2005) Long-distance plant dispersal and habitat fragmentation: identifying conservation targets for spatial landscape planning under climate change. Biol Conserv 123:389–401

Pena JCDC, Martello F, Ribeiro MC, Armitage RA, Young RJ, Rodriguez M (2017) Street trees reduce the negative effects of urbanization on birds. PLoS One 12:e0174484

Petit RJ, Hu FS, Dick CW (2008) Forests of the past: a window to future changes. Science 320:1450–1452

Pouzols FM, Moilanen A (2014) A method for building corridors in spatial conservation prioritization. Landsc Ecol 29:789–801

Rayfield B, Pelletier D, Dumitru M, Cardille JA, Gonzalez A, Travis J (2016) Multipurpose habitat networks for short-range and long-range connectivity: a new method combining graph and circuit connectivity. Methods Ecol Evol 7:222–231

Rehfeldt GE, Crookston NL, Warwell MV, Evans JS (2006) Empirical analyses of plant-climate relationships for the western United States. Int J Plant Sci 167:1123–1150

Rehfeldt GE, Worrall JJ, Marchetti SB, Crookston NL (2015) Adapting forest management to climate change using bioclimate models with topographic drivers. For Int J Forest Res 88:528–539

Robillard CM, Coristine LE, Soares RN, Kerr JT (2015) Facilitating climate-change-induced range shifts across continental land-use barriers. Conserv Biol 29:1586–1595

Saura S, Bodin Ö, Fortin M-J, Frair J (2014) Stepping stones are crucial for species' long-distance dispersal and range expansion through habitat networks. J Appl Ecol 51:171–182

Scholes RJ, Reyers B, Biggs R, Spierenburg MJ, Duriappah A (2013) Multi-scale and cross-scale assessments of social–ecological systems and their ecosystem services. Curr Opin Environ Sustain 5:16–25

Secretariat of the Convention on Biological Diversity (2012) Cities and Biodiversity Outlook. Montreal

Staddon P, Lindo Z, Crittenden PD, Gilbert F, Gonzalez A (2010) Connectivity, non-random extinction and ecosystem function in experimental metacommunities. Ecol Lett 13:543–552

Tomiolo S, Ward D (2018) Species migrations and range shifts: A synthesis of causes and consequences. Persp Plant Ecol Evol Systemat 33:62–77

United Nations (2012) World population prospects: The 2011 revision, highlights and advance tables. In: Department of Economic and Social Affairs, P.D. (Ed.). New York: United Nations Publications

Vimal R, Mathevet R, Thompson JD (2012) The changing landscape of ecological networks. J Nat Conserv 20:49–55

Walther GR, Post E, Convey P, Menzel A, Parmesan C, Beebee TJ, Fromentin JM, Hoegh-Guldberg O, Bairlein F (2002) Ecological responses to recent climate change. Nature 416:389–395

Walther GR, Berger S, Sykes MT (2005) An ecological 'footprint' of climate change. Proc R Soc B Biol Sci 272:1427–1432

Walther GR, Gritti ES, Berger S, Hickler T, Tang Z, Sykes MT (2007) Palms tracking climate change. Glob Ecol Biogeogr 16:801–809

Wessely J, Hülber K, Gattringer A, Kuttner M, Moser D, Rabitsch W, Schindler S, Dullinger S, Essl F (2017) Habitat-based conservation strategies cannot compensate for climate-change-induced range loss. Nat Clim Chang 7:823–827

Woodall CW, Oswalt CM, Westfall JA, Perry CH, Nelson MD, Finley AO (2009) An indicator of tree migration in forests of the eastern United States. For Ecol Manag 257:1434–1444

Wright H (2011) Understanding green infrastructure: the development of a contested concept in England. Local Environ 16:1003–1019

Wu J, Li H (2006) Concepts of scale and scaling. In: Wu J, Jones KB, Li H, Loucks OL (eds) Scaling and uncertainty analysis in ecology. Springer, Dordrecht, pp 3–15

Chapter 5
Landscape First! Nature-Driven Design for Sydney's Third City

Rob Roggema

Abstract Urbanization around the world has taken a flight towards rapid, sometimes uncontrolled growth. Megacities expanded, whilst erasing the developable area and adjusting the existing landscape to artificial water and nature systems. This rampant expansion often leads to monotonous new neighborhoods, often dominated by high rise, or extensive urban sprawl. The financial benefits often dominate the quality of the development. These widespread practices of urban development are hard to modify, to the detriment of sustainability. In this chapter the state of the art of urban development in Sydney and its associated problems are described first. An alternative approach, to take the landscape as the starting point of urbanization is then proposed, before conclusions are drawn.

Keywords Third city · Landscape · Ecosystem · Water system · Sydney

5.1 Introduction

Urbanization around the world has taken a flight towards rapid, sometimes uncontrolled growth. Megacities expanded, whilst erasing the developable area and adjusting the existing landscape to artificial water and nature systems. This rampant expansion often leads to monotonous new neighborhoods, often dominated by high rise, or extensive urban sprawl. The financial benefits often dominate the quality of the development. These widespread practices of urban development are hard to modify, to the detriment of sustainability. In this chapter the state of the art of urban development in Sydney and its associated problems are described first. An alternative approach, to take the landscape as the starting point of urbanization is then proposed, before conclusions are drawn.

R. Roggema (✉)
Research Centre for the Built Environment NoorderRuimte,
Hanze University of Applied Sciences, Groningen, The Netherlands

CITTA IDEALE, Office for Adaptive Planning, Wageningen, The Netherlands
e-mail: r.e.roggema@pl.hanze.nl; rob@cittaideale.eu

R. Roggema (ed.), *Nature Driven Urbanism*, Contemporary Urban Design Thinking, https://doi.org/10.1007/978-3-030-26717-9_5

5.2 Current Urbanism in the Sydney Region

Sydney is well known for its Opera House and Harbour Bridge, the beaches of Bondi and Manly. These icons of the largest city of Australia belong to the oldest parts of the city, constructed during the last two centuries and most of it is home of the Sydney City Council. The unbridled growth to the West has led to ongoing sprawl, build since the 1960s. This shapeless and borderless urbanization has brought the Greater Sydney Commission to define the Sydney Metropolitan Region as a city of three cities (GSC 2018): The Eastern Harbour city, the Central River City around Parramatta, and the new Western Parkland City around the new Western Sydney Airport (Fig. 5.1).

At metropolitan level, this vision has caused some inconvenience as it brought to the fore that the metropole was not very well planned, as a matter of fact it was a metropole of over 30 councils, each with their own growth ambitions and policies. In response other strategic plans were made to underpin the importance to integrate land use and infrastructure planning (Infrastructure NSW 2018) and the long-term and regional focus on the customer, making good places, services and the economy, beyond transportation as such (TfNSW 2018). The question remains whether the emphasis on integration and broadening of the policy from an infrastructure/transportation perspective will change the current practice of urbanization towards more sustainability.

Meanwhile the planning of the so-called Third City is on its way. The Western Sydney City Deal formulates as its main ambitions to deliver, around the newly to be developed Western Sydney airport, 'a 30-minute city by delivering the North-South rail link, Create 200.000 jobs capitalizing on the new Aerotropolis and the new agribusiness district, skilling residents in the region, respecting and building local character through a livability program, and coordinate and innovate through a planning partnership, continuing to consist of all three levels of government' (Australian Government and Government of NSW 2018). In the Deal, the attention for sustainability, environmental values and development of green and the landscape is relatively underwhelmed. Most of the attention is oriented towards the development of the new Aerotropolis, a development with and around the new airport, including new industries, homes and amenity. To put it at its most modest, the new airport will get two tartan runways, around which a large number of logistics will be realized. A lot of homes cannot be planned here because of safety and noise reasons, and the number of jobs for people from the region is quite limited as the majority of logistics functions will be robotized and most of the high-level jobs at the airport will be taken by employees living outside the region. The best opportunity for local jobs lies in the development of the agribusiness precinct, where R&D in agriculture, intensified agricultural production, and local fertile soils could be used to increase food productivity.

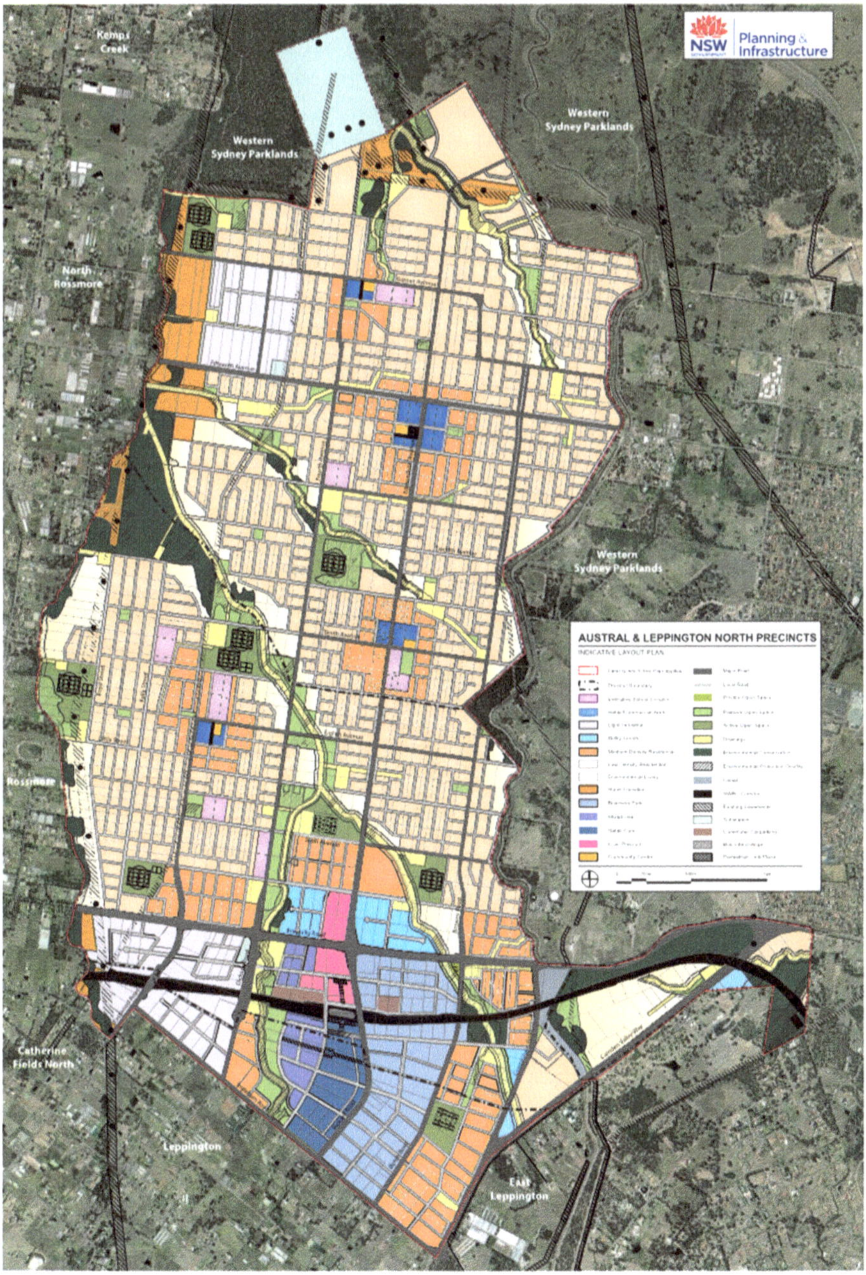

Fig. 5.1 Recent planning for the Western Sydney area: the plan for Austral & Leppington. (Source: Department of Planning and Infrastructure, State Government of NSW, 2017)

The fact this area will be home to over 1 million new residents, asks for a large-scale urban plan, in which future sustainability is the condition for newly built homes. In current practice however, short term practice is more dominant. In the Sydney urban context, it is very difficult to develop a long-term spatial strategy that is consistent for a longer period. Ad-hoc urbanism, driven by land-ownership, short term profits, a housing market under pressure, and the quick and efficient development process, leads to consistency but as an ever-continuing addition of new masses of housing. These new neighborhoods are mainly built in Sydney's west and consist of houses that are as large as possible on the plots they are built on, have black, heat-absorbing roofs, and are built according the 4-2-2- concept: four bedrooms, two bathrooms and a double garage. At the urban design level, the location of the station determines the lay-out of the neighborhood. The station is the center of the area and has all amenities, offices and multilevel multifunctional buildings next to it. The dogma to house most people within a radius of 500 meter from the station, means then that densities are highest here. It also means that the 4-2-2 housing finds its place in a convenient grid (80x40m or 80x80m) in the zones further away. The waterway and eventual riparian zones are seen as an obstructive stand in the way of more housing, which is then built with its back to the green zones, and in the lowest possible density. The plan for Austral/Leppington (Fig. 5.1) is one of the most recent plans designed according these principles.

In the end, when this form of urban development continues, and it does under the current market stress, new sprawl will be realized and a sea of identical houses in identical neighborhoods will be the result (Fig. 5.2). No problem if these neighborhoods were not so space consuming, eating green space and agricultural valuable grounds, were built sustainably and provided healthy environments for happy Sydneysiders. This Business as Usual model is based on a couple of principles (Fig. 5.3):

1. Circle around station and bus stop determines higher densities;
2. Outside the public transport circles other land-use, such as green is located, farthest away from the highest densities;
3. Road system forms a grid, as this is the most efficient way to cover as much space as possible with circles;
4. Coincidental waterways are shaped according the last row of circles, until the last meter used for housing;
5. Housing areas outside circles are left to being approached by car only.

The consequence of this way of developing the city is that a minimized number of people gets to enjoy the best quality: landscape amenity. In the urban designs the landscape has become the backyard of the city, while it is the place that holds most opportunities to enhance the quality of life and the value of real estate. And we don't offer these benefits to the majority of the people.

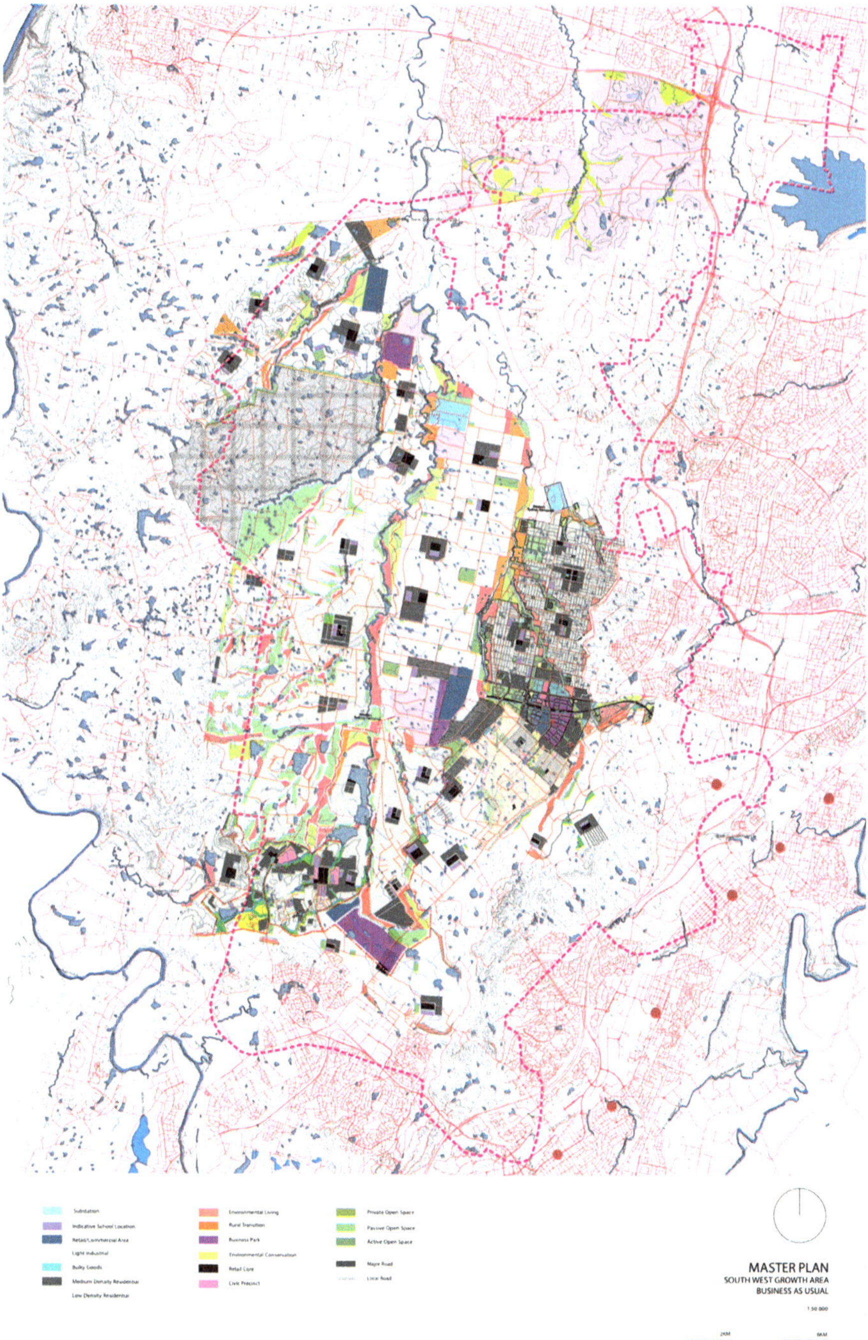

Fig. 5.2 An endlessly copied urban sprawl when policy is unchanged. (Source: Mushi et al. 2017)

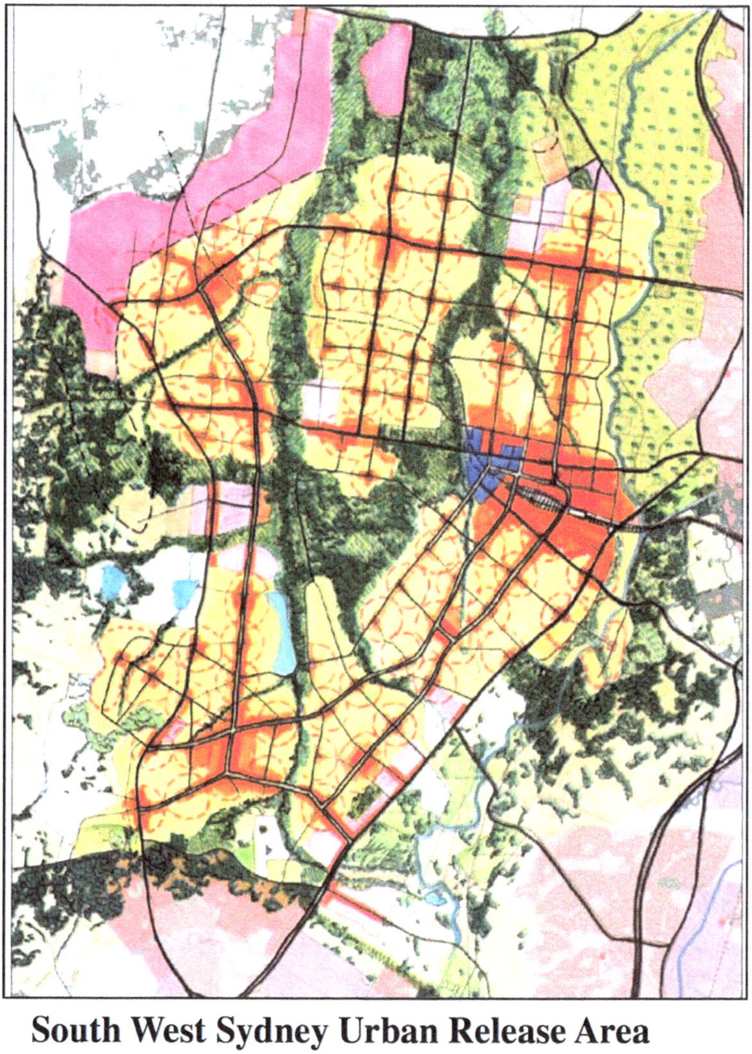

Fig. 5.3 Principles of business as usual urbanism. (Source: Department of Planning and Environment undated)

5.3 Problems Associated with Standard Urbanization

The Business as Usual urban model has a range of problems associated with it. In general, urbanization is now considered to have become one of the most important health challenges of the twenty-first century (World Health Organisation 2015), being associated with an increase in chronic and non-communicable conditions such as obesity, stress, poor mental health and a decline in physical activity (Dye

2008). It is not always clear whether these problems are solely determined by the urban form chosen, or that social aspects, price mechanisms and politics are more important causes of these problems. However, urban design, could it not prevent these problems only by design, the design of our urban environment, especially the distance to and accessibility of green spaces, could certainly contribute to shaping neighborhoods in which these problems are not exaggerated or positively influenced. A broad range of research has proven the role of green for many aspects of human life.

When a population becomes more urbanized, a higher green dose, both in frequency and duration, has a positive influence on mental health, social cohesion and physical behavior (Cox et al. 2018). People living in areas without access to nature were 1.27 times more likely to experience symptoms of depression (Gyeong-bok 2017).

Domestic gardens and parks have a positive influence on a range of health factors. The measured health-related outcomes of living close to green space (Jackson 2003; Gidlöf-Gunnarsson and Öhrström 2007; Kaplan 1995; Maas et al. 2006, 2009; Maller et al. 2006), having local access to naturalistic areas (Ward Thompson et al. 2014) and be able to undertake physical activity in nature (Bird 2007; Carrus et al. 2015; Marselle et al. 2014; Pretty et al. 2005; Tzoulas et al. 2007), include reduced levels of morbidity (Mitchell and Popham 2007). When people are living in areas with higher amounts of green spaces this reduces mortality by reducing cardiovascular disease (Gascon et al. 2016). Access to green space also reduces psychological stress (Mennis et al. 2018; Husqvarna Group 2013; Thompson et al. 2012). Greening a schoolyard for instance improves children's attention restoration during recess, but only after the schoolyard had already been greened for a longer period (Van Dijk-Wesselius et al. 2018) Spending time in a greener environment in general improves attention restoration (Li and Sullivan 2016; Sullivan and Chang 2011; Taylor et al. 2002). Green environments lead to increased social (Sullivan et al. 2004) and physical (Epstein et al. 2006) activity, in green schoolyards it is, in particular for girls, stimulating to become physically active (Van Dijk-Wesselius et al. 2018).

Patients with views of trees and greenery out their windows heal faster and with need for less medication (Cox et al. 2017). Moreover, in hospitals (Söderback et al. 2004; Marcus 2007), educational institutions (Lau and Yang 2009), places of work (Lottrup et al. 2013), conflict resolution (Krasny and Tidball 2015), in social education and transformation (Pudup 2008), and social cohesion (Kuo et al. 1998; Kuo and Sullivan 2001; Okvat and Zautra 2011) the restorative influence of gardens is made clear. What works in gardens and parks, might also be beneficial at a higher spatial scale, greening the city and creating greener urban systems would then extend these advantages to entire urban populations.

The influence of greenspace on children's spatial working memory and their cognitive functioning is positive. Spatial working memory is an important cognitive ability that is strongly related with academic achievement in children's, particularly mathematics performance (Flouri et al. 2018; Dovey 2018). The greening of schoolyards has a positive impact on children's appreciation of the schoolyards, and their cognitive and social well-being. Furthermore, it is and that it can support pro-social behavior amongst younger children. Children show greater appreciation of the schoolyard after their schoolyard has been greened, in particular younger children and girls. It also has a positive short-term impact on younger, but a negative impact on older children's prosocial orientation and greening a schoolyard is beneficial for children's social functioning, in particular for social support and self-reported peer problems (Van Dijk-Wesselius et al. 2018). Moreover, playing in green outdoor spaces fosters creative play and reduces symptoms of attention disorders in children (Shore 2017; Louv 2016).

Many of these neighborhoods are also susceptible to climate hazards. In the case of Western Sydney, the new to be developed neighborhoods have to suffer from heat waves in summer, face occasional floods due to heavy rainfall and flash flooding, and might be confronted with a high bushfire risk.

Finally, in the streets of these neighborhoods, trees are often absent or extremely small, and, due to maximization of the size of the house on the plot, front and backyards are small, paved or in use as a miniscule pool. A harsh environment, which increases, in conjunction with the black roofs, the urban heat island effect. The necessary aircon will be often in use, in itself adding to the household energy bills. Hence the non-natural environment increases the costs for residents, decreases their well-being and health, and devaluates the value of their real estate. A house in a tree-rich environment, close to green space is on average 50,000 Australian Dollar more worth (Swinbourne and Rozenwax 2018). Similarly, it has been calculated that one square kilometer of tree cover saves an average of US$83 per city resident (Endreny 2018).

So, there is abundant evidence to increase the role of nature and green spaces in the urban environment, as it benefits the economy, lowers the health insurance bills, reduces energy demand, improves the lives of residents, and mitigates climate impacts. Still, green is framed as expensive and relative useless space and therefore seen as the rest category when it comes to urban planning.

A radical different approach is therefore needed, in which the urban planning process is reversed (Roggema 2018), and green, landscape and ecology are taken as the starting point of the urban planning process. This is how the Urban Master Plan for the third city has been approached (Mushi et al. 2017).

5.4 Methodology

In order to allow for the landscape to play a crucial role in urban planning and design at the highest scales, that of the City Region Master Planning, a methodology is chosen in which all qualities of ecology, soil, and the water system can be brought to the fore, and the vulnerabilities can be brought to the surface. A regular urban development process would erase the existing landscape, because it is seen as just a bunch of green space, with low economic uses, such as agriculture or nature. The richness and value of different aspects of the landscape can only be seen when these are all separated in distinct maps of layers. This method is not new and has been developed in the 1960s (McHarg 1969) when ecological exhaustion and pollution of water systems and the landscape was widespread. A design-led approach in which all ecological values were separated put emphasis on their importance. In the Netherlands, learning from McHarg, the layer-approach was born (Van Schaick and Klaasen 2011), in which the landscape was subdivided in the underground/soil (abiotic), the ecology (biotic) and the occupation (urban/infrastructure) layers. Parallel to this thinking the value of ecology in the landscape was conceptualized in the so-called Casco-concept (Sijmons 1992). In this concept a distinction is made between the high-dynamic and low-dynamic land-uses. When combined the low-dynamic uses, such as water and ecology, always suffer from the high-dynamic ones. When these are spatially separated in the landscape the ecological and water networks could gain quality. The concept, mainly applied in rural areas, was later also transformed for urban environments. The strategy of the two networks (S2N) linked low-dynamic land-uses, such as water retention, ecology, organic agriculture, to the water network, while high-dynamic uses, such as heavy industry, energy plants, intensive agriculture, were connected to the traffic network. In between these sets of function urban living was positioned (Tjallingii 1995, 2015). Meanwhile the layer approach was enriched. First, a fourth layer was added, that of the sky, in which atmospheric functions and for instance solar energy were allocated (Tomásek 1979; Kristinsson 2012). Thinking in the same system the layer approach was then integrated and put in a design-oriented sequence. Design would start with the identification of networks (layer one), where these link one would find focal points (layer two), around which redundancy, or free space (unplanned) would provide the flexibility to react to unprecedented developments (layer three), before the natural resources could be located (layer four) and finally emergent occupation patterns were distinguished (layer five). This new layer approach was entitled Swarm Planning (Roggema 2012), and especially suited for dealing with climate adaptation in spatial planning and design. Recently, with the attention for smart cities, big data, IoT, Ai and VR, new layers seem to be required to understand our urban systems. Hence the cloud, IP-addresses, interfaces, users, networks are added to the set of layers (Van Timmeren and Henriquez 2015), which now have grown into an entire stack. Still, the basis of this approach lies in separation of in themselves valuable and vulnerable elements in the urban landscape, starting with the given and available functionalities of the landscape, in which precautionary spaces must be created to allow for space-requiring surprises that cannot be predicted, on top of which humankind is building its ever more advanced systems of infrastructure, technological and social. Both the Layer- and the Casco- approaches have been used, in combination, to develop a City Region Master Plan for the Badgerys Creek area in Western Sydney.

5.5 Landscape First

In order to create an alternative for the erasing way of urban growth currently happening in Western Sydney, the landscape needs to be torn apart in basic layers, which play a role in arranging future land use in a resilient way. This is an iterative process, in which the different layers are constantly adjusted, redesigned and revisited. As a first step, elevation, soil, vegetation, waterways, future flood-risk and ecological remnants are all mapped and used to first understand the nuanced sensitivities of life, and the reliability on available water, sunlight, shade, coolness and other factors that determine the chance the exiting plant- and animal life can coexist with future human occupation in the area. The landscape is there. An alternative way of developing takes the landscape as the basis and builds on an adaptive approach to decision making for planning. Taking decisions for the initial steps of development, without making future change impossible. Over time, decisions will then be made according the context in the future (technology, social capital, climate). The creeks and side streams, the contours and potential discharge of rainwater, and the natural vegetation form the basis for the Master Plan (Mushi et al. 2017). Therefore, the first step in the design process is to map the landscape, its elevation, water system and vegetation (Fig. 5.4).

Once the topography and the water system of the landscape is understood and mapped, its vulnerabilities can be explored. The vulnerability of the landscape is determined by the capacity of the landscape to adjust to new circumstances, and the external impacts on the landscape. In Western Sydney the landscape is vulnerable for external climate impacts. The creeks in the landscape occasionally flood, after heavy rains, and can reach levels up to nine meters above the normal. These floods occur mostly in the northern parts of the area. For the entire area the average flood, happening once in every one-hundred years is calculated and mapped, as is the so-called maximum potential flood (Bewsher 2004). These calculations, though still accepted as the prevailing ones, are reflecting the climate impacts as they were expected more than 15 years ago. If we want to sketch a realistic map of future flood impact in Western Sydney, we should exaggerate the space required to accommodate future flooding, because we know that heavy rainfall in a short period will occur more often and in a more severe fashion in the future. Therefore, the maps drawn for the flooding in Western Sydney take on average a twice as large area for potential maximum flood as the reports do (Fig. 5.5). This is an estimation only and should academically be investigated to get more accurate insights in expected dimensions.

Besides flooding other serious climate impacts in Western Sydney mainly regard to heat. With temperatures occasionally over 47 °C and expected to rise in the near future, the heat in summer, in combination with urbanization hence higher levels of concrete, rocks and asphalt, may lead to urban heat island effects, which raise the temperatures even further. It is evident these high temperatures will impact on human health, and increase the use of aircons, but it has also an influence on the growth of trees and plants, which becomes very difficult with these temperatures. This causes a downward loop, as less trees will increase the heat, which will decrease the growth of

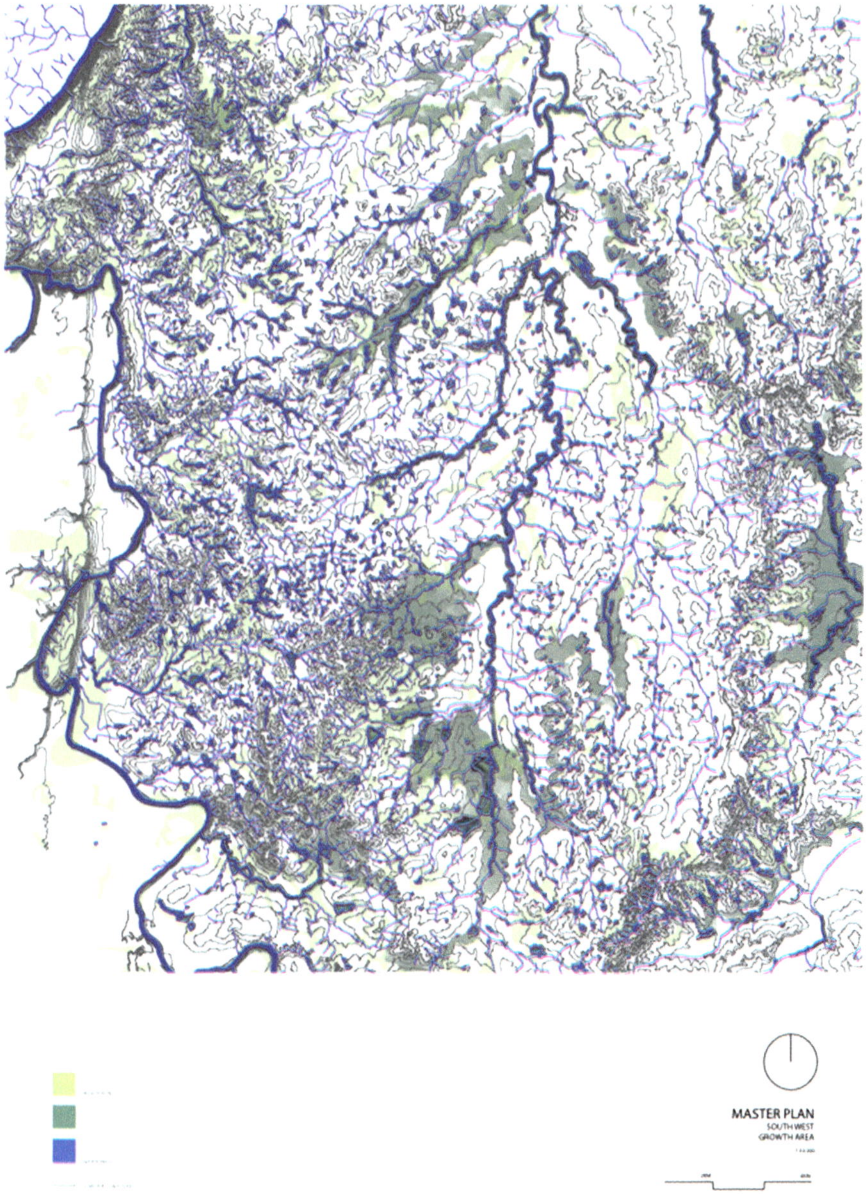

Fig. 5.4 Landscape with elevation, water and vegetation (Mushi et al. 2017)

trees, which, etcetera. The other impact of this is that drying vegetation increases the potential for bushfires in the area, especially during the dangerous summer period.

These vulnerabilities should be dealt with while designing Western Sydney. In order to give space to water and increase the cooling effect of vegetation (and water),

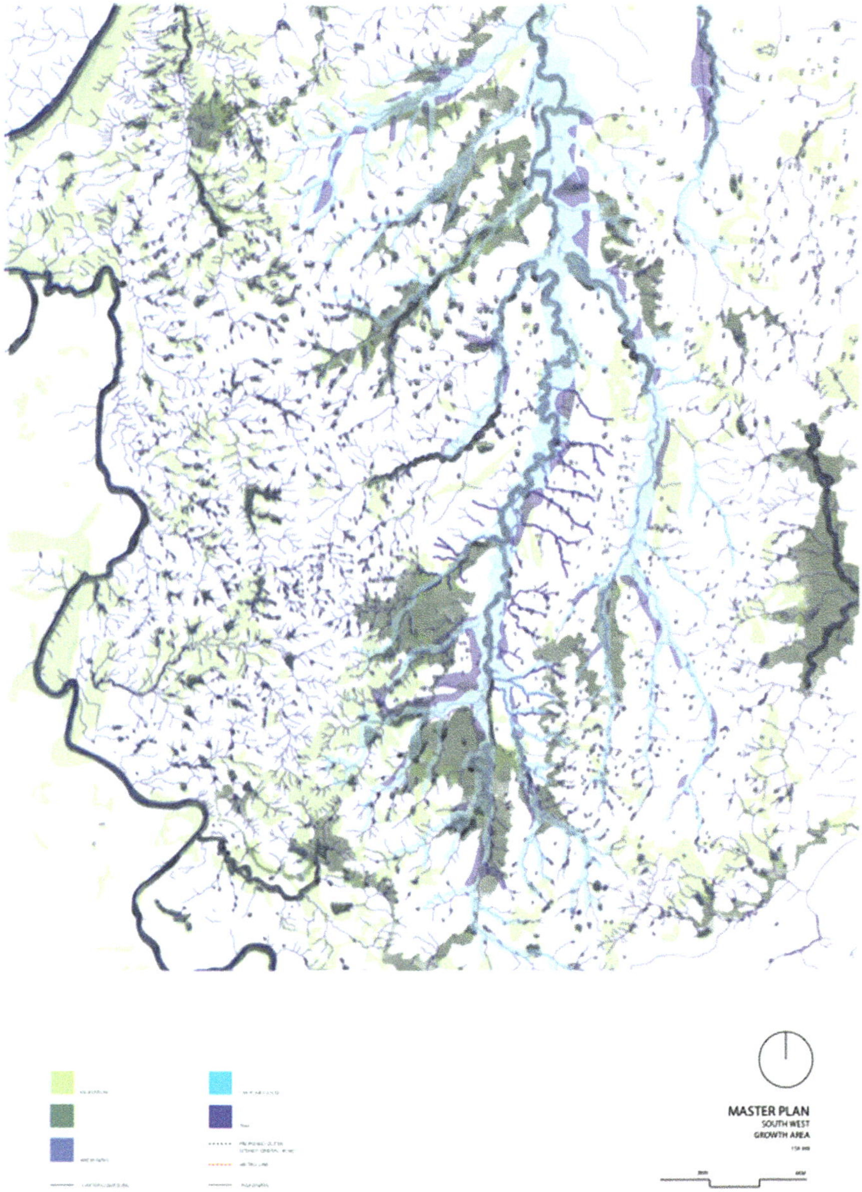

Fig. 5.5 Landscape with potential flood (Mushi et al. 2017)

the larger landscape system needs to be robust, and at the same time deeply intrude in the urban fabric. This is necessary to bring cool air from the surrounding landscape deeply into the urban precincts and makes it possible to create space for water after heavy rain, and also to discharge cleaned household water through the system.

The ambition to create a new lay-out for Western Sydney that intrinsically cools the living circumstances for future residents requires to accommodate flood water, store rainwater and keep cleaned household water in the creek system and smaller gullies. This way it determines the boundaries for urban development, the landscape offers the framework within which the future city can be embedded. The space needed for a satisfying cooling environment, in which vegetation, water and shade are large and rhizomically ramify itself in smaller branches of the landscape system, leaves the rest for urbanization in smaller and separated urban fields (Fig. 5.6), each approximately large enough for 10,000 houses.

In the initial phase of development these urban fields can be developed in a way that coincides with current delivery practice. This way existing land owners and developers, the market is not disturbed, and existing low-density neighborhoods are in first instance repeated. With a total area of 140km^2 and half of this space reserved for the cooling landscape, approximately 70 km^2 remains for urban development (e.g. 7000 ha). Applying an average low-density of 20 dwellings/ha there is a capacity of 140.000 dwellings or 350.000 people to be housed in the area (Fig. 5.7).

For every urban field a set of rules for good urbanism should be adhered. No matter where a new neighborhood is developed it should follow good urbanism ambitions. These ambitions can be reflected in six themes:

- Establish diversity. A lively city will emerge if housing types and densities are different, functions and land-use is mixed and urban spaces come in different sizes and atmospheres. This will attract a range of people with different backgrounds, a guarantee for a lively precinct.
- Create an equitable city. When the city offers equal access for everyone, no matter what background people are from, to housing and amenities the quality of life for everyone increases. It also does justice to the different lives people have, and ethnicity, level of education, income and gender do not matter.
- Strive for sustainability. A neighborhood needs to be safe for the hazards that climate change might bring, but also for criminality, it should be environmental-friendly with clean air, water and soil and no noise pollution. Moreover, transport and living should be energy efficient not using any fossil resources, and the neighborhood should be design in a water sensitive way, providing the space for locally growth of food. Finally, green spaces need to be located close to people and provide the conditions for a rich and biodiverse nature.
- Design a functional city. The city needs to work and be logical and understandable in its networks. It needs to be accessible for people with a range of transport means not the least public transport. There need to be sufficient schools of different types, shops and restaurants, and green space should be at a close distance where people live.
- Close cycles and become circular. In a circular city the flows of energy, water, nutrients and materials are minimized and when rest-flows are produced these should be recycled or reused within the urban environment. This way it is possible to create a city without waste, in which clean water, local energy and food is produced.

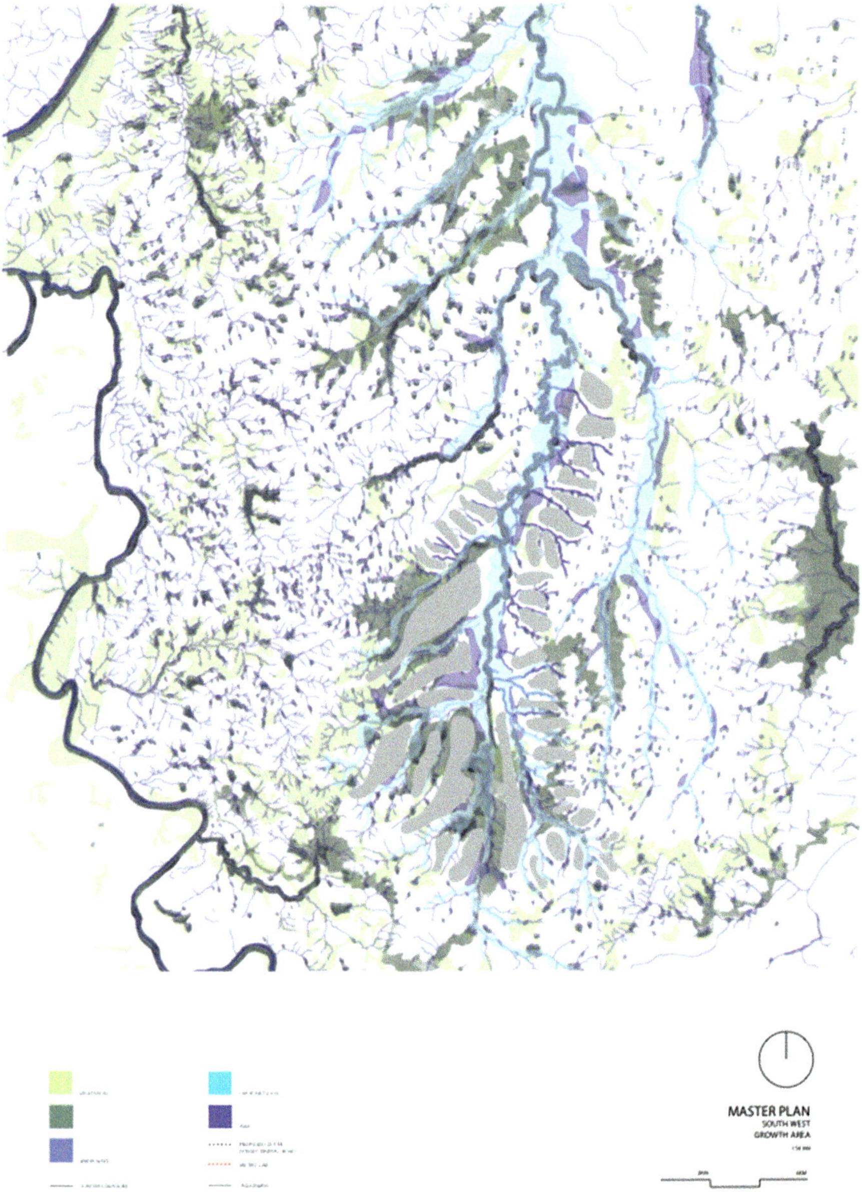

Fig. 5.6 Urban fields, determined by the space required to create a cooling landscape (Mushi et al. 2017)

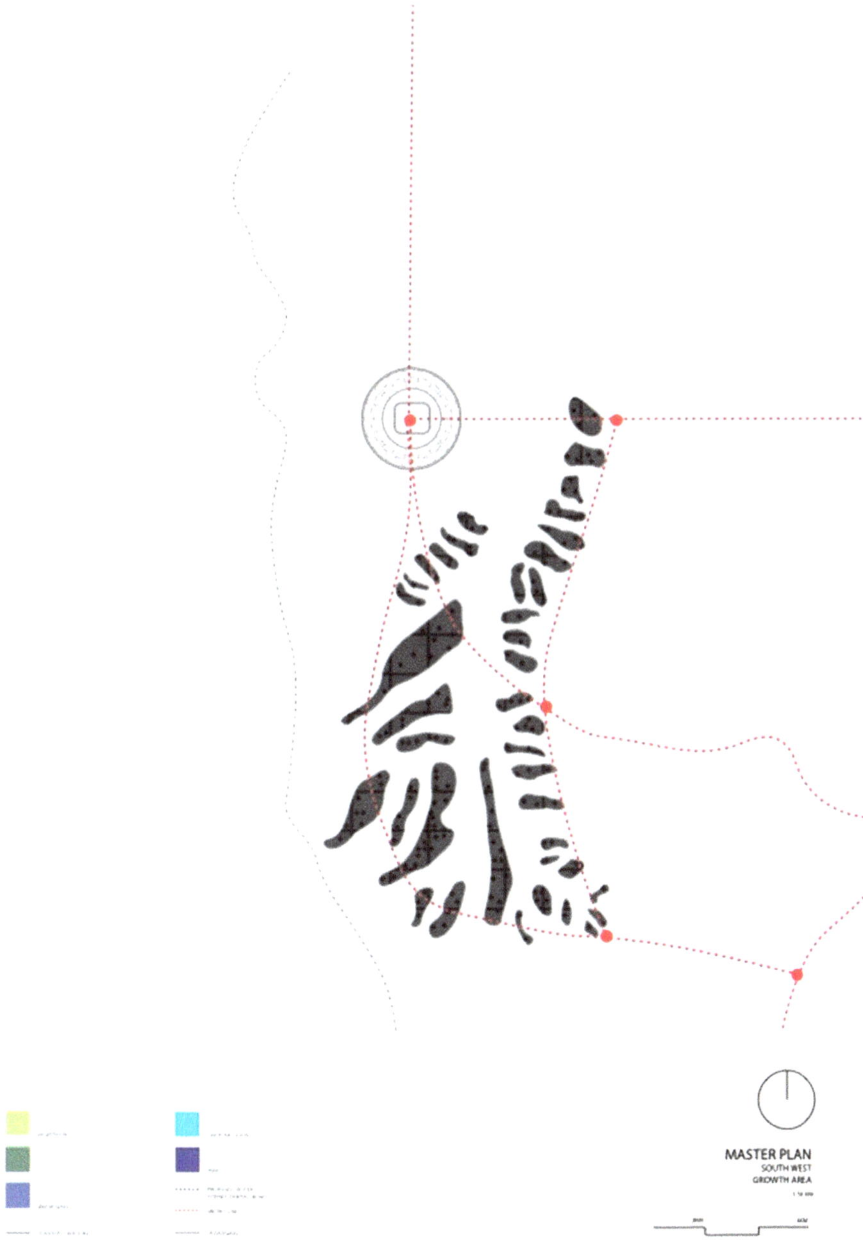

Fig. 5.7 Low-density urban fields (Mushi et al. 2017)

- Develop a beautiful city. Everyone enjoys living in a local environment that has paid attention to how it looks like, what the architecture is like, what materials are used and the local character is, it all determines the appreciation for the environment people will live in.

Each urban field in Western Sydney should be beautiful, diverse, equitable, sustainable, circular and functional.

At the larger scale these urban fields will then be connected to the broader area and each other. The primary way connectivity for the new city is provided through metro. These metro-lines will link each of the neighborhoods and connect the total area and the new airport to the existing network (Fig. 5.8) and stations such as Leppington, Parramatta and Campbelltown.

This basic Master Plan for Western Sydney, including the rules for good urbanism, will require adjustments over time. New problems and developments will emerge, and the urban development needs to be able to anticipate future needs. Therefore, the Master Plan, initially based on current market demands, is able to change. For instance, when growth continues densities can be increased. Only the future can tell if, and how future densification will take place, depending on changing technological, social and climatic contexts. It is very likely the final development will consist of a mix of densification typologies. In the Parkland City landscape three ways to create higher densities are distinguished: within the urban fields, in the valleys and on the ridges. The iterative process of finding optimal densification patterns (Fig. 5.9) starts with locating the urban fields within the boundaries of the cooling landscape, then exploring the three ways to increase density as rigid singular thought exercises, before the optimal mix is explored.

Three fundamental different densification strategies can be distinguished (Fig. 5.10).

The first strategy locates new housing all within the existing urban fields. This implies that plots and streets are intensified. This is possible if the promise of driverless cars will become reality. In that case no one will own a car and the space for parking on plots and in streets is no longer needed. Every double garage can be turned into a compact two to three level studio or apartment. In the backyard a similar addition is possible. Without increasing infrastructure or develop new greenfields, this way the density can be increased by three in an incremental way whenever time is ready. Should an apartment block replace the two houses at the top end of the street, the density could be increased four times. The total capacity in the Master plan would then rise to 560.000 dwellings or 1.4 million people.

In the second strategy the new housing is being located in the valleys, which separate the urban fields. As these areas are meant to offer space for flooding, discharging water after heavy rain and from the households through the gullies, where water is kept and a natural vegetation provides shade to cool the environment, the building types need to be adjusted in order to allow this to happen. Elegant and slender higher density buildings, such as apartment blocks of six to ten levels, placed on stilts in the valleys could raise the density of the entire Master Plan to 40 dwellings/ha, leading to a total of 280.000 dwellings or 700.000 people.

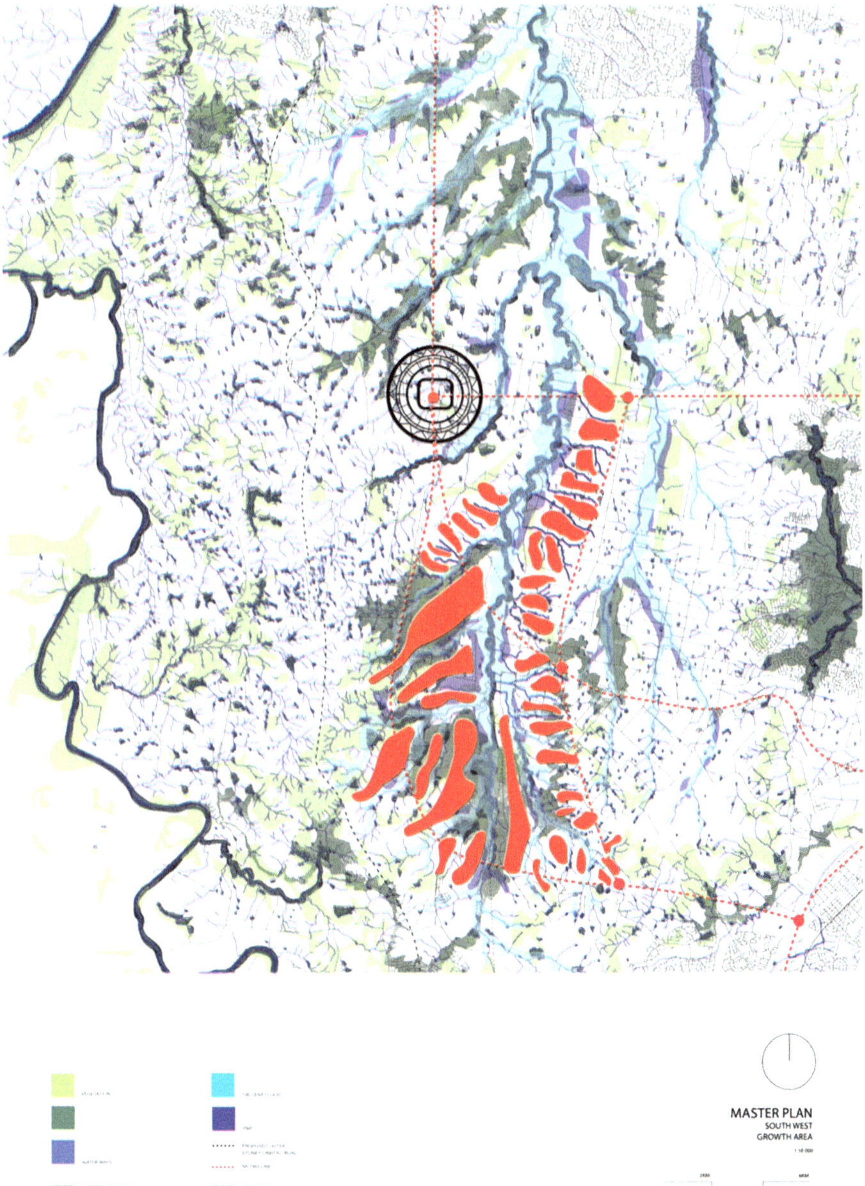

Fig. 5.8 Urban fields connected to the existing metro-network and Badgerys airport (Mushi et al. 2017)

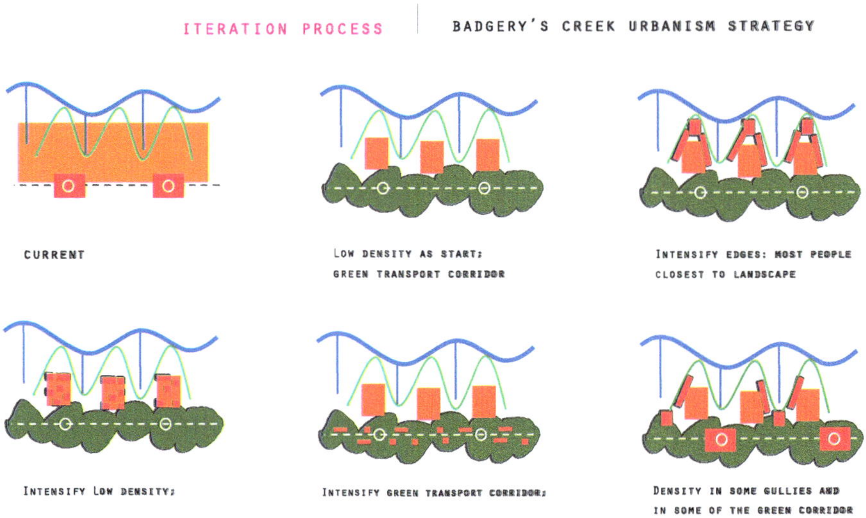

Fig. 5.9 Iterative process exploring the mix of densities (Mushi et al. 2017)

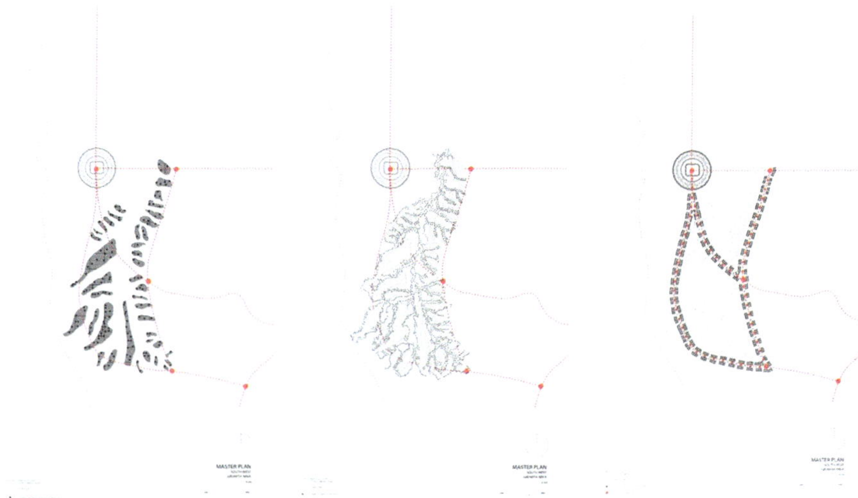

Fig. 5.10 Three strategies for increasing densities: within urban field (left), in the valleys (center) and on the ridges (right) (Mushi et al. 2017)

The third strategy locates all new housing on the ridges, where it is collocated with the metro stations in the form of mixed-use and apartment buildings of ten to fifteen levels. This strategy raises the density to 50 dwellings/ha, e.g. 350.000 dwellings or 875.000 people.

In reality the Western Parkland City will combine these typologies, maybe even include other strategies. Time will tell in what combination of realized typologies the new city will emerge, depending the context and decisions made along the way, but it is obvious future urbanization will contain a mix of strategies. It makes a big difference for the realized city what the mix will be. As a spatial exercise three typological mixes have been designed.

When the majority of new housing will be realized within the urban fields (60%) and 20% each in the valleys and on the ridges (Fig. 5.11), most of the landscape is still capable of functioning as a cooling machine, while connectivity is increased by densification around stations. This model could house 1,155,000 people.

In a second combination the new housing is evenly distributed, 33% each, over the three categories (Fig. 5.12). The landscape is relatively filled with development while two third of the urban field remain low density and the station areas are not fully exploited. The amount of people that can be housed in this model is 991,666.

In a third model the majority of new housing is located around the stations, 50%, while 30% is placed in the valleys, and 20% in the urban fields. In this option the station areas are well-used with most people living in higher densities around the metro-line. However, the urban fields remain very low density, and the valleys are relatively filled with buildings. The amount of people that can be housed in this variant is 927,500 (Fig. 5.13).

Evaluating the three models it becomes clear that most impact can be reached if most new housing is integrated in the urban fields, with support of intensification around stations. New developments in the valleys doesn't mean a lot in terms of numbers of people and disturbs the landscape functions. So new developments in valleys should be allowed incidentally and kept to a minimum. Should development in valleys be allowed at all, strict conditions should be adhered such as developing built structures loose from the underlying landscape.

5.6 Integrated Designs

Zooming in on a more detailed level the principles described before are applied at a lower scale, such as for the North Bringelly area (Fig. 5.14). In this urban design (Young 2017) the urban fields are located in between the gullies and valleys, leaving abundant space for the landscape to fulfill its cooling function. Within the urban field densities can be raised whenever appropriate.

The urban design is using landscape characteristics to create differences. There are five different street typologies, depending on their relation to the gullies and the wider landscape hence the edges are designed each in their own way. Every street that is angled towards the gullies is car free, so it becomes possible to walk down to the leisure areas connected to the valley and the transport and shops up the ridges, where the wider service roads and parking is located. Shared green space and private individual gardens are connected to the gully zone while the amenities such as schools and shops are located in the gully streets. The size of the gullies determines the size and shape of the neighborhoods and building lots. The highest densities are found near the gullies and inside the urban fields lower densities are realized. In the

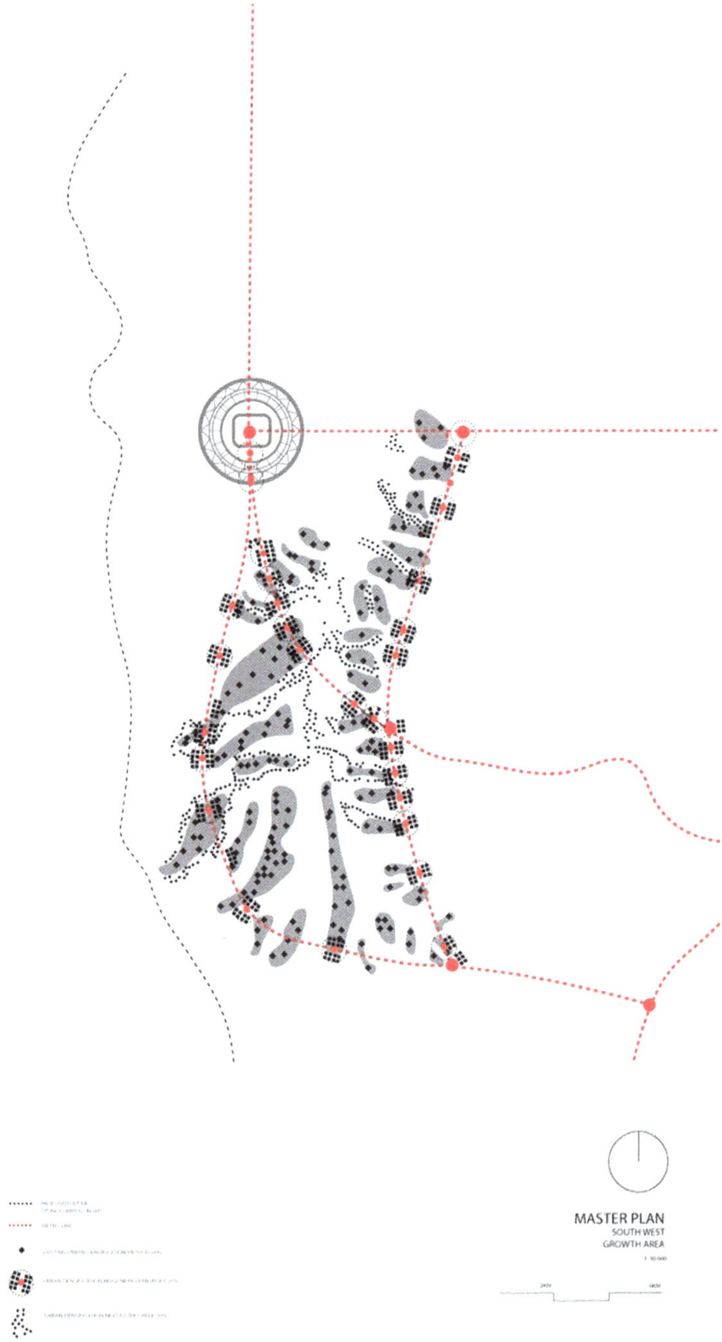

Fig. 5.11 Majority of new housing within the urban fields (Mushi et al. 2017)

Fig. 5.12 Evenly distributed densification in urban fields, valleys and ridges (Mushi et al. 2017)

Fig. 5.13 When the majority of new housing is located around metro stations (Mushi et al. 2017)

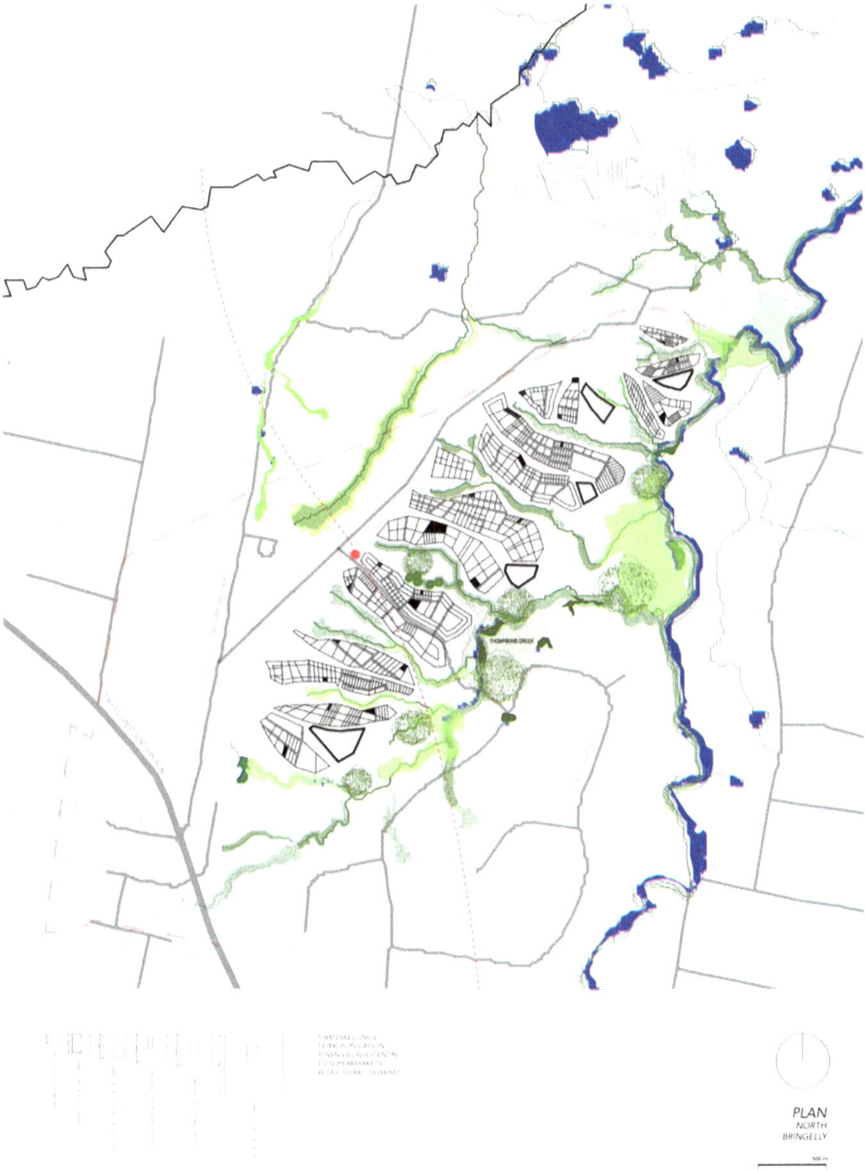

Fig. 5.14 Urban design North Bringelly (Young 2017)

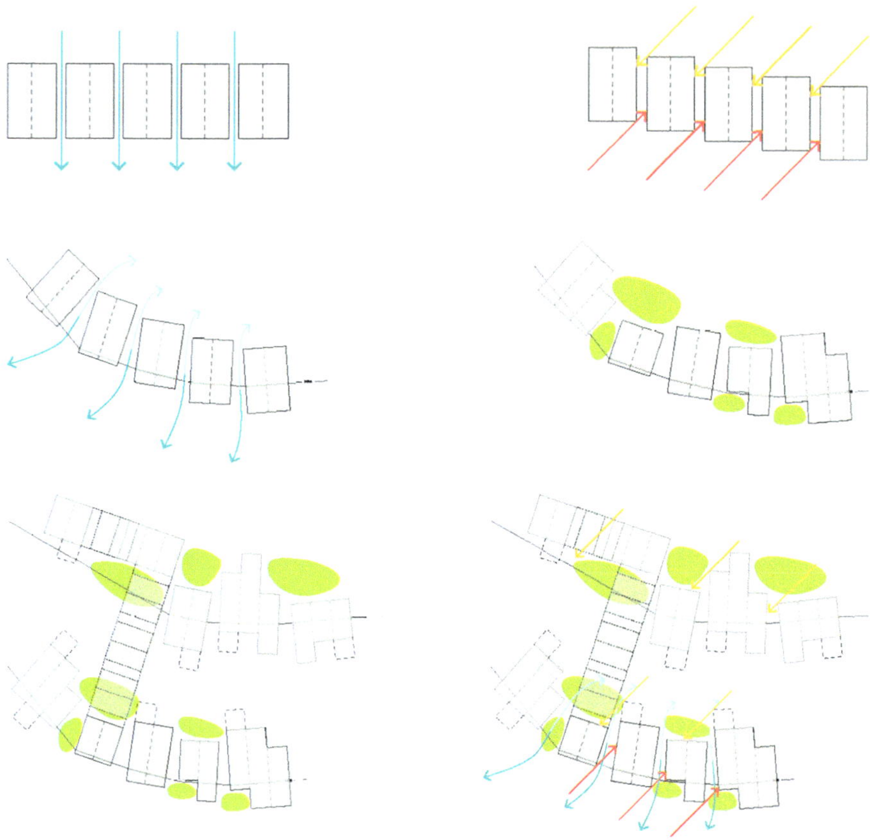

Fig. 5.15 Constructing the Hydraulic City (Veringa 2017)

future specific pockets at the edges of the ridges are selected to increase the densities further within the existing urban environment. The building lots can be subdivided built up, so the total density can potentially quadruple.

In the urban design 'Hydraulic City' (Veringa 2017) the landscape is also used as the driving force (Fig. 5.15). The discharge of rain and cleaned household water is taken as the starting point for subdividing the building blocks, which then are disconnected to allow summer and winter sunlight in the houses, at the same time providing shade during hot summer days. By orienting the blocks along the contours of the gully landscape, new spaces emerge which are used to integrate shady green spaces. The final design is connecting both sides of the gully by overbuilding the landscape and lifting the building blocks out of the underlying landscape, creating or safeguarding landscape functions such as green and waterflows to function undisturbed. By building upon the landscape relatively high densities (Fig. 5.16)

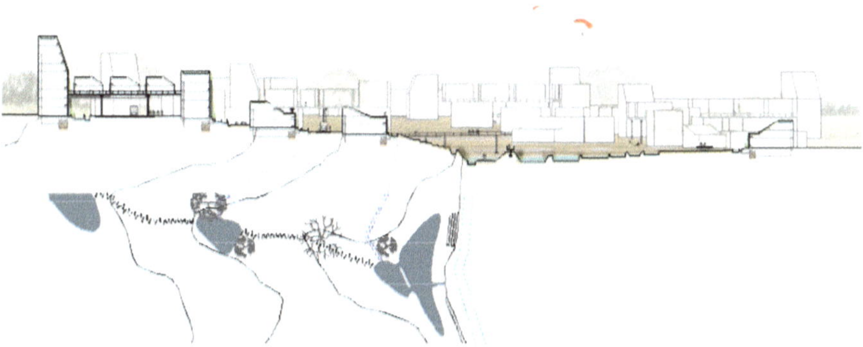

Fig. 5.16 Cross-section, showing high densities (Veringa 2017)

Fig. 5.17 Urban design Hydraulic City (Veringa 2017)

can be reached while the landscape is respected and keeps its natural and cultural heritage functionalities (Fig. 5.17).

Both designs show the use of landscape characteristics and the ambitions of realizing higher densities can be realized in a plan, which is very livable and walkable, close to nature, leisure and amenities.

5.7 Conclusions and Recommendations

In this chapter the planning and design of a large-scale new Master Plan for Western Sydney has been presented. The proposition takes the natural landscape as the entrance point hence the basis for thinking about the spatial future. It would be recommended to use the principle of 'landscape first' in Western Sydney development, but this is applicable in any new development. It places the landscape functions, such as ecology and water, and the impacts on the landscape such as heat, rain and other climate impacts at the center of the design process. Starting with the basic ingredients of what makes a landscape sustainable and ecological, the water system, elevation, soil and vegetation are the most relevant elements to keep the quality high and continue the functioning of the system. To start with these elements is de-facto reversing the regular planning process. In our current planning approaches thinking starts with the program that needs to be realized and where the land can be found to do so. Then the standard way of delivery is applied to the area that has been chosen, and average housing, i.e. as large as possible as cheap as possible, is proposed. The rest of the functionalities of the urban development are secondary and are woven in the main framework which is already determined by land ownership and economic reasoning. The landscape first approach firstly arranges the basic features of the landscape, including the way the area can deal with climate hazards, and then, based on the frameworks that are determined though this way of Master Planning make sure that the principles of good urbanism are applied to the neighborhoods which are planned within the landscape first framework. This requires thinking through the different spatial scales. The choice to take water and elevation as the entrance point for the Master Plan design should also be reflected at the scale of the urban designs, so the larger scale is supported by this. Additionally, the choices made at the higher scale of the landscape create the opportunities for high quality living and working and are supported by the individual urban designs, for instance by realizing higher densities at the cross-points of urban and landscape.

For the specific situation of Western Sydney, with its main problem of (over)heating in summer, the urban solutions should increase shade, and ventilation with cooling breezes using the water as the air-conditioning source. By doing so, an increase in densities is ideally realized inside the existing neighborhoods as it allows the cooling landscape to continue doing its work. When higher densities around the edges of the urban field are realized and combined with densification around metro trainstations, the majority of the new inhabitants will profit from the cooling landscape machine and will reach the highest number of new residents. Developments in the gullies, and valleys of the landscape are for the same reason not supported, or only incidentally with strict restrictions, such as that buildings do not touch the landscape.

Urban design at regional and local level are intrinsically connected through the continuing landscape that is everywhere, in between urban fields and underneath new urban developments, while connecting the urban with the rural, and making sure that flows of ecology and water can undisturbed run through the urban landscape. The health of plants and animals, and humans benefits from green, nature and landscape in their vicinity.

References

Australian Government and Government of NSW (2018) Western Sydney City Deal. Vision. Partnership. Delivery. NSW Government, Sydney

Bewsher Consulting PTY LTD (2004) South Creek floodplain risk management study and plan for the Liverpool local government area, Final report. Liverpool Council, Liverpool

Bird W (2007) Natural thinking: investigating the link between the natural environment, biodiversity and mental health. Royal Society for the Protection of Birds, London

Carrus G, Scopelliti M, Lafortezza R, Colangelo G, Ferrini F, Salbitano F et al (2015) Go greener, feel better? The positive effects of biodiversity on the well-being of individuals visiting urban and peri-urban green areas. Landsc Urban Plan 134:221–228

Cox DTC, Shanahan DF, Hudson HL, Plummer KE, Siriwardena GM, Fuller RA, Anderson K, Hancock S, Gaston KJ (2017) Doses of neighborhood nature: the benefits for mental health of living with nature. Bioscience:147–155. https://doi.org/10.1093/biosci/biw173

Cox DTC, Shanahanc DF, Hudson HL, Fuller RA, Gaston KJ (2018) The impact of urbanisation on nature dose and the implications for human health. Landsc Urban Plan 179:72–80

Department of Planning and Infrastructure, State Government of NSW (2017) Plan for Austral and Leppington. Government of NSW, Sydney

Department of Planning and Environment (undated) Planning for Sydney's South West. https://www.planning.nsw.gov.au/Plans-for-your-area/Priority-Growth-Areas-and-Precincts/South-West-Growth-Area

Dovey R (2018) Can parks make kids better at math? Next City https://nextcityorg/daily/entry/can-parks-make-kids-better-at-math. Published 11 September 2018

Dye C (2008) Health and urban living. Science 319:766–769. https://doi.org/10.1126/science.1150198

Endreny, T. (2018) We calculated how much money trees save for your city. *The Conversation* 27 April 2018. https://theconversation.com/we-calculated-how-much-money-trees-save-for-your-city-95198

Epstein L, Raja S, Gold S, Paluch R, Pak Y, Roemmich J (2006) Reducing sedentary behavior the relationship between park area and the physical activity of youth. Psychol Sci 17(8):654–659

Flouri E, Papachristou E, Midouhas E (2018) The role of neighbourhood greenspace in children's spatial working memory. Br J Educ Psychol 89:359–373. https://doi.org/10.1111/bjep.12243

Gascon M, Triguero-Mas M, Martínez D, Dadvand P, Rojas-Rueda D, Plasència A, Nieuwenhuijsen MJ (2016) Residential green spaces and mortality: a systematic review. Environ Int 86:60–67

Gidlöf-Gunnarsson A, Öhrström E (2007) Noise and well-being in urban residential environments: the potential role of perceived availability to nearby green areas. Landsc Urban Plan 83:115–126

Greater Sydney Commission (2018) A metropolis of three cities. Greater Sydney region plan – connecting people. State Government of New South Wales, Sydney

Gyeong-bok M (2017) Ministry of health and welfare

Husqvarna Group (2013) Global green space report 2013. Exploring our relationship to forests, parks and gardens around the globe. Husqvarna AB, Stockholm

Infrastructure NSW (2018) Building momentum. State infrastructure strategy 2018–2038. iNSW, Sydney

Jackson L (2003) The relationship of urban design to human health and condition. Landsc Urban Plan 64(4):191–200

Kaplan S (1995) The restorative benefits of nature: toward an integrative framework. J Environ Psychol 15(3):169–182

Krasny M, Tidball K (2015) Civic ecology: adaptation and transformation from the ground up. MIT Press, Cambridge, MA

Kristinsson J (2012) Integrated sustainable design. Delft Digital Press, Delft

Kuo F, Sullivan W (2001) Environment and crime in the inner city: does vegetation reduce crime? Environ Behav 33:343–367

Kuo F, Bacaicoa M, Sullivan W (1998) Transforming inner-city landscapes: trees, sense of safety and preference. Environ Behav 30(1):28–59
Lau S, Yang F (2009) Introducing healing gardens into a compact university campus: design natural space to create healthy and sustainable campuses. Landsc Res 34(1):55–81
Li D, Sullivan W (2016) Impact of views to school landscapes on recovery from stress and mental fatigue. Landsc Urban Plan 148:149–158
Lottrup L, Grahn P, Stigsdotter UK (2013) Workplace greenery and perceived level of stress: benefits of access to a green outdoor environment at the workplace. Landsc Urban Plan 110:5–11
Louv R (2016) Vitamin N. The essential guide to a nature-rich life: 500 ways to enrich your family's health & happiness. Atlantic Books, London
Maas J, Verheij R, Groenewegen P, De Vries S, Spreeuwenberg P (2006) Green space urbanity, and health: how strong is the relation? J Epidemiol Community Health 60(7):587–592
Maas J, Verheij R, de Vries S, Spreeuwenberg P, Schellevis F, Groenewegen P (2009) Morbidity is related to a green living environment. J Epidemiol Community Health 63(12):967–973
Maller C, Townsend M, Pryor A, Brown P (2006) Healthy nature, healthy people: 'contact with nature' as an upstream health promotion intervention for populations. Health Promot Int 21(1):45–54
Marcus C (2007) Healing gardens in hospitals. Interdiscip Design Res e-Journal 1(1):1–27
Marselle M, Irvine K, Warber S (2014) Examining group walks in nature and multiple aspects of well-being: a large-scale study. Ecopsychology 6(3):134–147
McHarg IL (1969) Design with nature. Natural History Press, New York
Mennis J, Mason M, Ambrus A (2018) Urban greenspace is associated with reduced psychological stress among adolescents: a Geographic Ecological Momentary Assessment (GEMA) analysis of activity space. Landsc Urban Plan 174:1–9. https://doi.org/10.1016/j.landurbplan.2018.02.008
Mitchell R, Popham F (2007) Greenspace, urbanity and health: relationships in England. J Epidemiol Community Health 61(8):681–683
Mushi M, Shalala R, Young J (2017) Badgerys urbanism. Elective Master of Architecture, UTS School of Architecture, July 2017, led by Rob Roggema and Rod Simpson
Okvat H, Zautra A (2011) Community gardening: a parsimonious path to individual, community, and environmental resilience. Am J Community Psychol 47(3–4):374–387
Pretty J, Peacock J, Sellens M, Griffin M (2005) The mental and physical health outcomes of green exercise. Int J Environ Health Res 15(5):319–337
Pudup M (2008) It takes a garden: cultivating citizen-subjects in organized garden projects. Geoforum 39(3):1228–1240
Roggema R (2012) Swarm planning: the development of a methodology to deal with climate adaptation. Delft, Wageningen: Delft University of Technology and Wageningen University and Research Centre. PhD-thesis
Roggema R (2018) Design with voids. How inverted urbanism increases urban resilience. Architectural Science Review (ASR). Special issue: time, place and architecture. https://doi.org/10.1080/00038628.2018.1502153
Shore R (2017) Kids need access to nature for mental health. Vancouver Sun. https://vancouversun.com/health/local-health/kids-need-access-to-nature-for-mental-health. Published 17 April 2017
Sijmons D (1992) Het casco-concept, een benaderingswijze voor de landschapsplanning. Ministerie van LNV, directive NBLF, Utrecht
Söderback I, Söderström M, Schälander E (2004) Horticultural therapy: the 'healing garden' and gardening in rehabilitation measures at Danderyd hospital rehabilitation clinic, Sweden. Pediatr Rehabil 7(4):245–260
Sullivan W, Chang C (2011) Mental health and the built environment. In: Dannenberg A, Frumkin H, Jackson R (eds) Making healthy places. Island Press/Center for Resource Economics, Washington, DC, pp 106–116
Sullivan W, Kuo F, DePooter S (2004) The fruit of urban nature –vital neighborhood spaces. Environ Behav 36(5):678–700

Swinbourne R, Rozenwax J (2018) Green Infrastructure. A vital step to brilliant Australian cities. AECOM, Sydney

Taylor A, Kuo F, Sullivan W (2002) Views of nature and self-discipline: evidence from inner city children. J Environ Psychol 22(1–2):49–63

Thompson C, Roe J, Aspinall P, Mitchell R, Clow A, Miller D (2012) More green space is linked to less stress in deprived communities: evidence from salivary cortisol patterns. Landsc Urban Plan 105(3):221–229

Tjallingii SP (1995) Ecopolis: strategies for ecologically sound urban development. Backhuys Publishers, Leiden

Tjallingii S (2015) Planning with water and traffic networks. Carrying structures of the urban landscape. Res Urban Ser 3(1):57–80. https://doi.org/10.7480/rius.3.832

Tomásek W (1979) Die Stadt als Oekosystem; Űberlegungen zum Vorentwurf Landschaftsplan Köln [The city as ecosystem. Considerations about the scheme of the landscape design in Cologne]. Landschaft+Stadt 11:51–60

Transport for NSW (2018) Future transport strategy 2056. State Government of New South Wales, Sydney

Tzoulas K, Korpela K, Venn S, Yli-Pelkonen V, Ka'zmierczak A, Niemelä J et al (2007) Promoting ecosystem and human health in urban areas using green infrastructure: a literature review. Landsc Urban Plan 81(3):167–178

Van Dijk-Wesselius JE, Maas J, Hovinga D, Van Vugt M, Van den Berg AE (2018) The impact of greening schoolyards on the appreciation, and physical, cognitive and social-emotional well-being of schoolchildren: A prospective intervention study. Landsc Urban Plan 180:15–26. https://doi.org/10.1016/j.landurbplan.2018.08.003

Van Schaick J, Klaasen IT (2011) The Dutch layers approach to spatial planning and design: a fruitful planning tool or a temporary phenomenon? Eur Plan Stud 19(10):1775–1796

Van Timmeren A, Henriquez L (2015) Ubikquity and the illuminated city. From smart to intelligent urban environments. TU Delft, Delft

Veringa G (2017) The Hydraulic City. Design Studio 'GO SOUTHWEST', Master of Architecture, UTS School of Architecture, Led by Rob Roggema and Rod Simpson August–November 2017

Ward Thompson C, Aspinall P, Roe J (2014) Access to green space in disadvantaged urban communities: evidence of salutogenic effects based on biomarker and self-report measures of wellbeing. Procedia Soc Behav Sci 153:10–22

World Health Organisation (2015) Urban health. www.who.int/topics/urban_health/en/. Accessed on 11th Nov 2016

Young J (2017) Urban design for North Bringelly. Master of Architecture Design Elective 'Badgerys Urbanism', led by Rob Roggema and Rod Simpson, School of Architecture, University of Technology Sydney, July 2017

Chapter 6
From Urban Green Structure to Tidal River in Rotterdam: Testing Grounds for Urban Ecology

Nico Tillie

Abstract The Rotterdam approach to nature-based urbanism is presented in this chapter. The urban green structure, the water vision 2035 and the tidal river park strategy, all part of the Rotterdam policies are the first testing grounds and form also the starting point for an urban ecology research agenda at Delft University of Technology (TUD). These examples show perfectly how urban ecology and Landscape Architecture and urbanism can go hand in hand. In fact, nature based urbanism also holds both of these aspects of ecology (natural) and design and planning (cultural).

Keywords Urban greenstructure · Ecology · Urban ecology · River · Rotterdam

6.1 Introduction

The Rotterdam approach to nature-based urbanism is presented in this chapter. The urban green structure, the water vision 2035 and the tidal river park strategy, all part of the Rotterdam policies are the first testing grounds and form also the starting point for an urban ecology research agenda at Delft University of Technology (TUD). These examples show perfectly how urban ecology and Landscape Architecture and urbanism can go hand in hand. In fact, nature based urbanism also holds both of these aspects of ecology (natural) and design and planning (cultural).

In his doctoral thesis, Tillie et al. (2018) describes the bridging role of landscape architecture and urbanism with urban ecology. Forman (2014) defines urban ecology as 'the interactions of organisms, built structures, and the physical environment, where people are concentrated'. Until recently, urban ecology has been predominantly approached from an ecological perspective. Alberti (2009) calls for other professions and research angles to enrich the urban ecology domain. When the main goal is to help the transition from cities to sustainable cities, modelled on cities as

N. Tillie (✉)
Delft University of Technology, Delft, The Netherlands
e-mail: n.m.j.d.tillie@tudelft.nl

R. Roggema (ed.), *Nature Driven Urbanism*, Contemporary Urban Design Thinking, https://doi.org/10.1007/978-3-030-26717-9_6

urban ecosystems, landscape architecture is one of the main domains that can provide an answer.

A main question in urban ecology related to the previous statement is: what is the role of humans in urban ecology? Forman (2014) discusses that 'it seems wiser to maintain and further build on the core strength of ecology, with its basic focus on plants, animals and microbes. Sister disciplines and professions will welcome and use principles developed by a strong Urban Ecology'. This might raise some questions about where humans stand in relation to nature. This debate, as Forman (2014) formulates, 'A human-as-inside-or-outside-of an ecosystem discussion is endless'.

From the perspective of a purely natural environment or conservation of these unspoilt areas without 'direct' human influence this is a most valuable approach. From a Landscape Architectural perspective in non-urbanised areas, this approach is a cornerstone for restoring and maintaining ecosystems and a better understanding of this. However, in urban landscapes, which are more influenced by humans, more knowledge is needed to understand relations and to design and anticipate for a reciprocal relationship. Sijmons (2014) specifically states that humans are part of nature and he uses this perspective to approach these phenomena, using urban metabolism (Wolman 1965; Newman 1999, Kennedy et al. 2010, Sijmons 2014; Tillie et al. 2014). Also, Alberti (2009) argues that taking a standpoint where humans are part of the ecosystem might gain new insights; she discusses the term cities as hybrid ecosystems and mentions that 'urban ecosystems are not different from other ecosystems simply because of the magnitude of the impact humans impose on ecosystem processes, nor are they so removed from nature that ecosystem processes become only a social construct in themselves'. If we conceptualize such systems purely in ecological or human system terms, we limit our ability to fully understand their functioning and dynamics (Collins et al. 2000; Alberti et al. 2003). As a result, we would also limit ourselves in identifying valuable synergies between these systems. Landscape Architecture is one of the disciplines well positioned to take a bridging role and combine the natural as well as human aspects of urban ecology.

6.2 Synergies in Improving the Quality of Life for People and Nature

The field of urban ecology is quickly diversifying, however there are still some interesting niches to be addressed especially when looking at how urban ecology can serve a broader agenda, such as improving the quality of life, focusing on health, sustainable urban development, urban water and energy issues, biodiversity and more. Although the concept of ecosystem services could be used for this, the problem is that in the end urban ecology is not automatically anchored in the urban planning and design disciplines involved in city making. For this, Bird Life Netherlands (Vogelbescherming Nederland) and the chair of Landscape Architecture at TU Delft have joined forces and set up an Urban Ecology chair with a more holistic point of view. The chair will collaborate with ecology specialists and make use of and extend the existing body of knowledge in urban ecology to be applied in Landscape Architecture, urban planning and design and architecture. As the first questions that arise are for instance: 'How can urban ecology contribute to a better quality of life in cities?', 'How can the environmental performance in cities improve?' Which urban strategies can be used to realise this?', 'What are the spatial urban ecological structures to strive for and what basic ecological conditions in (Dutch) cities are required for this?', 'Which ecological principles can be used how in planning and design?' In short: 'How to design with nature?'

Although most questions are part of the emerging research agenda for the coming years, some studies in the city of Rotterdam are conducted already and will be presented in this chapter.

6.3 The Urban Green Structures, a Template for Nature-Driven Urbanism

The City of Rotterdam and these examples can serve as starting point to test and project a hypothesis, design ideas and future findings against. Based on the natural system the fundamentals of an urban ecology are laid out in the urban green (or main ecological) structure of the city (De Greef et al. 2005a). Taking this as the basic framework, elaborations can be planned such as potential greening strategies for stony areas, urban water retention areas and using the river as a central park in the city. Rotterdam Water City 2035 (De Greef et al. 2005b) uses this structure, takes it a step further and uses the former peat bog landscape as a metaphor for its water storing capacity. From there links are established to projects in the city. For the tidal river parks (Gemeente Rotterdam 2018), design with nature is a cornerstone for the areas in the dynamic tidal river system of the city. In fact, the main axis of the urban green structure, the river, can be framed as zone. The river Maas as a Central Park!

6.3.1 Urban Green Plan

The 2005 urban green structure was based on the underground (soil, topography), rivers and ecological structures. As such, this induced nature-based urbanism on several scales. In urban environments, areas can change in a relatively short period of time, often drastically changing the conditions for certain urban ecological niches or habitats. Succession processes have to start all over. On the other hand, there are certain species (such as pioneers) which thrive in (constantly) changing environments (Odum and Barrett 2005). There are also less-dynamic areas in a city which can serve as a backbone for basic urban ecological qualities in a city. The Rotterdam urban green structure exists of a chain of rivers, parks, quays, lanes, open water and canals and is combined with routes, water storages and other synergies, it is an attempt to structure the city and future planning.

Over the years, the goals of the city of Rotterdam have become more specific and focused on improving quality of life as well as ecological qualities (Frantzeskaki and Tillie 2014).

In policy documents, such as the Urban Green Structure Plan 'Groenplan 2005' (Greef et al. 2005a) this process started with improving green qualities for citizens. In Rotterdam, sustainable growth document (Gemeente Rotterdam 2013) the densification and greening strategy is placed in a broader perspective. Also, densification and greening studies, such as 'People Make the Inner City' (Tillie et al. 2012, 2018) show this development. The sustainable growth document (Gemeente Rotterdam 2013) mentions the following goals for any development, which should:

1. Lead to a more child-friendly, green, clean and healthy living environment;
2. Create economic value today and on the long term and affordability for its citizens;
3. Contribute to a 50% carbon reduction and a 100% climate proof city.

Rotterdam has 117 public parks which total 1765 ha. The core ideas of the urban green structure plan are as follows: 'The Maas river is considered as the largest ecosystem in the region. The 'Groenplan' broke with previous green structure plans, which used an abstract concept such as the '4 green Corners' disconnected from natural processes. This plan is based on the natural conditions and (potential) flowscapes of rivers, biota and people. On the northern shore of the river Maas, the Schie and Rotte rivers and the Ringvaart canal are the major ribbons. They form a radial structure from the landscapes outside of the city to the city centre of Rotterdam. At a lower scale, there are different green structures, such as the 'Singels'(canals) which connect these three main ribbons. The Southern shore has a different structure. The green structure is not radial (north-south) but based on half rings (east-west) which follow the old dike structures of the sea clay polders along the Maas and Zuiderpark (Fig. 6.1) (Tillie and Van der Heijden 2016).

'In the city, the river landscape offers an attractive image right in the city center. The rivers are natural connections to the regional green areas. The structural support at city level – radials and half rings – are attractive green areas and form the chain that link different parks' (Tillie et al. 2016). These are ideal routes for recreational use but also for commuters by bike. For nature, they form ecological patches and a collection of corridors and stepping stones.

Over the years, other plans such as the 'Rotterdamse Stijl' (Gemeente Rotterdam 2009) (guidelines on quality, consistency and recognisability in the public space) have developed and new plans will come forward. However, as the 'Groenplan 2005' with the main urban green structure is based on primarily natural conditions (supplemented with cultural layers) it is not just one of many visions but in fact a representation of nature-driven urbanism of how the city developed and it can develop over time.

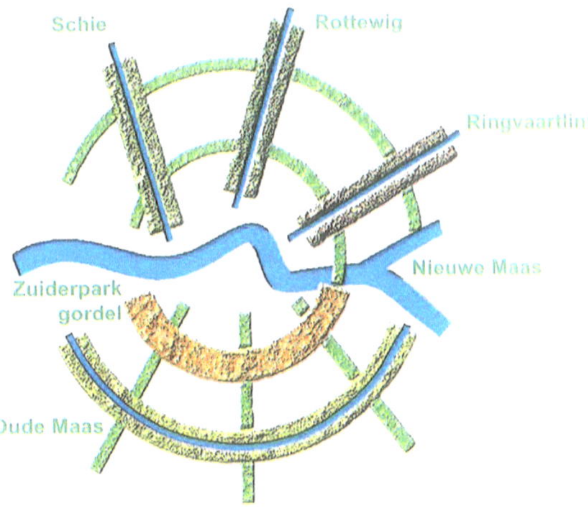

Fig. 6.1 Original schematic outline by Nico Tillie of the fundamental urban green structure in Rotterdam based on natural systems supplemented by cultural layers (De Greef et al. 2005a)

Like many cities in the world, Rotterdam is densifying. Many parts of the city are outside the main urban green structure. It is here that green qualities have to be introduced. Greening of neighbourhoods is a crucial task. This can be done by improving the quality of the main axis in the urban green structure as well as its accessibility (see Sect. 6.3.3 the River Maas as Cenral Park). Other strategies focus on the areas in the neighbourhood themselves. For the inner city of Rotterdam several 'greening strategies' have been formulated (Fig. 6.2) (Tillie et al. 2012, 2018).

The green roofs and facades, boulevard, quay, square, parks and playground strategies are all well-known typologies to work with in cities. They need to go hand in hand with densification. Other ideas for this are the introduction of urban roof landscapes such as Haaksma 2017 and Willemsen and Tillie 2018 propose. One of the challenges is to investigate what these typologies look like and need to be applied when urban ecology values need to be increased.

6.3.2 *Natural Peat Bogs, Sea Clay Landscapes and the Dynamic River as an Example in Watercity 2035 Vision*

'The area east of the coastal dunes, which nowadays is below sea level, used to be 10 meters above sea level (with the exception of Flevoland). Huge, often forested, peat beds fed with rainwater dominated this area and acted as a natural sponge storing the rainwater and releasing it slowly. As a way of access, early pioneers used the rivers such as Schie and Rotte, coming from these peat beds. In many areas, peat beds were drained for agriculture' (Steenbergen et al. 2009, Tillie 2018). Also, peat was dug away and dried, to be burned in homes for cooking and heating (Sijmons 2014). This resulted in peat polders or lakebed polders. The northern part of

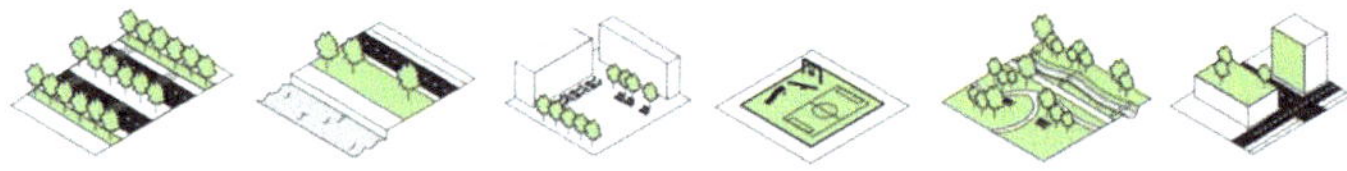

	BOULEVARDS	QUAYS	SQUARES	PLAYGROUNDS	PARKS	GREEN ROOFS & FACADES
CS-KWARTIER	10,400	0	34,900	2,500	0	39,500
OUDE WESTEN	13,600	0	0	0	0	172,600
LIJNBAANKWARTIER	7,700	0	21,300	4,500	2,300	119,300
LAURENSKWARTIER	35,900	48.900	42,700	4,500	5,100	155,200
HOBOKEN	6,500	11,900	6,000	0	0	113,200
COOL	8,300	0	4,600	4,500	7,000	41,200
WATERSTAD	30,000	26,300	0	0	25,800	122,900
NIEUWE WERK	11,300	59,100	0	2,500	0	59,900
KOP VAN ZUID	6,700	64,400	3,000	6,000	29,000	46,500
TOTAL IN 2040	130,400	210,600	112,500	24,500	69,200	870,300
TOTAL POTENTIAL NEW GREEN	1,417,500 M²/ 141,7 ha					

Fig. 6.2 Calculations per neighbourhood of potential new green per greening strategy which totals 1417 ha. (Courtesy of Doepel Strijkers Architects, Rotterdam)

Rotterdam is for a large part built in these polders. Over the centuries the water management adapted to reigning weather patterns. In the last decade peaks in rainfall have become higher which means that the existing water system cannot cope with every situation. More water storage is needed in the city.

The Rotterdam Water City 2035 consists of three images (Fig. 6.3): The River City (Rivierstad) in the centre, which is one of the most elevated area in the city, the Canal City (Vaartenstad) in the south, and the Singel City (Singelstad) in the north of Rotterdam (De Greef et al. 2005a, b). This last strategy was later reframed to 'sponge city' as it refers to the natural peat bogs working as a sponge (Tillie 2009).

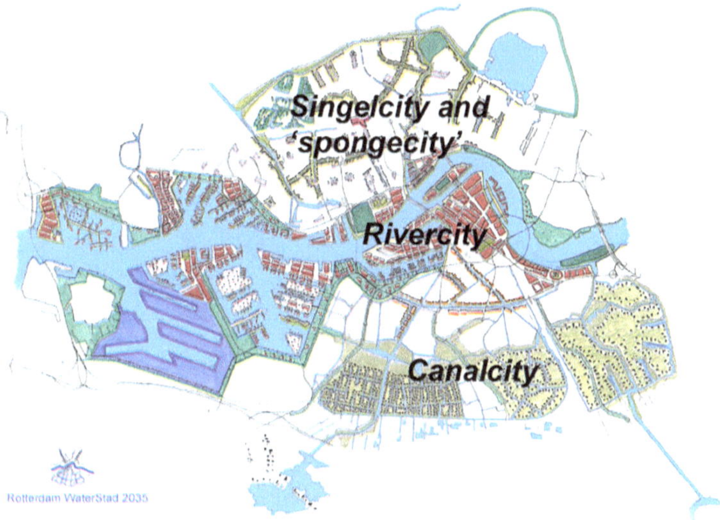

Fig. 6.3 Rotterdam Water City 2035 vision. Three images in different areas of the city

Nature-based urbanism in the Sponge City relates to the water storing capacity of the former peat bogs in this area. In the ideas mentioned 'Water will become tangible in the streets: it will be seen and heard. Rainwater will no longer be hidden' (Greef et al. 2005b, Tillie 2018). Here, the emphasis lies on combining an extended *singel,* which is a type of canal aligned with green embankments and housing, structure and realise a so-called dry solution for peak water storage. Also, an extended canal structure will store part of this peak. This is needed because, for a short period of time the level of the water in a Singel can rise too much. However, in any case a substantial part of peak storage has to be found where rainwater falls out of the sky, such as in green roofs and hard surfaces. Next to parks, bio-swales and semi-permeable surfaces, the water square (Tillie 2017) was introduced as a solution to slow down the discharge of rainwater. These dry-wet squares are linked to hidden gutters, and as soon as there is too much water for these gutters to handle, the squares will slowly fill up and store surplus of water (Fig. 6.4).

The southern areas of Rotterdam, the Canal City, is a part of city located behind the port infrastructures. It is made up of a single-sided population. Currently, the neighbourhoods offer insufficient living quality and the water system is inadequate. Nature-based urbanism can turn this around by taking a closer look at the sea clay landscapes, apparent underneath the city. 'The rain water will be almost directly drained to the surface water. The surface water will thus play a role in storing and draining rainfall from the area. Partly because the uninterrupted water network will be regarded as a quality of the south, it will be possible to create even more capacity than strictly necessary. Flooding will be prevented because there is ample storage space in the waterway system' (De Greef et al. 2005b; Tillie 2018). A connected network of waterbodies and canals for little boats, beautiful housing at the waterside

Fig. 6.4 Urban floodplain at Westersingel (left) and water squares Benthumplein Bellemyplein on the far right. (Images courtesy of city of Rotterdam and De Urbanisten)

and good recreational routes to the surrounding landscapes provides new urban qualities for this part of the city.

The River City is portrayed as the centre for further densification, bringing more dwellings and companies near and on the water, closer to the city centre. As the river is the main artery more public transport on water is promoted. The morphology of the uninterrupted dike structures can change, allowing it to respond to different rises in sea level. Rotterdam is planned as a city on the water. In open areas on the land side of the dikes there is enough space to reinforce dikes or to build new typologies of broadened dikes or flood mounts. This can create fascinating landscapes. The vision is that the city will grow over time in line with sea level rise and the dynamics of the tidal river.

6.3.3 The River Maas as a Central Park: Designing with Nature in a Dynamic Tidal River

At the mouth of the peat rivers such as the river Rotte the tidal influences from the North Sea occasionally pushed up water levels into the narrow river arms, making these areas as such unfit for habitation. In the thirteenth century a dam was constructed at the confluence of the small Rotte river and the Maas river to prevent tidal water coming upstream of the river Rotte and to collect toll from tradesmen who came from hinterlying rivers and needed to sail to the North Sea and vice versa (Pye 2015). Hence the name Rotter*dam* (as well as Schie*dam*, Amster*dam*, Appinge*dam*, etc.). Along these dams merchants settled and trade soon flourished. Tradespeople, fishermen and sailors created a settlement. The city grew and more ports were built. After the opening of the New Waterway (Nieuwe Waterweg) in 1872, which in essence was a canalisation of the main river creating a direct connection to the North Sea, the (port) economy of Rotterdam grew rapidly (Tillie et al. 2016). The river Maas became a highway for shipping. Ten years after the publication of the Urban Green Structure Plan (Greef et al. 2005a) in which the river is presented as the main axis, the river was slowly being rediscovered in a sense of green developments. Nowadays, the tidal river Maas is the key topic of an interesting strategy linking urban ecology, Landscape Architecture and urban planning.

6.3.3.1 Tidal Rivers; a Rare Ecosystem in Northwestern Europe

The Rotterdam region is situated in a dynamic delta (Fig. 6.5). For centuries, tides have shaped and influenced the landscape. With the development of the port as well as the building of the dikes and storm search barriers many tidal river areas have disappeared and along with it its natural and recreational qualities. Most tidal fresh (or brackish) water areas can be found in the National Park De Biesbosch and along the Oude Maas (Fig. 6.7). The Nieuwe Maas (or Meuse) serves as a shipping highway and there are only a few tidal river areas left. It is in fact this river that is the main green structure of Rotterdam. With a lot of densification in the city along the river there is a need for useful green areas for people.

A whole area along the river with new tidal river parks can serve more goals. It offers recreational value for citizens and it also improves the ecological values for the tidal river nature.

A coalition of NGOs, governments, companies and citizens in and around Rotterdam already started to work with natural processes in the tidal river to create new areas such as the 'Island of Van Brienenoord', others of a more recent date, such as the refurbishment of the Veerkade and Westerkade (Fig. 6.6) and the tidal area at Vijfsluizen.

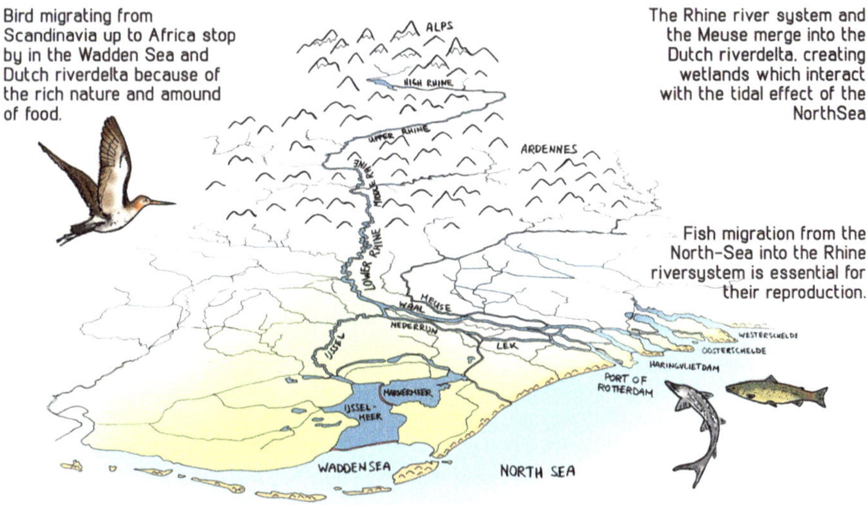

Fig. 6.5 The Netherlands as a delta (north is placed south) with the river Rhine and Meuse in Rotterdam being one of the few tidal non rivers open for fish migration. Many river mouths are closed off by dams or storm surge barriers for safety. (Image courtesy of Matthijs Hollanders)

Fig. 6.6 The Westerkade tidal river park is a urban tidal park with hard landscaping, trees and ornamental perennial planting. The steps and seating areas are designed is such a way that the visitor can experience the tide of the river. (Photos courtesy of Man-Chuan Lin)

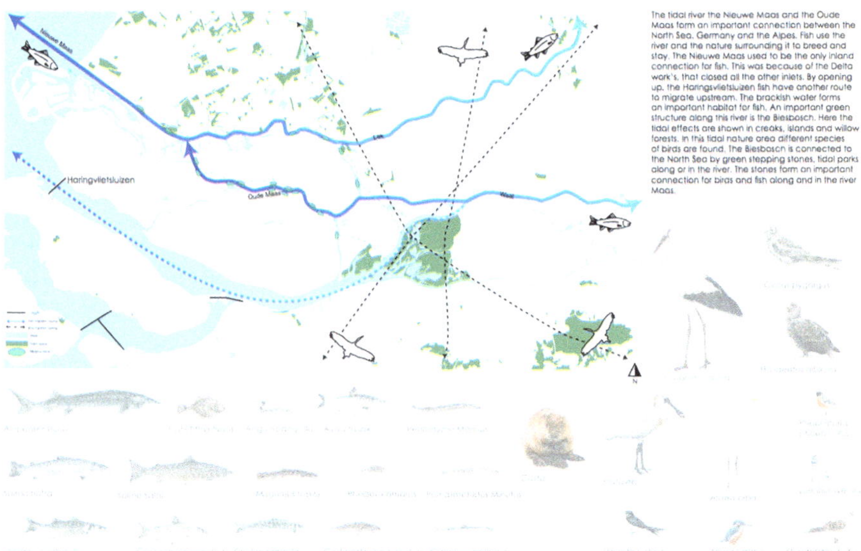

Fig. 6.7 The tidal river the Nieuwe Maas and the Oude Maas form an important connection between the North Sea and the hinterland. A main natural tidal area is the Biesbosch. Here the tidal effects are shown in creeks, islands and willow forests. In this tidal nature area many species of birds and are found. The Biesbosch is connected to the North Sea by green stepping stones along or in the river. (Image courtesy of Emma Kannekens)

The enthusiasm in the city for realizing new tidal parks has grown due to these examples and new tidal areas along the river. In fact, they show how city, port, nature and recreation can go together and lead to vital and attractive areas. This investment in the quality of the region is becoming increasingly important in retaining and attracting companies and people. Tidal parks also contribute to nature values, water quality, water safety and much more. Because of this wealth of objectives, its interlinkage and potential synergetic effects, the 'River as a Tidal Park'-program is of strategic importance to the region (Gemeente Rotterdam 2018).

6.3.3.2 Tidal Park Vision

In the so-called 'groeidocument' or growth document about the tidal river parks the vision is described. (Gemeente Rotterdam 2018) mentions, 'A tidal park is a park which makes the river more natural but also more attractive to be used in relation to the urban environment. A tidal park strengthens the relationship between water and land and between nature and the city. The tidal park tries to generate various added values for the city in terms of ecology and recreation, societal and economic values, and experience of natural dynamics.' As such it is also a testing ground where new knowledge is gathered about the combination of a safe shipping route, nature development and recreational values. A tidal park is a spatial intervention that uses different possibilities and connects multiple goals. These synergies increase the opportunity for realization because more sources of funding can be tapped into. In the document seven goals for tidal parks have been formulated (Gemeente Rotterdam 2018). Three of them are stressed here while other will be mentioned.

Create City and Nature

The tidal park can bring nature closer to the city. In fact, in the case of Rotterdam the nature can penetrate to the core of the city. Nowadays, in many places the river is difficult to reach from the city. A few of these places are being studied in the Waterscapes design studio at TU Delft. There are many (public and private) hard quays in the region which offer hardly any access to the water nor provide good conditions for natural gradients to develop. There is too little recreational and ecological value. Softening the banks creates conditions in which the water becomes more accessible and opportunities for recreation and nature development emerge. Looking at the river system as a whole from east to west, there is a need for more ecological and recreational continuity in the water, for example for migrating fish and for on-land movement of animals. The edges of the river are crucial to effectuate this (Fig. 6.7).

Increasing the Natural Values

Where high and low tide are effective, and where tidal nature has been lost through urbanization or other causes, a tidal park offers opportunities to restore and enhance natural values along the river. A landscape of gradients can be built, creating opportunities for different types of habitats and nature to develop. This makes the river in the longitudinal direction (east-west /west-east) more valuable for typical vegetation, as a migration route for species of fish, migratory birds and other animal species in search of nutrient-rich areas or resting places (Gemeente Rotterdam 2018).

Basis for Urban Development

The tidal parks can be seen as an investment in green living and working environments in the city. Especially with new densifications in the inner-city the demand for accessible green is high.

Other goals are:

- To use the tidal river parks as a learning environment. Already many fieldstrips and excursions for children in schools are organized to playfully increase the awareness of life in a river delta.
- To increase water safety. In many cases the nature development creates a wider foreland for the dikes which can lead to increased safety levels.
- To produce and experience local food (Niel 2015).
- To play an important role in regional recycling for different flows, an important one being sediments.

6.3.3.3 Exploration and Application of Natural and Ecological Principles

The natural conditions for tidal nature are not the same everywhere. In fact, the different conditions provide opportunities for many different habitats. Closer to the sea, the tidal range and with that the currents are in general much bigger having different effect on sedimentation. On the western side near the sea, the salt water gradually changes into brackish water. All this changes into a freshwater tidal plain land-inwards, further east. Gradients of different vegetation form in different conditions according to the salinity of the water as well as the tidal levels. All this results in gradient rich areas, safe areas for fish and birds to live and/or breed and so on. Species in these tidal environments settle, depending on the location and available space. The diversity of species, such as plants, insects, fish and birds increases.

The tidal parks are in the floodplains of the river because there the tides can be experienced. However, as the tidal river flows through a lowland area with (often urbanized) peat-, lakebed- and sea clay polders, lying below the water level of the river, the physical contact between (urban) landscapes and the river is often difficult to experience. Visual contact between the river and/or the tidal river parks is required in most cases to make it part of the urban system. However, each location demands a different relation to react to the genius locus of the place.

As stated in the urban green structure plan, there is a clear difference in characteristic between the north and south sides of the river. The primary dyke is much closer to the water on the northern side than on the southern side of the river. On the northern side a series of old creeks such as the Rotte and the Schie (medieval tidal environments) are connected by the primary dyke. Nowadays, these creeks have become regulated canals or serve as water storage bodies in the water system of the polder. From these canals, the surplus water is discharged into the river Maas. Gemeente Rotterdam (2018) mentions that: ‘The aim here is to design a gradient and bring the dike, the etching path and the river into contact with each other and as such strengthen its continuity (Fig. 6.8).

On the south side of the river the dyke is further away as the ports are in between the river and the city. The dike is usually located at a distance from the water. Here, the aim is to create easily accessible, special places near the river or the port inlets.

6.3.3.4 Extracting Principles for Building with Nature from Tidal River Island ‘De Zaag’

To encourage the development of tidal nature, the right conditions can be stimulated so different gradients and habitats can develop (Fig. 6.9). For instance, gradients from wet to dry, can be encouraged by design. Especially in a tidal river where different water levels influence the floodplains. Next to high- and low tide there is spring tide. This is the period of the tide in which the difference between high and low water is greatest. It occurs about every 14 days.

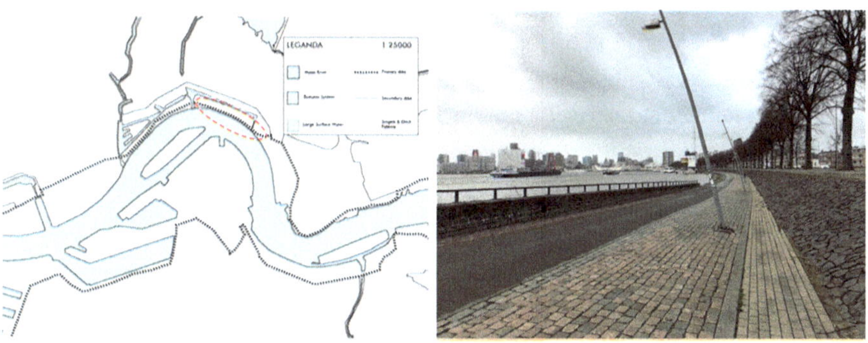

Fig. 6.8 The dike on the northern shore is closer to the river (map), nowadays its continuity is expressed as a water defense structure (right). In many cases there is space for adding natural qualities or planning a tidal river park that links the river to the city. (Map courtesy of Roberto Wijntje Santamaria, photo by Man-Chuan Lin)

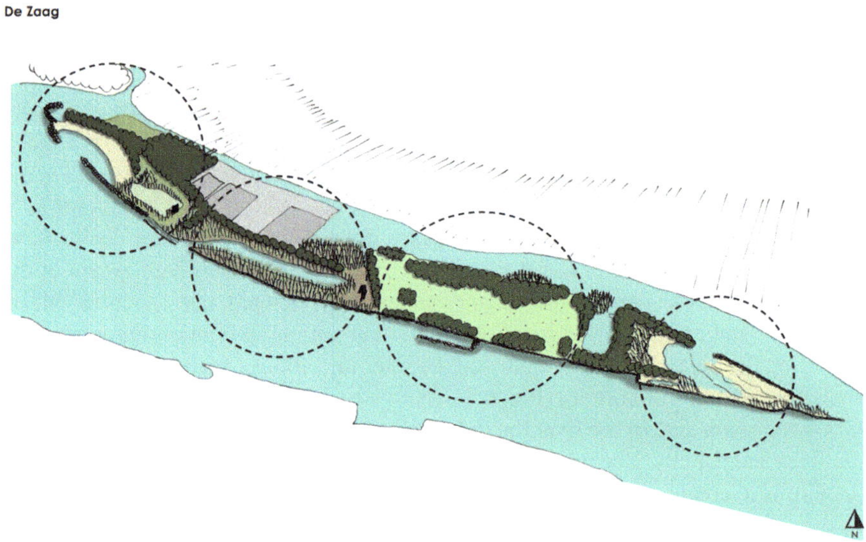

Fig. 6.9 The Zaag is an island in the river the Nieuwe Maas which for a large part was a brown field area. Different principles were applied to create conditions so a tidal river park with different habitats could develop over time. The height difference between high and low tide is still 1.3 m and the water is mild brackish. Apart from being a habitat and breeding ground for many species of fish it's also a habitat and green stepping stone for birds as well as beavers and other mamals. (Images courtesy of Emma Kannekens)

Not only is the high water higher than average, the low water is therefore lower than average.

Small interventions can be the difference between a flooded area or not, creating a variety of different habitats. Interventions can be raising slopes and lowering or strengthening dykes and groynes. Inside river bends show sedimentation and outside river bends show erosion. The speed of the river currents determines what material drops down. In high currents, stones and gravel drop down, next is sand. Clay particles only settle on the bottom of calm water. By responding well to the hydro-morphology of the river, one can play with these sedimentation processes.

6.3.3.5 Ongoing Projects

Along the river there are many projects in different phases of development.

Gemeente Rotterdam (2018) mentions the project near Rozenburg where the groynes have been strengthened and two longitudinal barriers (strekdammen) have been constructed in the water (Fig. 6.10). The effects are that tidal marshes start to develop and reed is growing and new habitats are forming.

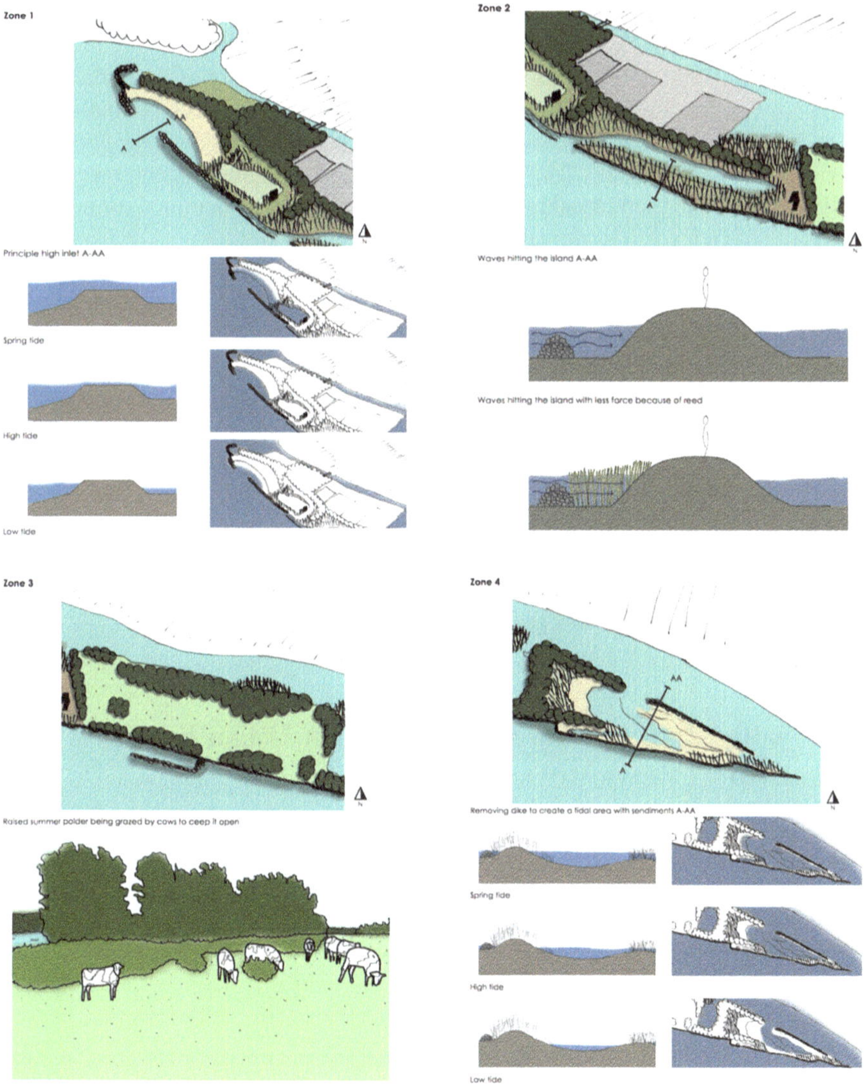

Fig. 6.10 Four principles to create a tidal river park are explained. The principles are; the height of the inlet, elongated dams, raised polder with grazing cows and the removing of the dike in some areas. Each principle is explained by sections or images and with each habitat the vegetation and animals are shown. (Images courtesy of Emma Kannekens)

The Wilhelminahaven (port inlet) in the city of Schiedam is in the design phase. This is one of the few places where the city of Schiedam and the port meet.

In Rotterdam at the south bank of the river Maas, the implementation of Mallegat tidal river park started in 2016. However due to erosion problems the project was put on hold. New research is required of the effects on the riverbanks before continuation is possible.

The World Wildlife Fund, Rijkswaterstaat, the municipality of Rotterdam, Zuid Hollands Landschap and ARK Nature Development also announced to realize tidal park 'Eiland van Brienenoord' in 2019. An important intervention is the creation of a river channel over the length of the island. Prior independent research in collaboration with users, local residents and companies have identified the possibilities for this (Gemeente Rotterdam 2018).

In the west of Rotterdam, north of the river Maas, the Keilehaven and the Schiehaven (two port inlets) offer opportunities for a tidal park in the long term. The quays are still in use.

A more large-scale project is the tidal park in the Maashaven. Here is an opportunity to create an urban park with recreational significance for the surrounding residential areas on the south bank. An area with few parks and green for recreation. The Municipality of Rotterdam and the Port of Rotterdam Authority are investigating the possibility to reuse sand that is released at the deepening of the Nieuwe Waterweg nearby (by dredging) for the development of a tidal park.

The Rotterdam 'River as a tidal park program' facilitates local planning and execution with design workshops, expertise, regional communication and the search for funds to build the planned projects.

6.4 A Research Agenda for the Design of Urban Ecology TU Delft

Several key aspects of urban ecology and landscape architecture will be explored. Both knowledge fields will be applied in research of urban and landscape planning and design and will focus on planning urban ecological structures, biodiversity in the context of sustainable cities.

1. Building a conceptual framework; how do urban ecology, landscape architecture and sustainable urban development interrelate?
2. Case study analysis: which recent programs and projects in (Dutch) cities provide examples and references for good practice on different scales?
3. Development and testing of design principles: which design methods and instruments can be formulated to improve urban ecological (as well quality of life) performance?
4. Strengthening the discipline: how can landscape architecture as well as urban ecology gain effectiveness in the context of urban sustainability goals?

Recent programs in Dutch cities such as described in paragraph three already lead to a new agenda for the urban green structure.

Although the urban green structure in 2005 was nature based, it was at the time not set up from an urban ecology perspective. To strengthen this urban ecology perspective some new challenges are addressed.

- What are *basic ecological conditions* in (Dutch) cities related to ambitions (from species to habitats), spatial arrangements & design, natural processes and maintenance.
- *Relating urban forestry, to urban ecology and landscape architecture.* In 2009 the Urban Green Structure Plan (Greef et al. 2005) was supplemented with an urban trees vision (Gemeente Rotterdam 2009). Planting trees and the choice of species can have a big influence over time as fully-grown specimen are not that easy replaced or planted. In the past decades, many Plane trees (*Platanus*) and Lime trees (*Tillia*) were planted for their majestic appearance. Nowadays one questions the ecological value of *Platanus* in cities. Other issues relate to the type of tree and how fast grows and its lifespan. To achieve quick results after 1945, many Poplars (*Populus*) were planted. Now many of them need replacing as old Poplars tend to lose large branches. This is an opportunity for a well-balanced mix in age, species, appearance and ecological benefits. Based on soil conditions and habitat knowledge the Tree Structure vision helps to locate where to plant what species. From an urban planning and design perspective one can argue that one species for lanes should be used. Is this the case everywhere? Where can different species be used?
- What will classical urban green typologies look like and how can they be enriched from object & ensemble level to strategy level?
- Identify principles for different (urban)ecosystems and habitats how design can create natural conditions so new habitats form. How can they become part of a landscape architects and urban planner's toolbox?

6.5 Conclusion: A Landscape Urbanism Perspective on Urban Ecology

To (re)build cities as sustainable ecosystems and find out what approach to follow, urban ecology is approached from a landscape architectural perspective. These two worlds meet at connected integrating themes (Fig. 6.11) as shown in the projects in this chapter. These connected integrating themes are key to potential synergies. A simple example is designing parkways to structure a city and regulate its water functions on a large scale while also make it accessible for biking and walking. This can be designed in a very technocratic way with concrete quays for waterways or in a way making use of natural elements such as soft water edges providing gradients for other forms of life. Also in the examples in paragraph three these relations are shown.

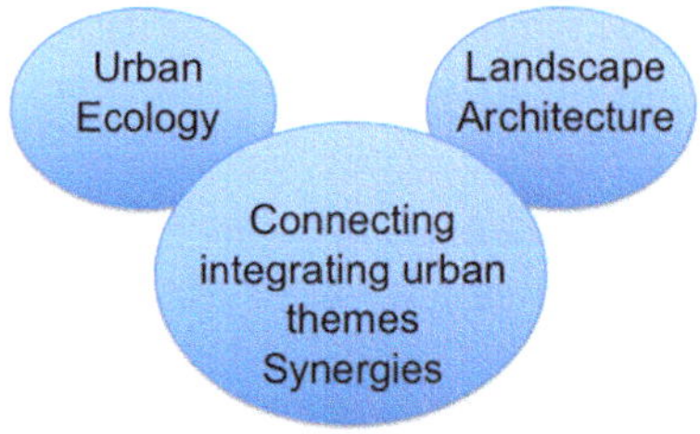

Fig. 6.11 Connecting integrating themes link the natural to the cultural and urban landscape environment and should deliver synergies. For instance, a Tidal Park can provide natural and recreational qualities as well as increased safety

The Urban Green Structure in Rotterdam based on natural systems can be used as a basic framework for the conservation and strengthening of existing (urban) ecological qualities as well as the development of new ecological qualities. Examples of this are the strengthening of the main axis of in the Urban Green Structure Plan (Greef et al. 2005), the river Maas with tidal river parks. The Tidal Parks have become a powerful metaphor used to convey the interdependencies between society and the natural systems on which it depends. It also shines light on the perspective spatial design disciplines have on urban ecology.

Also new qualities such as the seven green interventions in the city centre relate to the urban green structure.

Apart from the relation and links the urban green structure has to ecological qualities there is also a strong connection to other sectors such as recreational, health, transport and water management. The water structure in the Rotterdam Water City 2035 plan (Greef et al. 2005b) is directly linked to the urban green structure. The urban green structure is a crucial aspect in the sustainable development of the city. Within the context of sustainable urban development and synergetic urban landscape planning, Tillie (2018) talks about three perspectives: Urban ecology, landscape architecture and governance. Crucial elements to continue are the integrated point of view, using design and planning as well as a multi-disciplinary approach. Several overlapping frameworks use this. Tjallingi (1995) describes in the Ecopolis strategy **Flows** (urban ecology), **Areas** (Landscape architecture & urban landscape planning) and **Actors** (governance). These different elements will be used in a follow up studies as well as to work on an urban ecology research agenda at TU Delft.

Acknowledgements I would like to thank the city of Rotterdam for its cooperation, in particular Pieter de Greef who took a leading role in this. Also, from Bird Life Netherlands, Jip Louwe Kooijmans and Robert Kwak, for making the urban ecology research fellowship possible and exploring new grounds between urban ecology landscape architecture and urbanism. Last but not least the Landscape Architecture students at TU Delft for their input, drawings photos and many original design ideas in the waterscapes, tidal river park studio 2018–2019, in particular Emma Kannekens and Matthijs Hollanders.

References

Alberti M (2009) Advances in urban ecology, integrating humans and ecological processes in urban ecosystems. Springer, New York

Alberti M, Marzluff JM, Shulenberger E et al (2003) Integrating humans into ecology: opportunities and challenges for studying urban ecosystems. Bioscience 53:1169–1179

Collins JP, Kinzig A, Grimm NB, Fagan WF, Hope D, Wu J, Borer ET (2000) A new urban ecology. Am Sci 88:416–425

De Greef P, Den Heijer M, Tillie N, Soeterbroek M (2005a) Groenplan Rotterdam 2005. Gemeente Rotterdam, Rotterdam

De Greef P, Aarts M, Molenaar A, Oosterman J, Jacobs J (2005b) Rotterdam Waterstad 6. Episode Publishers, Rotterdam

Forman RTT (2014) Urban ecology, science of cities. Cambridge University Press, New York

Frantzeskaki N, Tillie N (2014) The dynamics of urban ecosystem governance in Rotterdam, the Netherlands. Ambio 43(4):542–555. https://doi.org/10.1007/s13280-014-0512-0

Gemeente Rotterdam (2009) Rotterdamse Stijl, Bomenstructuurvisie. Gemeente Rotterdam, Rotterdam

Gemeente Rotterdam (2013) Investeren in duurzame groei, Rotterdamse Duurzaamheidsmonitor. Gemeente Rotterdam, Rotterdam

Gemeente Rotterdam (2018) De Rivier als Getijdenpark, groeidocument 2018. Dienst Stadsontwikkeling, Rotterdam

Haaksma M (2017) Roofstructure Rotterdam: designing the fifth façade of the city Centre in Rotterdam. Master thesis. Delft University of Technology

Kennedy C, Pincetl S, Bunje P (2010) The study of urban metabolism and its application to urban planning and design. Environ Pollut 159:1965–1973

Newman P (1999) Sustainability and cities: extending the metabolism model. Landsc Urban Plan 44(1999):219±226

Niel L (2015) Productive Riverscapes as a tool to connect people and nature in the city of Rotterdam. Master thesis. Delft University of Technology

Odum EP, Barrett GW (2005) Fundamentals of ecology, 5th edn. Thompson Brook/Cole, Belmont

Pye M (2015) Aan de rand van de wereld, hoe de Noordzee ons vormde. De Bezige Bij, Amsterdam

Sijmons D (2014) Urban by nature. International architectural biennale Rotterdam 2014

Steenbergen C, Reh W, Nijhuis S, Pouderoijen M (2009) The polder atlas of the Netherlands. Uitgeverij THOTH, Bussum. ISBN:978-90-6868-519-0

Tillie N (2009) Climate adaptation strategies Rotterdam. Presentation at METREX, first US – European conference of metropolitan regions and areas, Alexandria, Virginia USA

Tillie N (2017) Redesigning urban water systems and exploring synergies lessons from an urban planning perspective on the 'Rotterdam Water City 2035, vision and follow ups 2005–2016. Eco web town. J Sustain Design 16:58–69

Tillie N (2018) Synergetic urban landscape planning in Rotterdam: Liveable Low-Carbon Cities. PhD-thesis. Delft University of Technology. https://doi.org/10.7480/abe.2018.24

Tillie N, van der Heijden R (2016) Advancing urban ecosystem governance in Rotterdam: from experimenting and evidence gathering to new ways for integrated planning. Environ Sci Pol 62:139–144. https://doi.org/10.1016/j.envsci.2016.04.016

Tillie N, Aarts M, Marijnissen M, Stenhuijs L, Borsboom J, Rietveld E, Visschers J, Lap S (2012) Rotterdam people make the inner city, densification plus greenification = sustainable city. Rotterdam Mediacenter, NL

Tillie N, Klijn O, Frijters E, Borsboom J, Looije M, Sijmons DF (2014) Urban metabolism, sustainable development in Rotterdam. IABR 2014 URBAN BY NATURE, Rotterdam

Tillie N, Dudok I, Pol P, Boot L, van der Heijden R (2016) Rotterdam case study: quality of life in remaking Rotterdam. In: Carter DK (ed) Remaking post-industrial cities: lessons from North America and Europe. Routledge – Taylor & Francis Group, New York

Tillie N, Borsboom-van Beurden J, Doepel D, Aarts M (2018) Exploring a stakeholder based urban densification and greening agenda for Rotterdam Inner City—accelerating the transition to a Liveable low Carbon City. Sustainability 10:1927. https://doi.org/10.3390/su10061927
Tjallingi SP (1995) Ecopolis, strategies for ecologically sound urban development. Backhuys Publishers, Leiden
Willemsen E, Tillie N (2018) Reconnecting green: towards a multi-dimensional biophilic city. In: Proceedings IIFLA conference 2018. IFLA, pp. 1130–1138, 55th IFLA world congress 2018, Singapore
Wolman, A. (1965) The metabolism of cities. Sci Am 213:179–190. https://doi.org/10.1038/scientificamerican0965-178

Chapter 7
From Urban Acupuncture to the Third Generation City

Marco Casagrande

Abstract The crisis of urbanism is analyzed as a vital phenomenon that prepares the Third Generation City—its connection with nature and its flesh. The industrial city is, on the contrary, fictitious. The example of the settlement of Treasure Hill, near Taipei, is given. As an organic ruin of the industrial city, Treasure Hill is a bio-urban site of resistance and an acupuncture point of Taipei, with its own design methodology based on Local Knowledge. This ruin is the matter from which parasite urbanism composts the modern city. Another example is offered by observing the daily life in Mumbai's unofficial settlements. Urban acupuncture, the Third Generation City, and the conceptual model of Paracity speak to the community that rests in the hands of its own people.

Keywords Urban acupuncture · Bio-urbanism · Third Generation City · Ruins · Parasite urbanism · Paracity · Local knowledge

7.1 Introduction

Missis Chen (Fig. 7.1) is 84 years old. She has lived together with the Xindian River all her life. Her family used to have a boat, like every Taipei family, and a water buffalo. Sometimes the kids would cross the river on the back of the buffalo. Sometimes an uncle might end up so drunk, that they hesitated, if they could put him back on his boat after an evening together. Children, vegetables, and laundry were washed in the river. The water was drinkable and the river was full of fish, crabs, snails, clams, shrimp, and frogs to eat.

Missis Chen used to work for sand harvesters, who dug sand out of the river bottom for making concrete. She made food for them. Many of the sand harvesters

M. Casagrande (✉)
Ruin Academy, Casagrande Laboratory, Helsinki, Finland

Bergen School of Architecture, Bergen, Norway

International Society of Bio-Urbanism, Artena, Italy
e-mail: marco.casagrande@casagrandelaboratory.com

R. Roggema (ed.), *Nature Driven Urbanism*, Contemporary Urban Design Thinking, https://doi.org/10.1007/978-3-030-26717-9_7

Fig. 7.1 Missis Chen drawing. (Source: Casagrande Marco 2013)

lived in the Treasure Hill settlement together with Missis Chen's family. In the past, the hill had been a Japanese Army anti-aircraft position, and it was rumored that the Japanese had hidden a treasure of gold somewhere in their bunker networks inside the hill—hence the name Treasure Hill.

Xindian River was flooding – like all Taipei rivers – when the frequent typhoons arrived in summers and autumns. The flood was not very high, though – the Taipei Basin is a vast flood plain and water has plenty of space to spread out. Houses were designed so that the knee-high flood would not come in, or in some places, the water was let into the ground floor while people continued to live on the upper floors. In Treasure Hill, the flood would also come into the piggeries and other light-weight structures on the river flood bank, but the houses with people were a bit higher up on the hill. All of the flood bank was farmed, and the farms and vegetable gardens were constructed so that they could live together with the flood. Flooding was normal. This pulse of nature was a source of life.

Missis Chen remembers when the river got polluted. "The pollution comes from upstream," she says, referring to the many illegal 'Made in Taiwan'—factories up on the mountains and river banks, which let all their industrial waste into the river. "Now not even the dogs eat the fish anymore." At some point, the river became so polluted that Taipei children were taught not to touch the water or they would go blind. The flood became poisonous for the emerging industrial city, which could no longer live together with the river nature. The city built a wall against the flooding river: a 12-meter high, reinforced concrete flood wall separating the built urban environment from nature (Fig. 7.2).

"One day, the flood came to Chiang Kai-shek's home and the Dictator got angry. He built the wall. We call it the Dictator's Wall," an elderly Jiantai fisherman recalls sitting in his bright blue boat with a painted white eye and red mouth and continues to tell his stories describing which fish disappeared which year, and when some of the migrating fishes ceased to return to the river. In one lifetime, the river has

Fig. 7.2 Taipei flood wall. (Photograph by the Author)

transformed from a treasure chest of seafood into an industrial sewer, which is once again being slowly restored towards a more natural condition. The wall hasn't moved anywhere. The generations of Taipei citizens born after the 1960s don't live in a river city. They live in an industrially-walled urban fiction separated from nature.

7.2 Treasure Hill

In 2003, the Taipei City Government decided to destroy the unofficial settlement of Treasure Hill (Kang 2006). By that time, the community consisted of some 200 households of, mainly elderly Kuomintang veterans and illegal migrant workers. The bulldozers had knocked down the first two layers of the houses of the terraced settlement on the hillside. After that, the houses were standing too high for the bulldozers to reach, and there were no drivable roads leading into the organically built settlement. Then the official city destroyed the farms and community gardens of Treasure Hill down by the Xindian River flood banks. Then they cut the circulation between the individual houses—small bridges, steps, stairs, and pathways. After that, Treasure Hill was left to rot, to die slowly, cut away from its life sources.

Roan Chin-Yueh of the WEAK! managed so that the City Government Department of Cultural Affairs invited me to Taipei. He introduced me to Treasure Hill's impressive organic settlement (Fig. 7.3) with a self-made root-cleaning

Fig. 7.3 Treasure Hill. (Photograph by Stephen Wilde)

system of gray waters through patches of jungle on the hillside. Treasure Hill was composting organic waste into fertilizer for the farms and using minimum amounts of electricity, which was stolen from the official grid. There was even a central radio system through which Missis Chen could transmit important messages to the community, such as inviting them to watch old black and white movies in the open-air cinema in front of her house.

At that point, the city had stopped to collect trash from Treasure Hill, and there were lots of garbage bags in the alleys. I started to collect these garbage bags and carried them down the hill into a pile close to a point that you could reach with a truck. The residents did not speak to me, but instead they hid inside their houses. One could feel their eyes on one's back, though. Some houses were abandoned when I entered them. The interiors and the atmospheres were as if the owners had left all of a sudden. Even photo albums were there and tiny altars with small gods with long beards. In one of the houses, I could not help looking at the photo album. The small tinted black and white photos started in mainland China, and all the guys wore Kuomintang military uniforms. Different landscapes in different parts of China, and then at some point the photos turned to color prints. The same guys were in Taiwan. Then there was a woman, and an elderly gentleman posed with her in civil clothes by a fountain. Photos of children and young people. Civil clothes, but the Kuomintang flag of Taiwan everywhere. A similar flag was inside the room. Behind me, somebody enters the house, which is only one room with the altar on one end and a bed on the other. The old man is looking at me. He is calm and observant, somehow sad. He speaks and shows with his hand at the altar. "Do not touch" – I understand. I look the old man in the eyes and he looks into mine. I feel like looking at the photo album. The owner of the house must have been his friend. They have travelled together a long way from the civil war of China to Taiwan. Literally, they have built their houses on top of Japanese concrete bunkers and created themselves a life in Treasure Hill. His friend passed away. There is a suitcase and inside is the absent owner's trousers and his shirt, both in khaki color. I continue collecting the garbage bags and carry the old man's bag around the village. The next day the residents start helping me with collecting the garbage. Professor Kang Min-Jay organizes a truck to take the bags away. After a couple of days, we organize a public ceremony together with some volunteer students and Treasure Hill veterans and declare war on the official city: Treasure Hill will fight back and it is here to stay. I'm wearing the dead man's clothes (Figs. 7.4 and 7.5).

We have a long talk with Professor Roan about Treasure Hill and how to stop the destruction. He suggests that Hsieh Ying-Chun (Atelier 3, WEAK!) will join us with his aboriginal Thao tribe crew of self-learned construction workers. I start touring at local universities giving speeches about the situation and try to recruit students for construction work. In the end, we have 200 students from Tamkang University Department of Architecture, Chinese Cultural University, and National Taiwan University. A team of girl students manage to make a deal with the neighboring bridge construction site workers, and they start offloading some of the construction material cargo to us from the trucks passing us by. Mainly we receive timber and bamboo; they use mahogany for the concrete molds.

With the manpower and simple construction material, we start reconstructing the connections between the houses of the settlement, but most importantly, we also restart the farms. The bridge construction workers even help us with a digging machine. Missis Chen comes to advise us about the farming and offers us food and

Fig. 7.4 Reconstructed steps in Treasure Hill. (Photograph by Stephen Wilde)

Fig. 7.5 Collective farm in Treasure Hill. (Photograph by Stephen Wilde)

Chinese medicine. I am invited to her house every evening after the workday with an interpreter. She tells the story of her life and I see how she is sending food to many houses whose inhabitants are very old. Children from somewhere come to share our dinners as well. Her house is the heart of the community. Treasure Hill veterans join us in the farming and construction work. Rumors start spreading in Taipei: things are cooking in Treasure Hill. More people volunteer for the work, and after enough urban rumors, suddenly the media arrives. After the media, the politicians follow. Commissioner Liao from the City Government Cultural Bureau comes to recite poems. Later Mayor Ma Ying-Jeou comes jogging by with TV-crews in his slip-stream and gives us his blessings. The City Government officially agrees that this is exactly why they had invited me from Finland to work on the issue of Treasure Hill. The same government had been bulldozing the settlement away 3 weeks earlier.

One can design whole cities simply with rumors.

Working in Treasure Hill had pressed an acupuncture point of the industrial Taipei City. Our humble construction work was the needle that had penetrated through the thin layer of official control and touched the original ground of Taipei—collective topsoil where Local Knowledge is rooting. Treasure Hill is an urban compost, which was considered a smelly corner of the city, but after some turning is now providing the most fertile topsoil for future development. The Taiwanese would refer to this organic energy as "Chi."

7.3 Urban Acupuncture

After this initial discovery in Treasure Hill, the research of Urban Acupuncture continued at the Tamkang University Department of Architecture, where Chairman Chen Cheng-Chen, under my professorship, added it to the curriculum in the autumn of 2004 (Casagrande and Ross 2004; Epifanio 3 2005). In 2009, the Finnish Aalto University's Sustainable Global Technologies research center with Professor Olli Varis (Casagrande 2009) joined in to further develop the multidisciplinary working methods of Urban Acupuncture in Taipei, with focus on urban ecological restoration through punctual interventions (Casagrande 2011a). In 2010, the Ruin Academy was launched in Taipei with the help of the JUT Foundation for Arts & Architecture (Harrison 2012). The Academy operated as an independent multidisciplinary research center moving freely in between the different disciplines of art and science within the general framework of built human environment. The focus was on Urban Acupuncture and the theory of the Third Generation City.

Third Genereation City is the organic ruin of the industrial city, an organic machine and open form of the mechanical urbanism which has learned to become biological. Ruin is when man-made has become part of nature. The industrial control has been opened up in order for the nature to step in. (Dudareva 2015). The seeds of the third generation city are coexisting together with the current industrial urbanism – for example the illegal collective urban farms and settlements of Taipei. These are the punctual interventions which are penetrating through the thin layers of asphalt and concrete and reach the original soil – tuning the city towards the organic. The process towards the third generation city is happening all the time and in all the scales. Nature has only one rule: existence maximum. It wants the city to be part of the life-providing process. Now our cities are anti-acupuncture needles in the life-providing tissue.

Ruin Academy collaborated with the Tamkang University Department of Architecture, the National Taiwan University Department of Sociology, Aalto University SGT, the Taipei City Government Department of Urban Development, and the International Society of Bio-urbanism.

Urban Acupuncture is a bio-urban theory (Casagrande 2013), which combines sociology and urban design with the traditional Chinese medical theory of acupuncture. As a design methodology, it is focused on tactical, small-scale interventions on the urban fabric, aiming in ripple effects and transformation on the larger urban organism (Kaye 2011). Through the acupuncture points, Urban Acupuncture seeks to be in contact with the site-specific Local Knowledge. By its nature, Urban Acupuncture is pliant, organic, and relieves stress and industrial tension in the urban environment, thus directing the city towards the organic—urban nature as part of nature. Urban Acupuncture produces small-scale, but ecologically and socially catalytic development on the built human environment (Kim 2010).

Urban Acupuncture is not an academic innovation. It refers to common collective Local Knowledge practices that already exist in Taipei and other cities, self-organizing practices that are tuning the industrial city towards the organic machine—the Third Generation City.

In Taipei, the citizens ruin the centrally governed, official mechanical city with unofficial networks of urban farms and community gardens. They occupy streets for night markets and second-hand markets, and activate idle urban spaces for karaoke, gambling, and collective exercises (dancing, Tai-Chi, Chi-Gong, et cetera). They build illegal extensions to apartment buildings and dominate the urban no man's land by self-organized, unofficial settlements, such as Treasure Hill. The official city is the source of pollution, while the self-organized activities are more humble in terms of material energy-flows and more tied with nature through the traditions of Local Knowledge. There is a natural resistance towards the official city. It is viewed as an abstract entity that seems to threaten people's sense of community and separates them from the biological circulations.

Urban Acupuncture is Local Knowledge in Taipei, which on a larger scale, keeps the official city alive (Fig. 7.6). The unofficial is the biological tissue of the mechanical city. Urban Acupuncture is a form of bio-urban healing and a development process connecting modern man with nature.

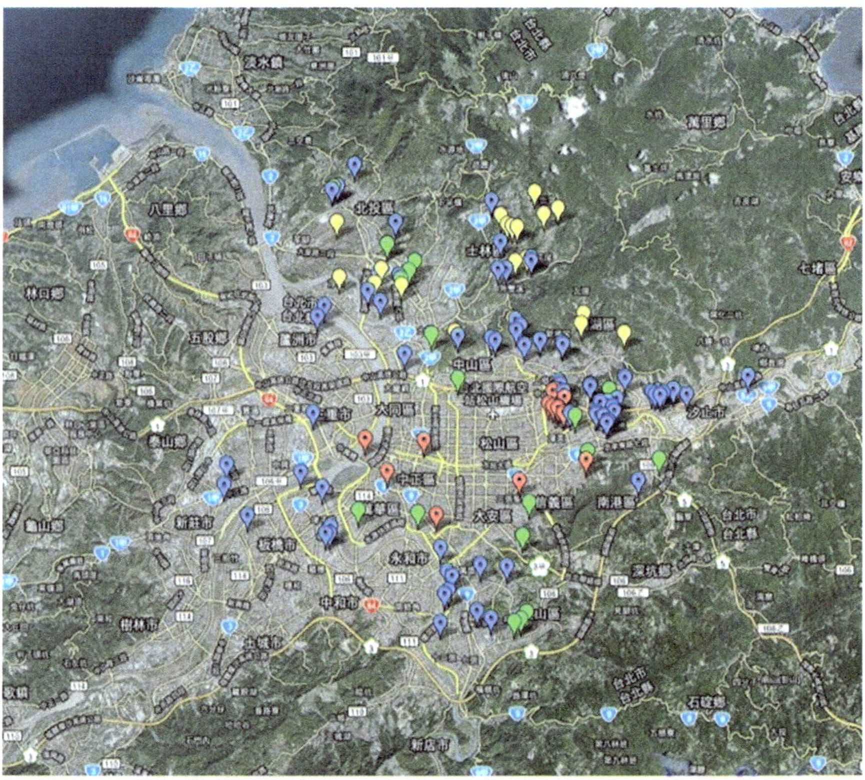

Fig. 7.6 Unofficial community gardens and urban farms of the Taipei Basin, the real map of Urban Acupuncture. (Image by the Ruin Academy)

7.4 Third Generation City

The first-generation city is the one where the human settlements are in straight connection with nature and dependent on nature. The fertile and rich Taipei Basin provided a fruitful environment for such a settlement (Casagrande 2011b). The rivers were full of fish and good for transportation, with the mountains protecting the farmed plains from the straightest hits of the frequent typhoons.

The second-generation city is the industrial city. Industrialism granted the citizens independence from nature—a mechanical environment could provide everything humans need. Nature was seen as something unnecessary or as something hostile—it was walled away from the mechanical reality (Casagrande 2011b).

The Third Generation City is the organic ruin of the industrial city, an open form, an organic machine tied with Local Knowledge and self-organized community actions (Yudina 2018). The community gardens of Taipei are fragments of third generation urbanism when they exist together with their industrial surroundings. Local Knowledge is present in the city, and this is where Urban Acupuncture is rooting. Among the anarchist gardeners are the Local Knowledge professors of Taipei.

The Third Generation City is a city of cracks (Casagrande 2016). The thin mechanical surface of the industrial city is shattered, and from these cracks the new bio-urban growth emerges, which will ruin the second generation city. Human-industrial control is opened up in order for nature to step in. A ruin is when the manmade has become part of nature. In the Third Generation City, we aim at designing ruins (Mik 2018). The Third Generation City is true when the city recognizes its local knowledge and allows itself to be part of nature.

To find a form that accommodates the mess, that is the task of the artist now. (Samuel Beckett).

7.5 Parasite Urbanism

The emerging bio-urban cities are not homogeneous platforms for single cultures, races, economical doctrines, timelines, or other ways of life or being. They are urban composts where organic knowledge is floating into the cracks of the industrially developed surroundings. This organic knowledge has the ability to treat and heal the surrounding city as a positive parasite. It can suck in and treat urban and even industrial waste, and it is able to build bridges between modern man and nature. It can grow to places where the industrial city cannot go to and, through punctual interventions, it can tune the entire urban development towards the organic; a built human nature as part of nature (Revedin 2015).

This symbiotic coexistence between the "official, developed" city, and unofficial, self-built and organic parasite bio-urbanism has been existing already for a long time with slums, favelas, camps of migrating workers, unofficial settlements, urban enclaves of resistance, community gardens and urban farms, and even refugee camps. These strongholds of urban nomads are harvesting the surrounding city from what it calls waste, surplus material streams of industrial life. Without these urban nomads, these material streams will end up in nature as, what we call, pollution. The unofficial is the buffer zone between development and nature—trying to save the city from itself.

This parasite urbanism should be encouraged to grow on the expense of industrial efficiency. It should eat the urban industrialism away up until a point, where the city is in tune with the life-providing systems of nature. Within this new bio-urban human mangrove, the relicts of the industrial hardness will emerge as islands, ridges or hills, maybe even as volcanoes. This urban compost is the Third Generation City. It already exists in many places and on many scales, from Jakarta to Rio, and from the collective urban farms of Taipei (Bauwens 2010) to the buffalo sheds of Mumbai (Casagrande 2006). It is not a utopia, but a way in which the different material cycles of cities have coexisted for much longer than industrialism.

For example, in Mumbai there have always been countless buffalo sheds along the monsoon floodwater streams. The respected animal gives fuel (dung-cakes) and milk to the surrounding city. Here, the river or stream is an essential part of this symbiosis. The buffalo dung is pushed to the low water stream, where women mix it by foot with straw before it gets transported back to the sheds for the making and drying of the dung-cakes. The buffaloes also need to get washed every day. The buffalo caretakers are living on decks above the animals (Fig. 7.7).

Fig. 7.7 Buffalo shed in Oshiwara, Mumbai. (Photo by Author 2006)

People have always brought their household waste from the surrounding city to the buffalo sheds in exchange for the milk and energy. The first one to eat from this organic waste is the buffalo, which will pick up the best parts. Then comes the goat, which can even eat paper. After the goat comes the dog, who goes through the possible small remnants of bones, skins, and meat. The last one in the chain is the pig, who will eat even rotten meat and already digested material (Casagrande 2006). The surrounding city cannot live without the buffalo sheds. This chain of animals worked perfectly before the age of industrial materials. Then, materials started to appear in the trash bags that even a pig could not consume—plastics, aluminum, et cetera. The city became in need of a new animal: man.

The slums of Mumbai have grown around the buffalo sheds. Millions of people have been transported from the poorest areas of India to take care of the developed city. Only in the Owhiwara River chain of slums is there estimated to live some 700,000 inhabitants. The recycling stations and illegal factories (Figs. 7.8 and 7.9) are situated here, just next door to Bollywood. What cannot be recycled or treated ends up in the river, just like in Jakarta it ends up in the bay. Monsoon will flush the toilet.

The buffalo sheds are the original acupuncture needles of Mumbai. Now, together with slums, they present a strong culture of parasite urbanism. The harvesting, processing, and recycling of urban waste is harmful for the people who do it and for nature. The Third Generation City is looking towards a situation where the parasite urbanism has reached another level presenting a bio-urban balance between the rivers, slums, and the surrounding city.

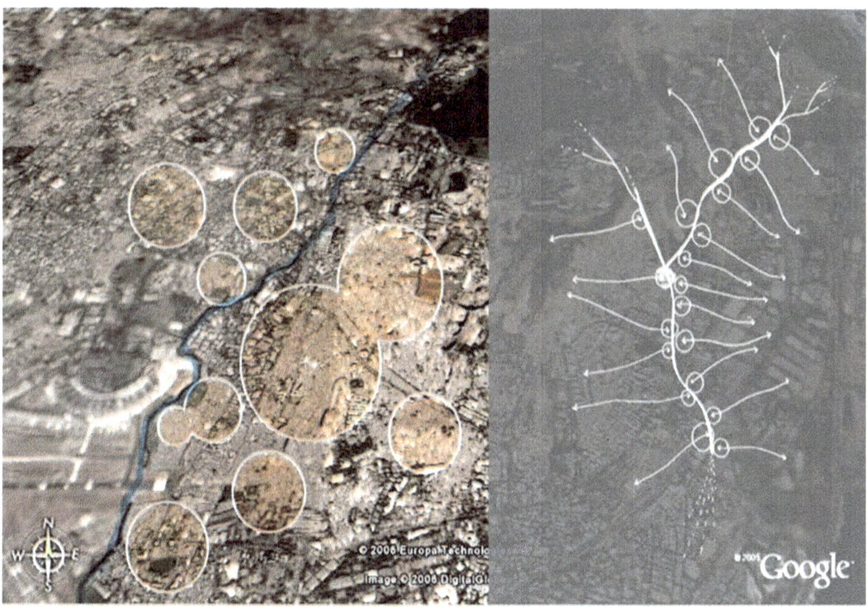

Fig. 7.8 Location of recycling stations in the Oshiwara chain of sluims, Mumbai. (Casagrande workshop, Urban Flashes Mumbai, 2006)

Fig. 7.9 Workers in one of the recycling stations separating aluminium from bottles. (Photo by Author, 2006)

7.6 Paracity

Learning from the cases of Taipei and Mumbai, we have developed a conceptual model to further study the possibilities of parasite urbanism: Paracity (Casagrande 2015a; Taipei City Government Department of Urban Development and Chinese Central Art Academy 2014). Paracity is a bio-urban organism that is growing on the principles of Open Form (Casagrande 2015b): individual design-built actions generating spontaneous communicative reactions on the surrounding built human environment. This organic constructivist dialogue leads to self-organized community structures, sustainable development, and knowledge building. Open Form is close to the original Taiwanese ways of developing the self-organized and often "illegal" communities. These micro-urban settlements contain a high volume of Local Knowledge, which we believe will start composting in Paracity, once the development of the community is in the hands of the citizens.

The agritectural organism of the Paracity is based on a primary wooden three-dimensional structure, an organic grid with spatial modules of 6 x 6 x 6 meters, constructed out of CLT (cross-laminated timber) beams, and columns (Fig. 7.10).

This simple structure can be modified and developed by the community members. The primary structure can grow even in neglected urban areas such as flood plains, hillsides, abandoned industrial areas, storm water channels, and slums. Paracity is perfectly suited for flooding and tsunami risk areas and the CLT primary structure is highly fire-resistant and capable of withstanding earthquakes (HolzBuild 2009).

Paracity (Fig. 7.11) provides the skeleton, but the citizens create the flesh. Design should not replace reality—'Flesh is More'. Paracitizens will attach their individual, self-made architectural solutions, gardens, and farms on the primary structure, which will offer a three-dimensional building grid for Do-It-Yourself (DIY) architecture. The primary structure also provides the main arteries of water and human circulation, but the finer Local Knowledge nervous networks are weaved in by the inhabitants. Large parts of Paracity is occupied by wild and cultivated nature following the example of Treasure Hill and other unofficial communities in Taipei.

Paracity's self-sustainable bio-urban growth is backed up by off-the-grid modular environmental technological solutions, providing methods for water purification, energy production, organic waste treatment, waste water purification, and sludge recycling. These modular plug-in components can be adjusted according to the

Fig. 7.10 High-end version of the Paracity in Taipei. (Source: Casagrande Laboratory 2010)

Fig. 7.11 Paracity model. (Photograph by the Author)

Fig. 7.12 Schematic section of a fragment of the Paracity Taipei. (Image sourced by the Author)

growth of the Paracity, and moreover, the whole Paracity is designed not only to treat and circulate its own material streams, but to start leeching waste from its host city and thus becoming a positive urban parasite following the similar kinds of symbiosis as in-between slums and the surrounding city. In a sense, Paracity is a high-tech slum, which can start tuning the industrial city towards an ecologically more sustainable direction. Paracity is a Third Generation City, an organic machine urban compost, which assists the industrial city to transform itself into being part of nature (Fig. 7.12).

The pilot project of the Paracity grows on an urban farming island of Danshui River, Taipei City (Figs. 7.13 and 7.14).

The island is located between the Zhongxing and Zhonxiao bridges and is around 1000 meters long and 300 meters wide. Paracity Taipei celebrates the original

Fig. 7.13 Paracity organic grid on a flooded island in Taipei. (Source: Casagrande Laboratory 2010)

first-generation Taipei urbanism with a high level of "illegal" architecture, self-organized communities, urban farms, community gardens, urban nomads, and constructive anarchy.

After the Paracity has reached critical mass, the life-providing system of the CLT-structure will start escalating. It will cross the river and start taking root on the flood plains. It will then cross the 12-meters high Taipei flood wall and gradually grow into the city (Fig. 7.15) The flood wall will remain in the guts of the Paracity, but the new structure enables Taipei citizens to fluently reach the river. Paracity will reunite the river reality and the industrial urban fiction. Paracity is a mediator between the modern city and nature. Seeds of the Paracity will start taking root within the urban acupuncture points of Taipei: illegal community gardens, urban farms, abandoned cemeteries, and wastelands (Fig. 7.16).

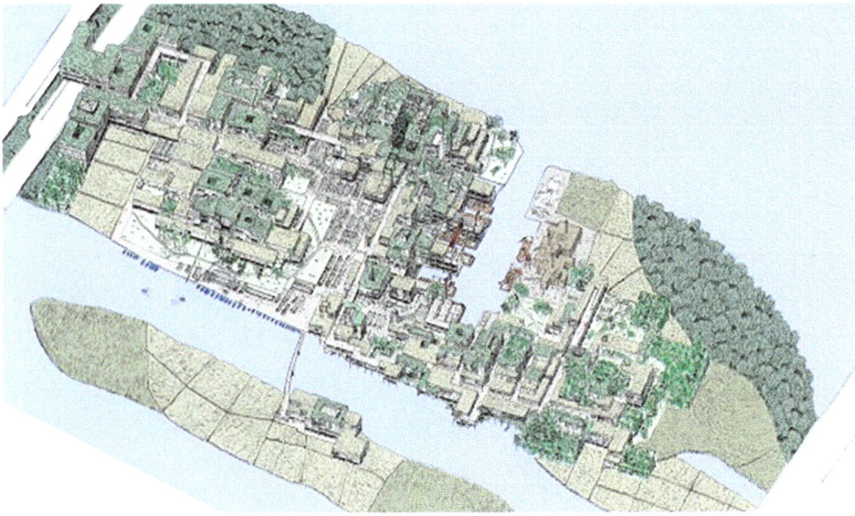

Fig. 7.14 The green roofs and collective farms of the Paracity Taipei. (Source: Casagrande Laboratory 2010)

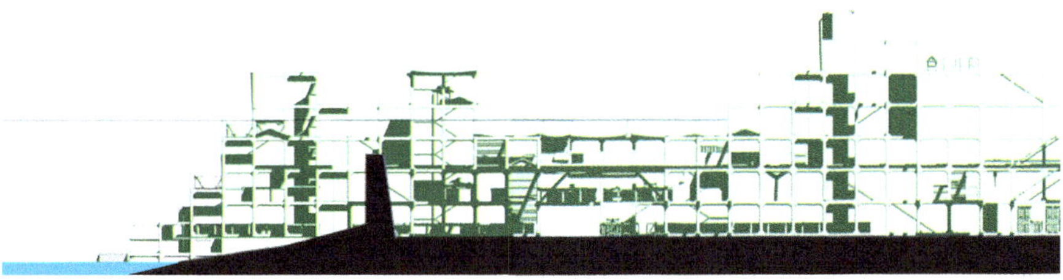

Fig. 7.15 Paracity growing over the concrete flood wall. (Source: Casagrande Laboratory 2010)

From these acupuncture points, Paracity will start growing by following the covered irrigation systems such as the Liukong Channel, and eventually the bio-urban organism and the static city will find a balance—the Third Generation Taipei. Paracity has a lot of holes, gaps, and nature between houses (Fig. 7.17).

This is a city of cracks. The system ventilates itself like a large-scale beehive of post-industrial insects. The different temperatures of the roofs, gardens, bodies of water and shaded platforms will generate small winds between them, and the hot roofs will start sucking in breeze from the cooler river. The individual houses should also follow the traditional principles of bioclimatic architecture and not rely on mechanical air-conditioning (Fig. 7.18).

The bio-urbanism of the Paracity is as much landscape as it is architecture (Fredirckson 2014). The all-encompassing landscape-architecture of Paracity includes organic layers for natural water purification and treatment, community gardening, farming, and biomass production as an energy source (Fig. 7.19).

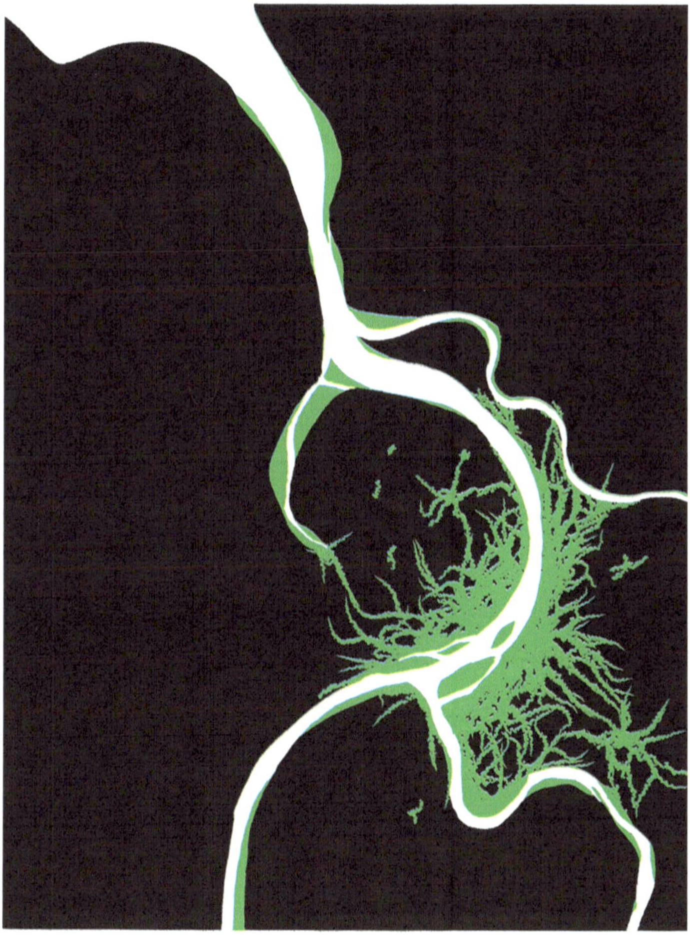

Fig. 7.16 Paracity growing into the surrounding city as a positive cancer. (Source: Casagrande Laboratory 2010)

Infrastructure and irrigation-water originates from the polluted Danshui River and will be both chemically (bacteria-based) and biologically purified before being used in the farms, gardens and the houses of the community (Fig. 7.20). The bacteria/chemically purified water gets pumped up to the roof parks on the top level of the Paracity, from where it will, using gravity, start to circulate into the three-dimensional irrigation systems (Fig. 7.21).

Paracity is based on free flooding. The whole city stands on stilts (Fig. 7.22), allowing the river to pulsate freely with the frequent typhoons and storm waters. The Paracity is actually an organic architectural flood itself, ready to cross the flood wall of Taipei and spread into the mechanical city.

Paracity Taipei will be powered mostly by bioenergy that uses the organic waste, including sludge, taken from the surrounding industrial city and by farming

Fig. 7.17 Paracity Taipei with the CLT primary structure supporting local life. (Source: Casagrande Laboratory 2010)

Fig. 7.18 First-generation Taipei. (Drawing by Niilo Tenkanen/Casagrande Laboratory)

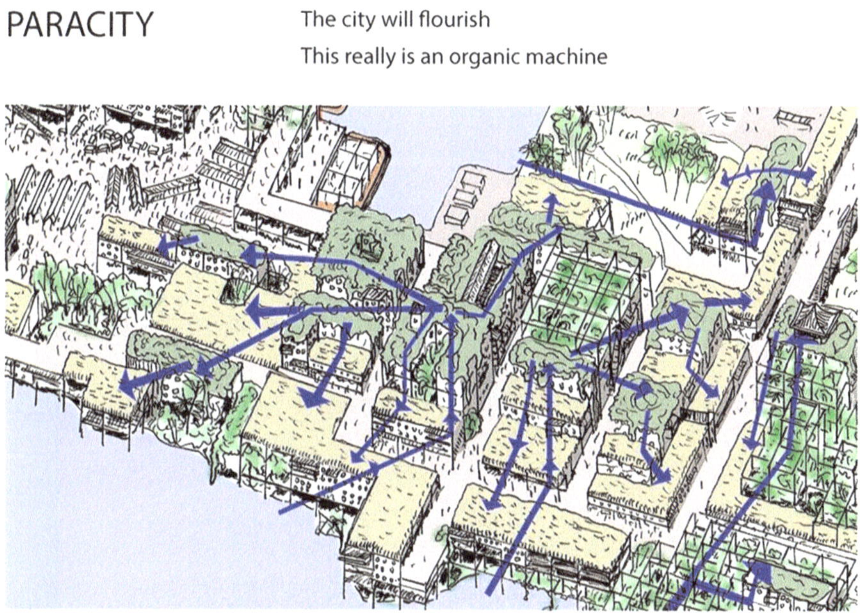

Fig. 7.19 Water circulaton of the Paracity originated from the polluted river and root-cleansed by the green roofs. (Source: Casagrande Laboratory 2010)

Fig. 7.20 4-story Paracity structure in the Stalker factory in Tallinn. (Photo by the Author 2015)

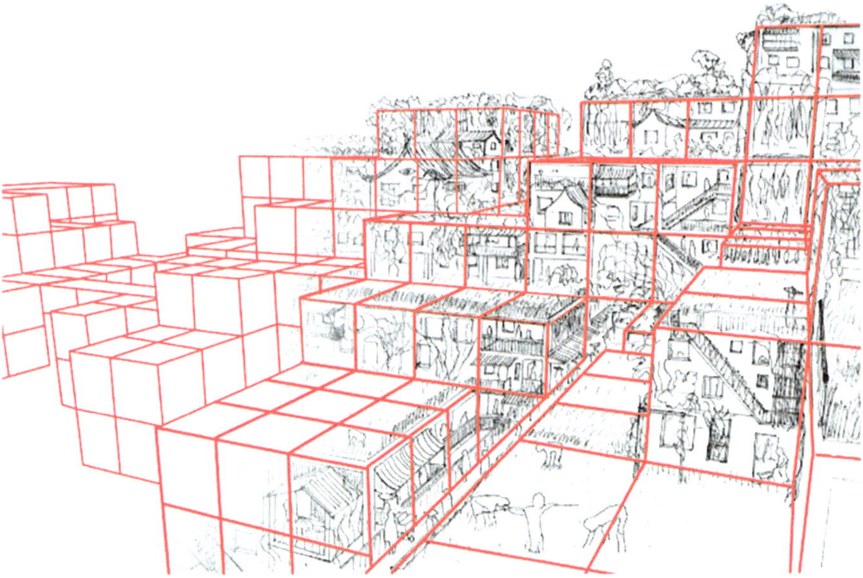

Fig. 7.21 Agritecture of the Paracity (Drawing by Niilo Tenkanen / Casagrande Laboratory)

Fig. 7.22 Paracity, flood-water scenario. (Image sourced by the Author)

Fig. 7.23 Paracity CLT-module, 6 x 6 x 6 m. (Photograph by Jan Feichtinger/Casagrande Laboratory)

fast-growing biomass on the flood banks of the Taipei river system. Paracity Taipei will construct itself through impacts of collective consciousness, and it is estimated to have 15,000–25,000 inhabitants.

The wooden primary structure (Fig. 7.23) and the environmental technology solutions will remain mostly the same, no matter in which culture the Paracity starts to grow, but the real human layer of self-made architecture and farming will follow the Local Knowledge of the respective culture and site. Paracity is always site-specific and it is always local. Other Paracities are emerging in North Fukushima in Japan and the Baluchistan Coast in Pakistan.

7.7 Conclusion

The way towards the Third Generation City is a process of becoming a collective learning and healing organism and of reconnecting the urbanized collective consciousness with nature. In Taipei, the wall between the city and the river must go. This requires a total transformation from the city infrastructure and from the centralized power control. Otherwise, the real development will be unofficial. Citizens on their behalf are ready and are already breaking the industrial city apart by themselves. Local knowledge is operating independently from the official city and is providing punctual third generation surroundings within the industrial city: urban acupuncture for the stiff official mechanism.

The weak signals of the unofficial collective consciousness should be recognized as the futures' emerging issues; futures that are already present in Taipei. The official city should learn how to enjoy acupuncture, how to give up industrial control in order to let nature step in.

The Local Knowledge-based transformation layer of Taipei is happening from inside the city, and it is happening through self-organized punctual interventions. These interventions are driven by small-scale businesses and alternative economies benefiting from the fertile land of the Taipei Basin, and of leeching the material and energy streams of the official city. This acupuncture makes the city weaker, softer, and readier for a larger change.

The city is a manifest of human-centered systems—economical, industrial, philosophical, political, and religious power structures. Bio-urbanism is an animist system regulated by nature. Human nature as part of nature, also within the urban conditions. The era of pollution is the era of industrial urbanism – the second generation city. The next era has always been surviving within the industrial city, like a positive cancer. The first-generation city never died, it went underground, but the bio-urban processes are still surviving. The seeds of the Third Generation City are present. Architecture is not an art of human control; it is an art of reality. There is no other reality than nature.

References

Bauwens, M. (2010) The community gardens of Taipei. Taipei Organic Acupuncture, P2P Foundation. https://blog.p2pfoundation.net/the-community-gardens-of-taipei/2010/12/04. Accessed: 3 Jan 2019

Casagrande M (2006) Can't push a river. Taiwan architect issue 4/2006

Casagrande M (2009) Guandu: river urbanism Taiwan Architect issue 10/2009. http://casagrandetext.blogspot.com/2009/10/guandu-river-urbanism.html

Casagrande M (2011a) Urban Ecopuncture. La Vie Issue 90, p 136–145. http://casagrandetext.blogspot.com/2011/10/urban-ecopuncture.html

Casagrande M (2011b) Taipei from the river. International Society of Biourbanism. www.biourbanism.org/taipei-from-the-river. Accessed 3 Jan 2019

Casagrande M (2013) Biourban acupuncture—from Treasure Hill of Taipei to Artena. International Society of Biourbanism, Rome

Casagrande M (2015a) Paracity: urban acupuncture. Oil Forest League, Netherlands

Casagrande M (2015b) De l'acupuncture urbain á la ville de 3ieme gènèration. In: Revedin J (ed) La Ville Rebelle. Democratiser le Projet Urbain, pp 38–49

Casagrande M (2016) From urban acupuncture to the third generation city. Urbanista 3. www.urbanista.org/issues/sustainable-urbanism/features/from-urban-acupuncture-to-the-third-generation-city-marco-casagrande. Accessed 3 Jan 2019

Casagrande Laboratory (2010) Anarchist gardener. Taipei: Ruin Academy. https://issuu.com/ruinacademy/docs/anarchist_gardener_issue_one. Accessed 28 Aug 2018

Casagrande M, Ross M (2004) Trojan Rocking Horse Taipei. www.youtube.com/watch?v=_aVAs9QH1lA. Accessed 3 Jan 2019

Dudareva L (2015) Ruins of the future. Strelka Institute, Moscow. https://strelka.com/ru/magazine/2015/06/01/interview-marco-casagrande. Transl: https://www.casagrandelaboratory.com/2014/05/14/ruins-of-the-future/

Epifanio 3 (2005) Tulevikupaviljon, Taivani Disaini Expo 2005. www.epifanio.eu/nr3/est/tulevikupaviljon.html. Accessed 3 Jan 2019

Fredirckson T (2014) Marco Casagrande presents modular paracity for habitare in Helsinki. designboom. www.designboom.com/architecture/marco-casagrande-paracity-habitare-helsinki-08-31-2014/. Accessed 3 Jan 2019

Harrison AL (2012) Architectural theories of the environment: post human territory. Routledge, New York

HolzBuild (2009) X-Lam Earthquake Test. www.youtube.com/watch?v=T08KRyVhyeo. Accessed 3 Jan 2019

Kang MJ (2006) Altered space: squatting and legitimizing treasure hill, Taipei. Cultural Development Network's Forum 'Activism: the role of arts in regeneration', 23 June 2006. www.culturaldevelopment.net.au/downloads/KangMinJay.pdf. Accessed 2 Jan 2019

Kaye L (2011, July 21) Could cities' problems be solved by urban acupuncture? The Guardian. http://www.guardian.co.uk/sustainable-business/urban-acupuncture-community-localised-renewal-projects. Accessed 28 Aug 2018

Kim J (2010) An anarchitect and an archetist have a talk. Ar2com.de: 1 March 2010. http://www.ar2com.de/radiofavela-blog/marco-casagrande/. Accessed 3 Jan 2019

Mik E (2018) If you are not connected with nature, you produce pollution. Archidea, 57. http://casagrandetext.blogspot.com/2018/11/if-you-are-not-connected-with-nature.html

Revedin J (2015) La Ville Rebelle. Paris: Gallimard. http://www.gallimard.fr/Catalogue/GALLIMARD/Alternatives/Manifesto-Alternatives/La-ville-rebelle

Yudina A (2018) Marco Casagrande: who cares, wins the third generation city. JUT Foundation for Arts & Architecture, Taipei

Yudina A. ~~it's anarchical it's acupunctural~~, well it's both/Marco casagrande. Monitor, 68. http://casagrandetext.blogspot.fi/2012/09/its-anarchical-its-acupunctural-well.html. Accessed 31 August 2018

Chapter 8
Urbanism on Water and Ecology: The Early Example of Westerpark, Breda

Rob Roggema

Abstract In this chapter we move back in time, to when it was not an usance to base our city designs on the natural systems of water and ecology. By the end of the 1980s the dogma of separation of functions, and dividing the city in areas for working, living, leisure and traffic was slowly abandoned and especially the focus on the traffic system, more in particular the car, was leading to uproar. In this timeframe an alternative to apply the principles of nature in urban design was very new and, in the beginning, needed to be conquered on the traditionalists who would pertain using their old-school design standards. In this chapter the development story of Westerpark, and Heilaar-Steenakker is presented. This area in the western outskirts of the city of Breda, in the south of the Netherlands, was one of the first, maybe even the first to use knowledge about the water system, ecological typologies and nature as the basis for urban planning. This article starts with a description in sections two and three of the policy context at national level to illustrate the momentum of change from rationalism towards ecological planning. In section four the policy context in Breda in the early nineties is presented as the context within which the planning of Heilaar-Steenakker (Sect. 8.5) and Westerpark (Sect. 8.6) could be based in a strong sense of the natural processes of ecology and water that formed the landscape in history.

Keywords Breda · Water · Westerpark · Natural system · Heilaar-Steenakker · Sustainable urbanism

R. Roggema (✉)
Research Centre for the Built Environment NoorderRuimte, Hanze University of Applied Sciences, Groningen, The Netherlands

CITTA IDEALE, Office for Adaptive Planning, Wageningen, The Netherlands
e-mail: r.e.roggema@pl.hanze.nl; rob@cittaideale.eu

R. Roggema (ed.), *Nature Driven Urbanism*, Contemporary Urban Design Thinking, https://doi.org/10.1007/978-3-030-26717-9_8

8.1 Introduction

In this chapter we move back in time, to when it was not an usance to base our city designs on the natural systems of water and ecology. By the end of the 1980s the dogma of separation of functions, and dividing the city in areas for working, living, leisure and traffic was slowly abandoned and especially the focus on the traffic system, more in particular the car, was leading to uproar. In this timeframe an alternative to apply the principles of nature in urban design was very new and, in the beginning, needed to be conquered on the traditionalists who would pertain using their old-school design standards. In this chapter the development story of Westerpark, and Heilaar-Steenakker is presented. This area in the western outskirts of the city of Breda, in the south of the Netherlands, was one of the first, maybe even the first to use knowledge about the water system, ecological typologies and nature as the basis for urban planning. This article starts with a description in sections two and three of the policy context at national level to illustrate the momentum of change from rationalism towards ecological planning. In section four the policy context in Breda in the early nineties is presented as the context within which the planning of Heilaar-Steenakker (Sect. 8.5) and Westerpark (Sect. 8.6) could be based in a strong sense of the natural processes of ecology and water that formed the landscape in history.

8.2 Vinex-Policy

The Dutch national government presented its national spatial policy in the so-called VINEX (Vierde Nota Ruimteijke Ordening Extra) in the early 1990s (Ministerie van VROM 1992). As a reaction on the former policies, as depicted in the Second and Third Policy Document on Spatial Planning in the Netherlands, in which urbanization occurred in so-called bundled deconcentration (Ministerie van Volkshuisvesting en Ruimtelijke Ordening 1966; Ministerie van Volkshuisvesting en Ruimtelijke Ordening, Raad van Advies voor de Ruimtelijke Ordening 1976), implying that large chunks of urban developments were grouped together in ‘growing cores’ (groeikernen), visualized on the so-called ‘Blokjeskaart’ (Fig. 8.1). The VINEX policy therefore wanted to develop greenfield locations closer and attached to existing urban boundaries. The main characteristics of these Vinex locations were a relative high density (at least in the Dutch context), of 35 dw/ha mainly low-rise dwellings, a range of building typologies, direct public transport to the city center and a general notice of sustainability as an ambition to for instance realize sustainable building principles and perform energy efficiency. Every region in the Netherlands had have the task to realize a certain amount of Vinex-housing in, by the central government, pre-determined very specifically defined locations (Fig. 8.2). With everyone living close to a train station, it was expected this way of urbanization would limit car traffic hence congestion. Within the neighborhoods, schools, amenities, and green spaces were foreseen according carefully calculated amounts.

Fig. 8.1 Blokjeskaart, Third National Spatial Plan

8.3 Urbanism Pre-Vinex: From High-Rise to Haagse Beemden

In the period before the Vinex, urbanism took place through a combination of restructuring old urban neighborhoods, and development of new towns, autonomous cities and towns at a distance from the central city, such as Almere, Purmerend, Zoetermeer, etc. At the time the typology of new neighborhoods was such that high rise and separation of land use in living, working, leisure and traffic was the common thought. People would live safely and healthy because confrontations with polluting industries, or traffic was prevented. In these towns, the complete amenity structure of a regular city would be developed, as such

Fig. 8.2 So-called Vinex locations in the region Amsterdam (Fourth National Spatial Plan 'Extra')

providing everything the people would need to stay in their new towns. However, what happened was that people, mainly of the social lower classes, moved from the larger cities, such as Amsterdam, The Hague and Rotterdam, but kept their jobs in the city. This leads to serious increase of congestion in and around the central city, with all the environmental impacts this brought. At the same time, because lower income groups gathered in these areas, criminality and unsafety became a relatively large problem. Some of these neighborhoods started to become the problem-areas in cities.

In Breda this problem was less significant due to the size of the city. In this relatively small urban conglomeration distances were not so large that entire new towns were realized. The main new neighborhoods were planned to the north, the so-called Haagse Beemden. The history of this area is quite interesting, as when planning started the common discourse was the high-rise, and plans were made to erase the landscape, prepare a tabula rasa, and develop a serious amount of high-rise buildings. In the 1970s people started to reject the separation of land-use, with its negative imaginery of unsafe high-rise 'ghetto's'. For Haagse Beemden the turn came on time, as the existing plans were not adopted by the municipal council and a renewed 180 degrees adverse planning process was started, in which the existing landscape, with its farms, meadows, forests and waterways was respected and taken as the basis for the neighborhoods that have been realized ever since in this area. This also gave rise to the appreciation of and support for thinking about sustainability.

8.4 Sustainable Urban Development Breda

The municipality of Breda was one of the first councils to adopt its first Environmental Policy Plan of the Netherlands (Milieudienst Gemeente Breda 1990). Moreover, this plan was not only discussing environmental issues, such as the water quality improvements, noise reduction, soil regeneration or the increase of ecological qualities, it also pointed at the spatial planning and design of existing and new urban neighborhoods as the platforms to increase overall sustainability. As follow-up and partially parallel to the development of the environmental policy plan, the attention for landscape and green was brought into concrete policy with the Municipal Green Structure Plan (Bureau voor Ruimtelijke Ordening Van Heesewijk en Milieudienst gemeente Breda 1986), and not much later the Policy Brief on urban parks (Buys and Van Vliet 1992; Sector Natuur en Landschap, gemeente Breda 1992).

This process was supported by national policies and thinking about the external integration of environmental topics in spatial planning (Ministerie van VROM 1993). Breda, seen as an innovative local government, became a pilot in the national DOSS-program, which stands for Duurzame Ontwikkeling Stedelijke Systemen (Sustainable Development of Urban Systems). Within this program research was conducted how to increase the sustainability of urban systems, such as the energy, water and transportation systems (Rijksplanologische Dienst 1994, 1995). In the EVSO-study (Tjallingii 1992; Rijksplanologische Dienst 1996), in which Breda was taken as one of the example areas, the so-called Strategy of the Two Networks (S2N, Fig. 8.3), was developed in

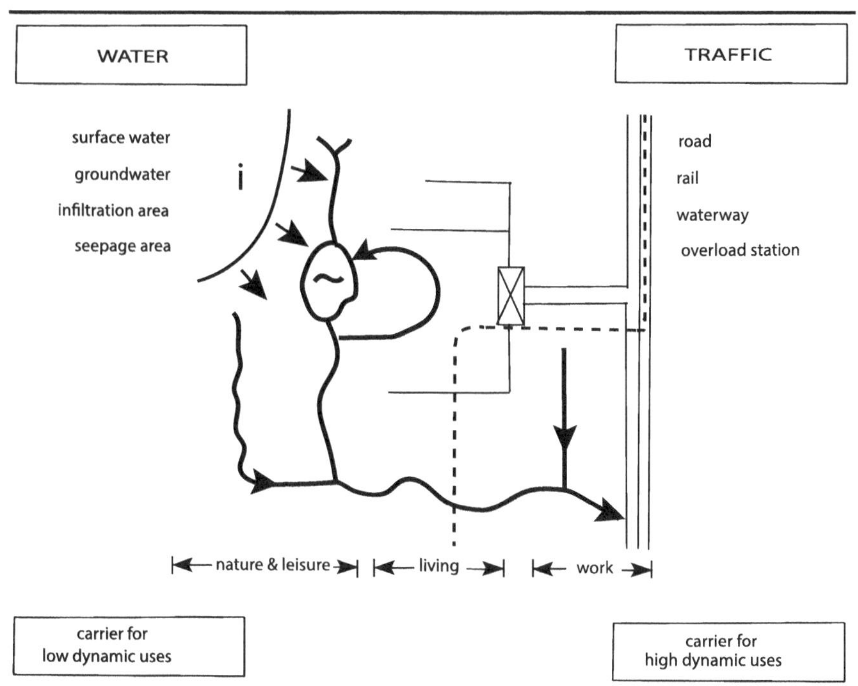

Fig. 8.3 Strategy of the two networks (Tjallingii 1992)

which higher dynamic land uses, such as intensive agriculture, industries and offices, intensive traffic, linked through the traffic system and energy grids, were separated from lower dynamic uses, such as water retention and purification, ecology, low-dynamic traffic, which were linked with each other by the water and ecological networks. In between these two networks housing would find its place (Tjallingii 1995, 2015).

In the early nineties the program of Sustainable Urban Development Breda started, which was the first program in the country paying attention to sustainability and urban design as two integrated objectives, hence it was a cross sectoral, cross disciplinary, discipline-overarching ambition. The program consisted of pilot projects, such as Westerpark, IJpelaar, Teteringen northeast, Bavelse Leij-Molenlei and others. These projects were identified as spatial development projects with a high sustainability ambition. Sustainability was defined in terms of an integrated environmental quality, consisting of the grey (noise, soil, air quality, waste), green (ecology, green space, sustainable building) and blue (water system, sewage system, civil engineering) environmental factors, and were connected with the design, the spatial quality of the development. In each of these projects a couple of young ambitious staff members of the environmental department (Milieudienst) would suggest spatial propositions to accommodate the highest possible environmental quality, e.g. sustainable urban development. Besides this on the ground work, the program organized lecture and knowledge development for municipal staff members and published the results in a series of yearbooks Sustainable Urban Development Breda (Roggema et al. 1994a, 1994b; Van Ginkel et al. 1995, 1997).

The planning process of Heilaar-Steenakker, of which Westerpark was an integral part, was a logical exemplar of these policy efforts and the coincidentally gathered young, ambitious members of staff who believed in the need to make urban environments more sustainable.

8.5 Plan Heilaar-Steenakker

When the Development Vision (Fig. 8.3) for Heilaar-Steenakker was conceived a traditional program directed the content. The size of the industrial area, mainly consisting of logistics, the number of greenhouses, offices and dwellings, and the way the road system would organize the accessibility was guiding the plan. (Dienst Ruimtelijke Ordening 1990; Fig. 8.4). Green connections were established between the circular boulevard (Singel) of the Breda City Centre and the ecologically valuable Liesbos. The water system did not play a significant role in the spatial lay-out of the area.

Under influence of the Municipal Environmental Policy Plan (Milieudienst Gemeente Breda 1990) thinking in natural systems, such as the ecological and water systems, became more and more important. For the development of the Land-Use plan for Heilaar Steenakker (1991, Fig. 8.8), these systems were taken as the starting point of design: even if the original creeks are no longer visible in the field, the

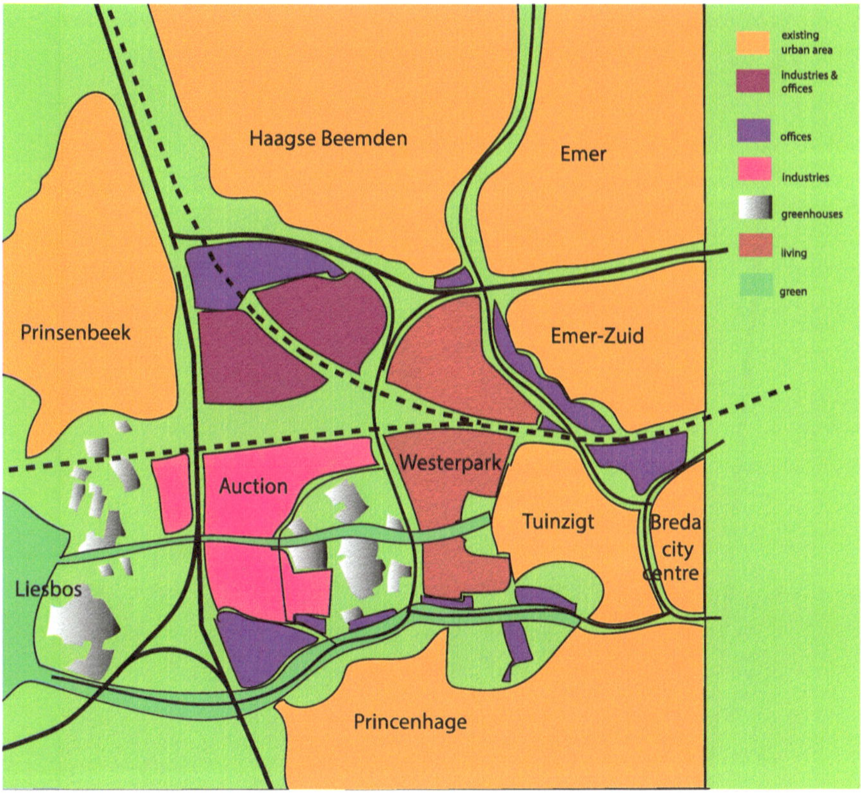

Fig. 8.4 Development Vision for Heilaar-Steenakker, 1990 (Roggema 1993)

underlying system is still functional. In order to identify these basic systems, historic, ecological and water analyses were carried out.

A first analysis of natural circumstances is the eco-typology (Roggema and Brekelmans 1996). In this analysis ecotypes determine where coherent groups of plants appear, based on the type of soil, humidity and nutrients. The eco-typology for Heilaar-Steenakker (Okhuysen and Ten Horn 1995) shows that humid and nutrient rich soils are found in clear and well-structured parts of the area (Fig. 8.5). These areas are the exact places where the historic creeks created their valleys.

The three creeks, which flow through the area from the southwest to the northeast, have in history been running dry for certain periods of the year. These parts are ecologically the upstream of the creeks, with strong alternating water levels and higher dynamics in flow rate. Typical ecological communities that are found upstream are heather-like vegetation, and seepage dependent ones downstream. The creeks have always been fed by surface rainwater and local and regional seepage (Van Acht et al. 1995).

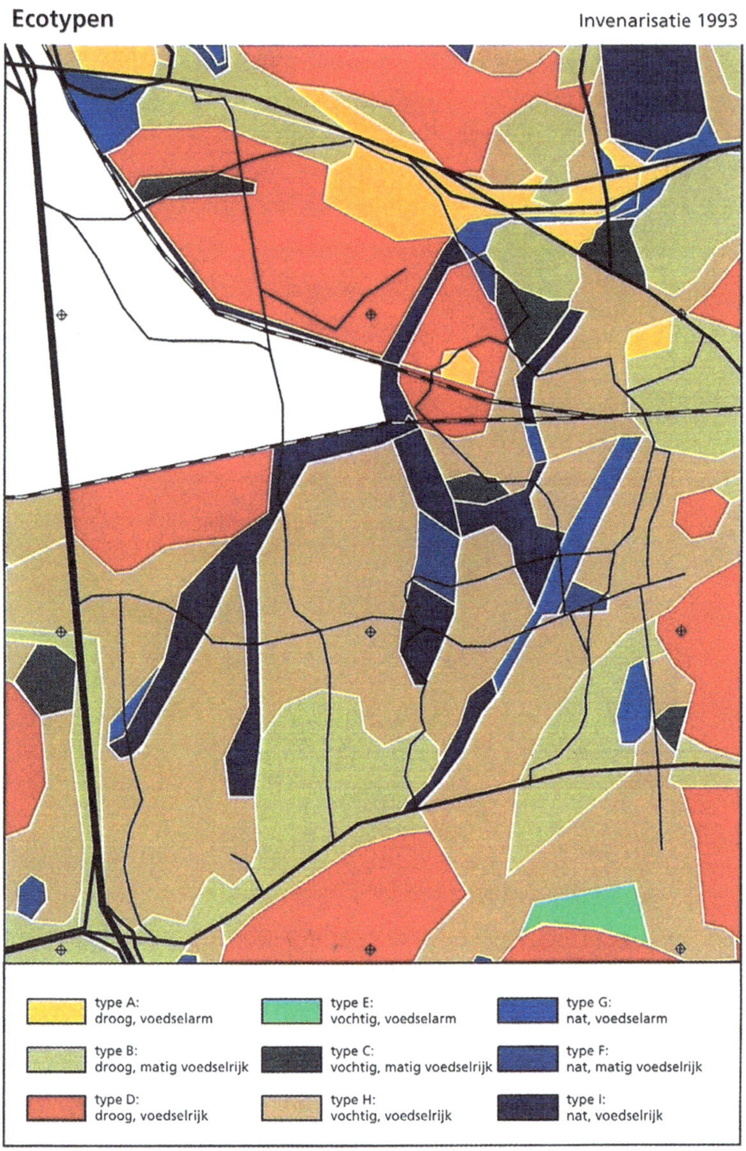

Fig. 8.5 Eco-typology Heilaar-Steenakker (Okhuysen and Ten Horn 1995)

These historic creek valleys are not congruent with the current hydrology in the field (see Fig. 8.6). This means that adjustments in the hydrological system have been executed to accommodate the intensive agriculture that was the predominant land-use before the urban development. However, the eco-typology does not lie, as it shows the original location of creek valleys, originating from the Holocene. It also shows that wherever in the area green plantation is desired, the eco-typology gives the complex of species that will thrive in that location. And based on these specific

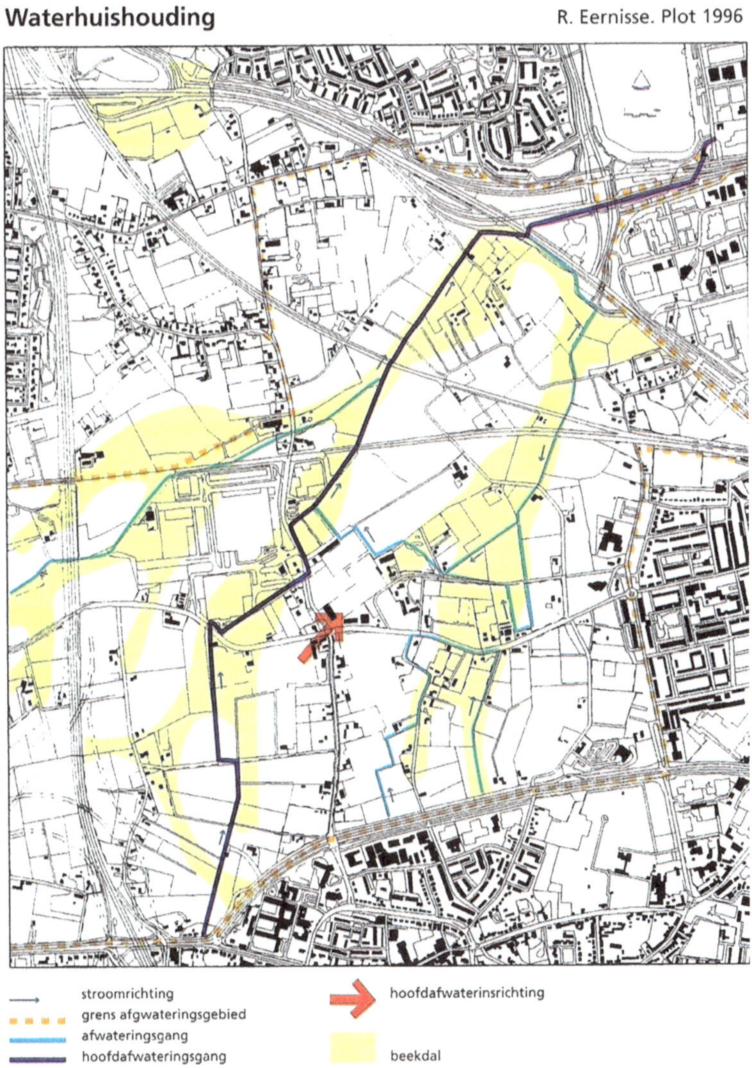

Fig. 8.6 Creek valleys and existing hydrology (Okhuysen and Ten Horn 1995)

plant species the habitat for animals, birds and insects is created. The eco-typology is therefore an area-covering method to determine the best development possibilities for a coherent ecosystem in every part of the area.

The planned water system (Heidemij Advies 1992) for Heilaar-Steenakker is based on ecological design-principles.

1. The water system is based on the historic and natural creeks:
2. Pollution is to be prevented at the source. Groundwater will be prevented through preventing to connect polluted surfaces to the sewage system, use sustainable materials for rooftops and preventing pollution of surface water. Surface water

itself will be prevented by installing an improved separated sewage system, which discharges only in very low frequencies, leaving only the cleanest water in the system

3. Storage of seepage and rainwater. As much local water as possible will be kept as long as possible in the area. In order to let hardly any clean rainwater flow towards the sewage plant, so-called clean surfaces are disconnected from the sewage system. The disconnected water will be stored in infiltration- and other green spaces and open water (the red dots in Fig. 8.7). This is dependent on the specifics of soil type, groundwater levels, and vegetation. Additionally, a level

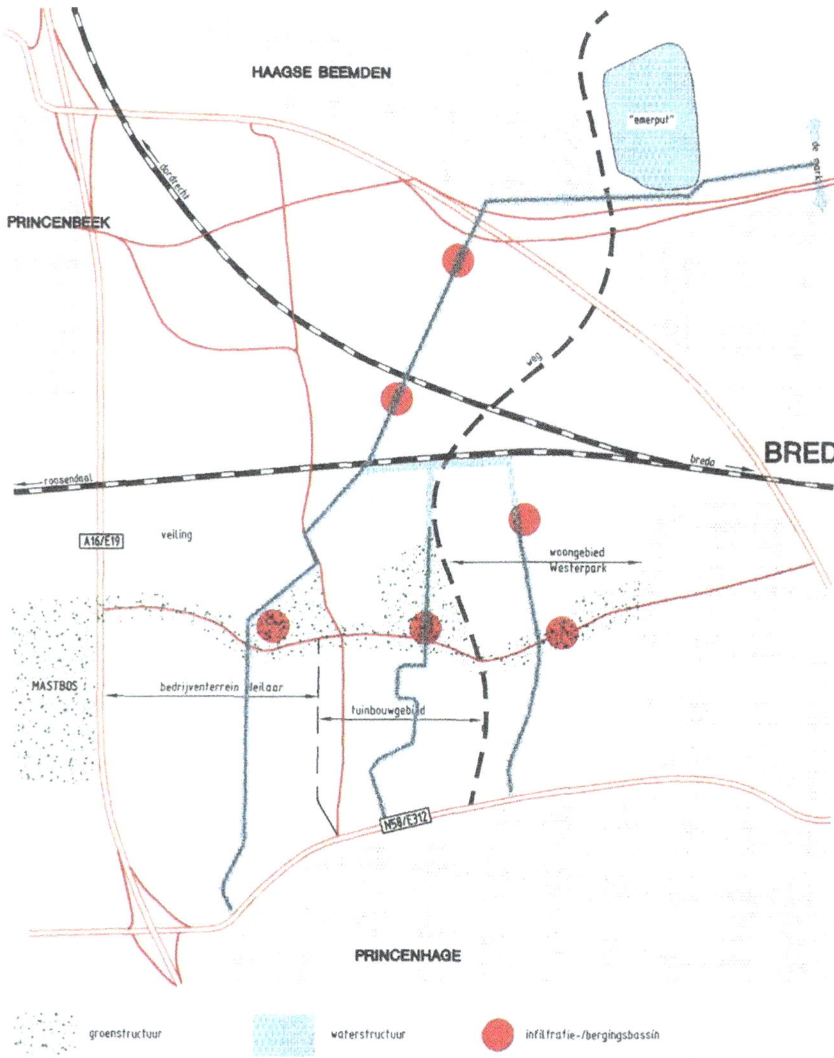

Fig. 8.7 Abstract design for the water system Heilaar-Steenakker (Van Acht et al. 1995)

fluctuation in the creeks of 60 centimeters is installed to support the desired dynamics of the typical natural water system of the area;
4. Let water flow from clean to polluted. The aim here is to create parts in the area as large as possible with relatively clean surface and groundwater. Polluted water from streets and roofs will therefore be discharged downstream as far as possible. This way a sequence of poor to rich water environments will be established, with accompanying typical plant species and animals. As a result of this principle almost no clean water is mixed with polluted water and the costs of cleansing in sewage plants will be minimized;
5. Create chances for water-based nature. These chances are increased if connections are established with neighboring water systems, surpluses of rainfall are infiltrated in the soil, waters' edges are constructed in a varied and curved way, inflow of extraneous water is prevented, the dynamics of discharge and fluctuations in water levels is congruous with the natural dynamics of upstream creeks, and the pollution levels from sewage discharge is minimized. Further to this the plantations are planted in a differentiated structure and materials used are sustainable.
6. The area required for water is minimized, as per the natural system, for instance through application of dry buffer zones, which will fill up quickly in wetter periods, hence store the surpluses of (rain)water. In summer, these areas can be used as playing areas for children, while in winter when freezing it easily transforms into an ice-rink. The water buffers function as the storage space for peaks in rainfall.

The water and ecological system were designed in order to allow them to behave like a natural ecological system. Large overflowing basins were connected through straightened creeks, which were designed with extended slow slopes as ecological edges. Plants would occupy these slopes and create the conditions for easy migration of a suite of species. The upstream, sandy, soils that form the upper parts of the creeks could dry up in period with low rainfall, as this is the natural habitat in this landscape. Species that can withstand these dynamics would survive even thrive and form the origins of the ecosystem.

In this approach water and ecology or green were planned as an integral system. Before, the land use of 'water' was seen as a different category as 'ecology/green'. Where in former days the square meters of these categories were added up it lead to a lower amount of land that could be sold. In Heilaar-Steenakker the integration of these land-use categories gained space, hence a higher percentage of the land could be sold, and likewise, the financial result was higher (Roggema et al. 1994a, b).

As shown in the zoning plan for Heilaar-Steenakker, 1991 (Fig. 8.8), the historic creek structure plays a dominant role in the final lay-out of the area and gives space to the dynamics of the typical water- and ecological system. These structures, together with the discharging clean water from the neighborhoods, and industrial areas, form the lower dynamic part of the Strategy of the Two Networks, while intensive land uses, such as industries and logistics are connected to the intensive road system. Housing was in this case integrated in the low-dynamic network.

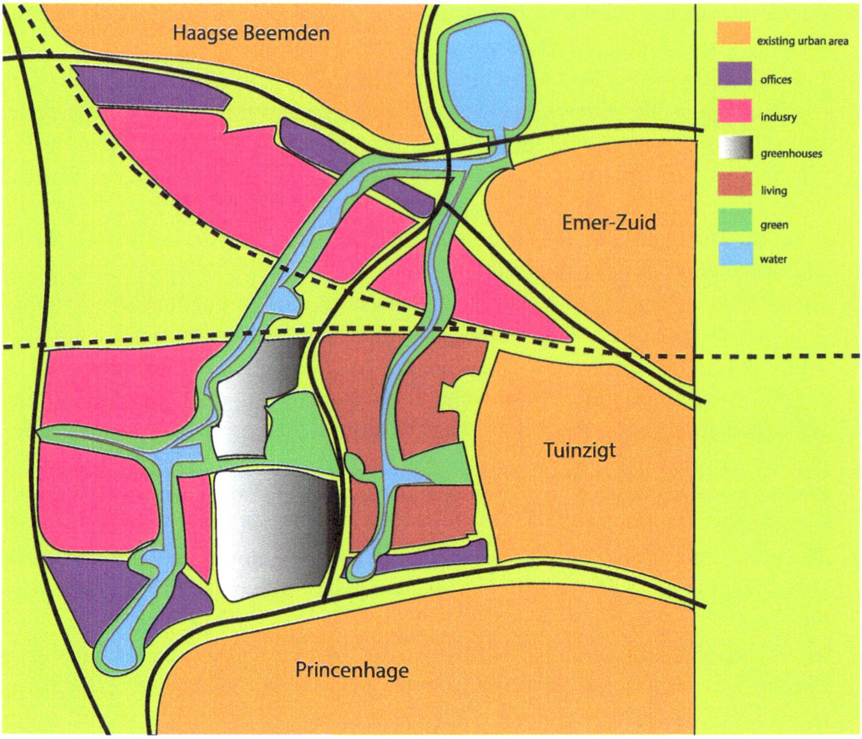

Fig. 8.8 Zoning plan for Heilaar Steenakker (Roggema 1993)

On the basis of these design principles a detailed urban plan was made for the industrial area (Fig. 8.9), in which the principles were further designed in concrete measures, such as the exact capacities, sizes of waterbodies, slope grades, and the way water was transported from the roofs to the water bodies.

8.6 Plan Westerpark

Westerpark is a living area, and part of the overall development of Heilaar-Steenakker. Its location is to the eastern side of the Development Vision of Heilaar-Steenakker, which positions it in between the industrial area and the city center and connects it directly to the neighborhood of Tuinzigt (Fig. 8.10). It forms also a stepping stone for urban ecology in between the city center of Breda and the Liesbos area to the west. This makes Westerpak contributing to the compact city of Breda, with short and good connections by bike, as pedestrian and by public transport, exactly the required qualities demanded in the Vinex.

The program for Westerpark is to realize 1400 dwellings on approx.. 52 ha, of which 48% (25 ha) is public space and 52% (27 ha) land-ownership. The density is

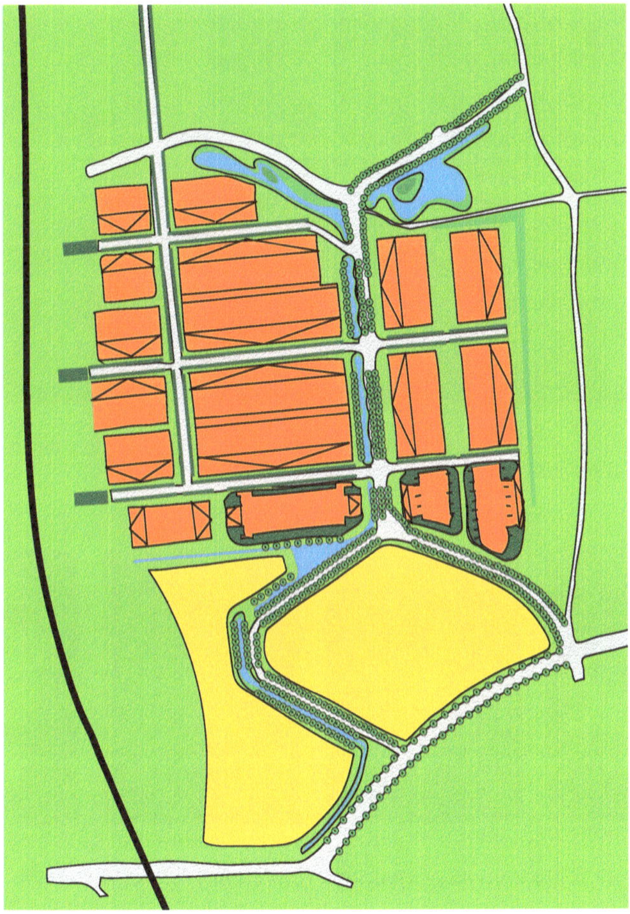

Fig. 8.9 Urban Plan of the Industrial zone, Heilaar-Steenakker (Roggema 1993)

net 50 dwellings/ha. 88% of the houses are land-based homes and the rest apartment dwellings.

This program is used to make an urban design for Westerpark with the following ambitions:

1. Realizing an urban quarter in Breda-west with smooth connections to its surrounding neighborhoods;
2. Realization of a high urban design and architectural quality;
3. Realization of a highly qualitative green space and -structure, with a leisure value for the inhabitants of Tuinzigt;
4. Realization of a high and integrated spatial and environmental quality.

The waterways, the park, the boulevard and the built accents form the main spatial structure (Fig. 8.11). These spatial elements define the shapes of the different

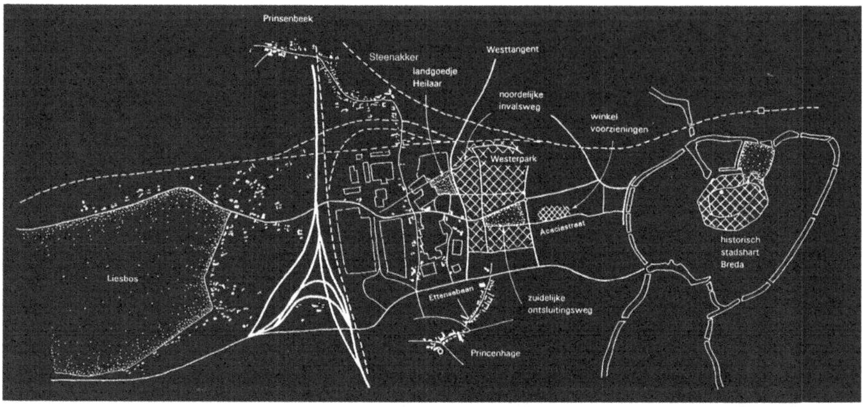

Fig. 8.10 Position of Westerpark in between Liesbos and City Center, and the industrial site of Heilaar-Steenakker and Tuinzigt (Van Ginkel et al. 1995)

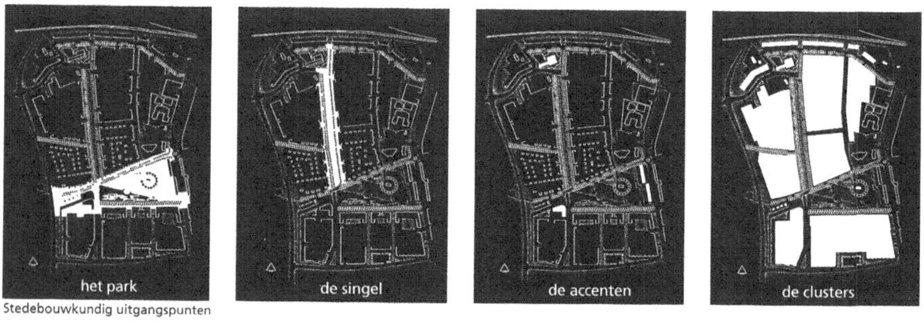

Fig. 8.11 Urban design elements of Westerpark (Brekelmans and Roggema 1995)

neighborhoods in Westerpark. Because these are determined by natural boundaries of the creek valleys, each of them has their own specific natural conditions (see also the eco-typology).

8.6.1 Water and Ecology

As mentioned before and similar to the plan for the entire Heilaar-Steenakker area, the original water system was taken as the starting point for the design of Westerpak. With the eco-typology analyzed, the creeks in the Westerpak area have only periodically water in it especially upstream, namely when it rains. During these wet periods the surplus water, then there is a surplus which needs to be captured and stored in the ponds in the park. The regulated dams where the waterway 'leaves' the housing development, keep sufficient water levels in the main parts of the development.

However, upstream the creeks will dry out to give space for the dynamic nature that belongs to this landscape, species, such as plants, and animals that can deal with wet, then dry conditions flourish.

The neighborhoods are connected with the ecological framework through a series of measures. There are several options to treat the water from roofs and clean hard surfaces. The water can be infiltrated in the soil, only when the soil is permeable enough and no excess water stagnates on top of the soil. Only clean water can be infiltrated, as most of the traffic space is too much polluted to be infiltrated in a safe way. A second option is to transport water to surface water, such as ponds and other water surfaces, using gutters and through streets. Water can also be discharged through wadis. These systems, consisting of depressions of grass in the public green space including a sandy infiltration bed underneath the grass, allow the water to slowly sink in the soil impaired as it is through the soil and sand beds. The rainwater falling on roofs can be collected and stored in buffers on the roof before it is used for toilet flushing in the house. This technique leads to less water use. Another way to store water on roofs is through grass roofs. This water helps the grass to grow, evaporates and adds to the microclimate at neighborhood level. All these options work better under specific conditions, determines by the locale circumstances.

Every part of the neighborhood has own characteristics which determine the type of measure that can be used best, on the basis of the following criteria:

- The amount of water that needs to be discharged is determined by the neighborhood; the amount of hard surfaces, the type of houses, the way the traffic system is designed, and the number of parking spaces;
- The future ground level, as not all parts of the neighborhood will be suppleted with sand. The height of the ground determines whether it is possible to let the water flow to surface water of buffers in a natural way;
- The permeability of the soil determines whether infiltration is possible Some of the soils in certain parts of Westerpark are less perneable than others.

Each of the parts of Westerpark have been analyzed using these criteria and the most suitable measures, which are subsequently used in the designs, have been determined on that analysis.

8.6.2 *Other Sustainability Measures*

In Westerpark not only water and ecology have led to the urban plan and urban design (Fig. 8.12). Different other themes had their influence on the final design. The desire to use as little energy as possible influenced the orientation of the building blocks to predominantly east-west, allowing the sun to heat the houses optimally in winter and preventing the sun from shining in the house in summer periods. Apart from the orientation the houses were all very well insulated to reduce energy demand further. Westerpark has also the ambition to become a quiet and noise reduced neighborhood. In order to achieve this the main roads are planned outside

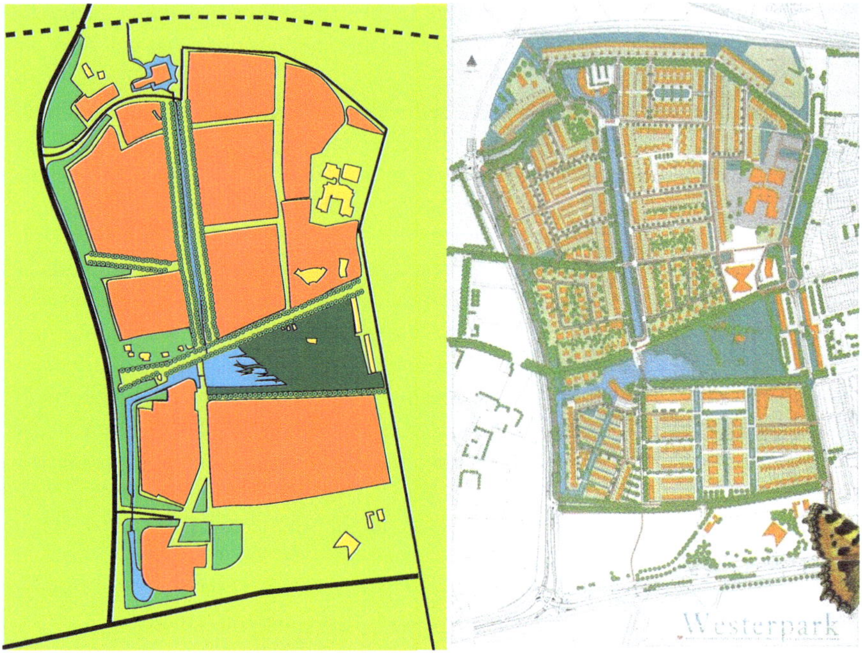

Fig. 8.12 Westerpark urban plan (Roggema 1993) and final urban design (Van Ginkel et al. 1995)

the actual neighborhood, and the distance taken into account is sufficient to reduce the noise to pleasant levels. Furthermore, to the north, where the trainline between Breda and Roosendaal runs, the noise is captured by implementing dwellings as noise-wall. The traffic plan is based on the principle not to allow any through roads through the living areas. Therefore, the road system is disconnected where it would cross the park, and the main traffic is guided around the area. From this road single delivery road provide the access to the homes. Finally, and as a joint effort with the architectonic quality, architects have to realize a very high standard in sustainable building. All these measures have been integrated in the urban design for Westerpark (Fig. 8.12).

These ambitions were laid down in the so-called Note on Quality and Realization (Kwaliteits- en Uitvoeringsnota, Gemeente Breda 1994). In this document integration schemes were developed for each of the parts of the plan. In these schemes the demanded spatial and environmental quality was described in detail at two levels of scale: the scale at which the government itself is responsible for realizing the ambitions, and the level of the building envelope at which the realizing party (developer, builder, architect) is responsible. This made sure the ambitions were aligned both in theme and scale (Van Hal and Roggema 1995).

8.6.3 Wester-Wetering

A special zone was safeguarded from developers and their architects, because a group of inhabitants of Breda wanted to jointly develop an extra high sustainability standard and presented their plans to the municipality. This initiative, called Wester-Wetering (Fig. 8.13), adapted the urban design and increased the sustainability level by introducing higher standards for sustainable building (materials), they introduced extra measures to stimulate ecological developments (breeding boxes, plantation), and they proposed to fully recycle all waste they produce in their part of Westerpark. Furthermore, this initiative successfully disconnected their water system from the municipal sewage systems and made sure to recycle and reuse water

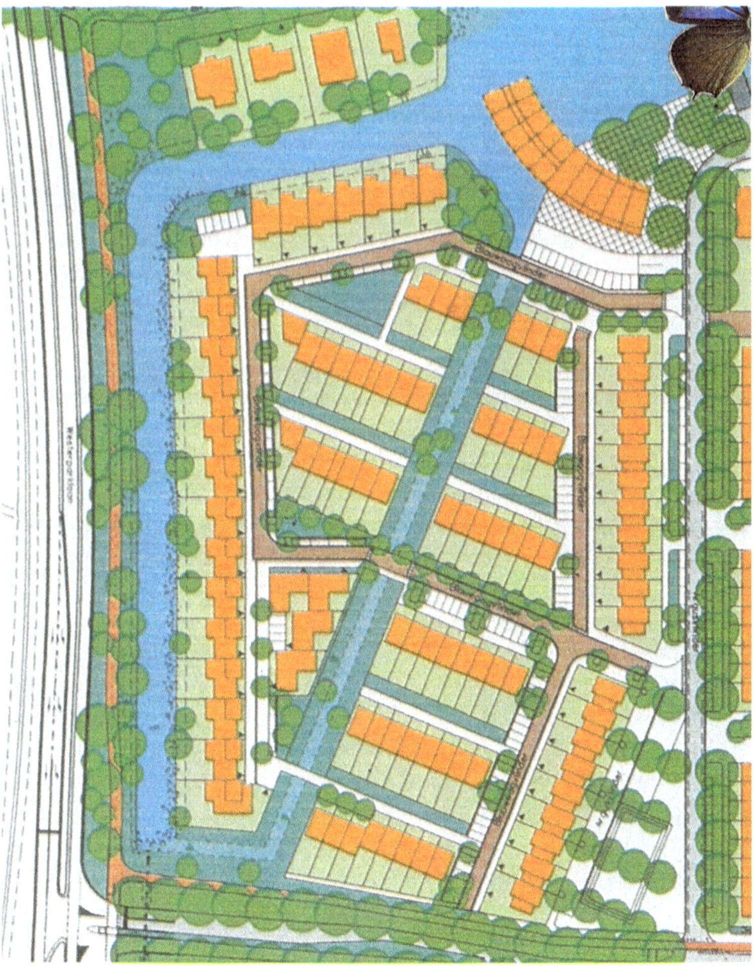

Fig. 8.13 Urban design for Wester-Wetering (Van Ginkel et al. 1995)

wherever possible and discharge rainwater in the natural water system. The water quality in their section of the plan is very good.

The involvement of citizens in achieving their own goals was initially not easy, as the municipal staff was used to determining what should be realized and was not extremely open for alternative ideas. After a while, however, when the initiators became more serious and could also show their financial soundness, the municipality not only accepted their proposals, but also embraced the measures and ambitions they represented, and even used this in negotiations with developers about their ambitions.

8.7 Conclusions and Recommendations

It sounds so simple, still so difficult to achieve: take water and ecology as the basis for the development of our cities. One of the main reasons seems to be that there are no direct earnings from including water or ecological features in an urban design. The appearance of water and green may raise the value of real estate in the vicinity a bit, and it may improve the health of people living near it, but in itself, the square meters of green and water eating up space in the urban plan only cost money. During the planning process there is often only few people representing the interests of green and water. Without real, mostly indirect, financial reasons the role of water and ecology/green in the negotiations with urbanists, real estate developers, the municipal financial department, housing department and traffic engineers, is still relatively weak. This will continue to be an irrational argument, as it has been proven over and over again that water and ecology have financial benefits for the real estate owners, as well as the costs of care hence the health insurers.

In the planning process itself, urban designers, without being fully indulged in ecological principles, have not always been able to give ecology and water the attention, and space it requires. Even if the designer on duty is convinced of the importance of green and water for the quality of his/her urban design, it could well be located in places that were economically the least interesting and generally not planned with an ecosystems' view in mind.

In this chapter however, it has been shown that to incorporate ecological and water systems in urban development it is not too difficult, but some basic principles need to be taken into account. First and foremost, the role of the landscape, in which the ecological and water systems determine the origin hence are the driving forces, should play a dominant role in discussions about spatial transformation of an area from the earliest beginnings of planning. Subsequently, a methodology that consequently implements a sustainable water and ecological system, should be applied throughout the planning process, from inception through to the maintenance phase. This methodology could be a very simple step-by-step approach. This article has shown which steps in the planning process are essential to realize the best results.

1. Analyze the history of water and ecology in the landscape. What was the historic topography and the contours, what geomorphological specific elements can be distinguished and which main waterbodies and -courses dominate(d) the landscape;
2. Use the eco-topology as analytical method to identify existing, maybe hidden, ecological systems, and the potential for certain plant species in the future;
3. Design water and ecology first, e.g. before urban uses are considered. This way the most optimal ecological and water system can be planned, and it becomes clear which trade-offs occur when urban uses are introduced;
4. Integrate urbanization with the demands of the natural system. What are the contributions of the urban areas to enhancing the natural quality of the entire area? Does it 'produce' clean water, does it host habitats for certain species, could it be an area where temporary nature can be located or where water can be stored temporary?
5. Design water and ecology from the system level through to the detail of the architecture and design of the public spaces.

This easy to follow way of working can be used to apply ecological and water principles to any planning process, no matter whether it is an industrial zone or a living area, no matter the spatial scale. Even though the first case studies based on this approach are more than 25 years old, it is still not very common. Often, urban designers do see ecology as exchangeable space for other green space, not necessarily with ecological qualities nor based in the ecological systems. This leads to absence, maybe even ignorance when it comes to the qualities of nature, but also the potential beauty of incorporating these systems in the urban designs of our cities, neighborhoods and houses.

In times of climate change this is even more important, because these systems contain a natural resilience when external climate impacts threaten the city. An integrated ecological water system provides the opportunity to urban areas to recuperate because water and ecological zones have the flexibility to deal with heat, drought and heavy rains at the same time. A feature that every urban area needs to have to be able to survive in the (near) future.

References

Brekelmans H, en Roggema R (1995) De ontwikkeling van het stedebouwkundig plan. In: Van Ginkel, M., Van den Elshout, A. & Timmermans, W. Jaarboek Duurzame Stadsontwikkeling Breda. Breda: Gemeente Breda. pp. 26–31

Bureau voor Ruimtelijke Ordening Van Heesewijk, Milieudienst Gemeente Breda (1986) *Gemeente Breda: Groenstructuurplan* 1986. Gemeente Breda, Breda

Buys, Van der Vliet (1992) Parkennota Breda. De ontwikkeling van een beleidsvisie. gemeente Breda, Breda

Dienst Ruimtelijke Ordening (1990) Ontwikkelingsvisie Heilaar-Steenaaker. Gemeente Breda, Breda

Gemeente Breda (1994) Kwaliteits- en Uitvoeringsnota Westerpark. Gemeente Breda, Breda

Heidemij Advies (1992) Integraal waterbeheerplan Heilaar-Steenakker. Gemeente Breda, Breda
Milieudienst Gemeente Breda (1990) *Gemeentelijk Milieubeleidsplan 1990–1992*. Gemeente Breda, Breda
Ministerie van Volkshuisvesting en Ruimtelijke Ordening (1966) Tweede nota over de ruimtelijke ordening in Nederland. Staatsuitgeverij, s-Gravenhage
Ministerie van Volkshuisvesting en Ruimtelijke Ordening, Raad van Advies voor de Ruimtelijke Ordening (1976) Derde nota over de ruimtelijke ordening Verstedelijkingsnota : deel A: beleidsvoornemens over spreiding, verstedelijking en mobiliteit. Staatsuitgeverij, s-Gravenhage
Ministerie van VROM (1992) Vierde nota over de ruimtelijke ordening Extra. Staatsuitgeverij, s-Gravenhage
Ministerie van VROM (1993) Nationaal Milieubeleidsplan (NMP2). Staatsuitgeverij, s-Gravenhage
Okhuysen M, Ten Horn L (1995) Ecologie en Vormgeving. In: Van Ginkel M, Van den Elshout A, Timmermans W (eds) Jaarboek Duurzame Stadsontwikkeling Breda. Gemeente Breda, Breda, pp 40–45
Rijksplanologische Dienst (1994) Duurzame Ontwikkeling Stedelijke Systemen (DOSS). Ministerie van VROM, Den Haag
Rijksplanologische Dienst (1995) Werkdocument DOSS. Ministerie van VROM, RPD, afdeling Thematische Planvorming, Den Haag
Rijksplanologische Dienst (1996) Visie Ecopolis. De Strategie van de Twee Netwerken. Ministerie van VROM, Den Haag
Roggema R (1993) The influence of sustainable planning on spatial planning in the municipality of Breda. *Proceedings Congress Ecological Strategies for Cities*, Dordrecht, 30 October 1993
Roggema R, Brekelmans H (1996) Milieu als inspiratiebron voor stedebouw. Duurzaam Bouwen 1-1996:18–21
Roggema R, Offenberg N, Okhuysen M, Van Ginkel M (1994a) Jaarboek duurzame Stadsontwikkeling Breda. Milieudienst Gemeente Breda, Breda
Roggema RE, Wiersinga WA, en Zonderland HGF. (1994b) Ecologische sturing van de ruimtelijke planvorming. Denken en doen in de Bredase praktijk. Groen 4: 12–17
Sector Natuur en Landschap, Milieudienst Gemeente Breda (1992) Beleidsvisie Parken. Milieudienst gemeente Breda, Breda
Tjallingii S (1992) In: Instituut voor Bos- en Natuuronderzoek (ed) Ecologisch Verantwoorde Stedelijke Ontwikkeling, Wageningen
Tjallingii SP (1995) Ecopolis: strategies for ecologically sound urban development. Backhuys Publishers, Leiden
Tjallingii S (2015) Planning with water and traffic networks. Carrying structures of the urban landscape. Res Urban Ser 3(1):57–80. https://doi.org/10.7480/rius.3.832
Van Acht, W., Figge, H., Timmermans, W. en Spanjers, J. (1995) Waterketen basis voor waterbeheersplan Heilaar-Steenakker Breda. H_2O 28(1): 22–25
Van Ginkel M, Van den Elshout A, Timmermans W (1995) Jaarboek Duurzame Stadsontwikkeling Breda. Gemeente Breda, Breda
Van Ginkel M, Kater G, Noordegraaf M (1997) *Milieu en Economie, Jaarboek Duurzame Stadsontwikkeling Breda*. Gemeente Breda, Breda
Van Hal A, Roggema R (1995) Breda zet hoog in. Duurzame Stedebouw in Westerpark. Bouw 50(11):52–54

Chapter 9
Blue Design for Urban Resilience in Drylands: The Case of Qatar

Anna Grichting

Abstract Gulf cities are generally characterized by hot, arid climates and extremely rapid developments with resulting demographic increases and accompanying environmental degradation. Until now, resources, technology, and capital have allowed expansion without limits—into the ocean with landfills and artificial islands, into the sky with tall buildings, and into the desert with Zero Energy Cities (Grichting, Int J Middle East Stud, 50(3), 580–585, 2018) With an economy largely focused on non-renewable energy sources, many of these Gulf countries are now looking to develop new postcarbon identities and to improve the sustainability and livability of their cities (Grichting, Int J Middle East Stud, 50(3), 580–585, 2018). Climate change and extreme weather conditions are also affecting the region, with more frequent and intense flooding (Al Jazeera, https://www.aljazeera.com/news/2018/11/raining-qatar-season-181112125930905.html. Accessed 17 Dec 2018). In October 2018, Qatar experienced unprecedented Flash Floods and the capital city of Doha received 84 mm of rain (Floodlist, http://floodlist.com/asia/qatar-floods-october-2018. Accessed 17 Dec 2018, 2018) the equivalent of the average rainfall of Qatar for one whole year. Buildings, roads, tunnels and coastal areas were severely flooded. In 1990, Qatar's only sources of water were groundwater abstraction (65%) and desalinated seawater (35%) (MDPS Ministry of Development, Planning and Statistics, Water statistics in the State of Qatar 2015. https://www.mdps.gov.qa/en/statistics/Statistical%20Releases/Environmental/Water/2015/Water-Statistics-2015-En.pdf, 2017). Treated wastewater for agriculture and green spaces irrigation purposes (5%) came into use in 2004, and increased to 11% in 2014 (MDPS Ministry of Development, Planning and Statistics, Water statistics in the State of Qatar 2015. https://www.mdps.gov.qa/en/statistics/Statistical%20Releases/Environmental/Water/2015/Water-Statistics-2015-En.pdf, 2017). Today, municipal potable water is obtained 99% from desalination and 1% from groundwater (Government of Qatar Web Portal, http://portal.www.gov.qa/wps/portal/topics/Environment+and+Agriculture/wateranddesalination. Accessed 13 Oct 2018,

A. Grichting (✉)
Institute for Environmental Diplomacy and Security, University of Vermont, Burlington, VT, USA
e-mail: anna.grichting@post.harvard.edu

R. Roggema (ed.), *Nature Driven Urbanism*, Contemporary Urban Design Thinking, https://doi.org/10.1007/978-3-030-26717-9_9

undated)). At the same time, the waters of the Gulf are becoming increasingly polluted due to desalination, over-fishing and pollution from hydro-carbon industries and shipping. Researchers believe that the Gulf marine environment has already surpassed its buffering capacity, with very poor circulation of water and extreme pollution (Nadim et al., Ocean Coast Manag 51:556–65, 2008). In the face of climate change, rising sea levels, food, water, energy insecurity, and loss of biodiversity - citizens, communities, and nations must work on common visions to address these future challenges by working with, and not against nature.

Keywords Drylands · Blue Networks · Green to Blue · Blue Urbanism · Resiliant Cities · Coastal Zone Management · Regenerative Systems · Symbiosis, Biodiversity · Rising Seas · Transboundary Planning · Ecological Networks · Saline Agriculture · Desalination · Food Water Energy Waste Nexus · Water Treatment · Constructed Wetlands · Urban Forestry · Landscape Urbanismt · Climate Change

9.1 Introduction

Gulf cities are generally characterized by hot, arid climates and extremely rapid developments with resulting demographic increases and accompanying environmental degradation. Until now, resources, technology, and capital have allowed expansion without limits—into the ocean with landfills and artificial islands, into the sky with tall buildings, and into the desert with Zero Energy Cities (Grichting 2018) With an economy largely focused on non-renewable energy sources, many of these Gulf countries are now looking to develop new postcarbon identities and to improve the sustainability and livability of their cities (Grichting 2018). Climate change and extreme weather conditions are also affecting the region, with more frequent and intense flooding (Al Jazeera 2018). In October 2018, Qatar experienced unprecedented Flash Floods and the capital city of Doha received 84 mm of rain (Floodlist 2018) the equivalent of the average rainfall of Qatar for one whole year. Buildings, roads, tunnels and coastal areas were severely flooded. In 1990, Qatar's only sources of water were groundwater abstraction (65%) and desalinated seawater (35%) (MDPS Ministry of Development, Planning and Statistics 2017). Treated wastewater for agriculture and green spaces irrigation purposes (5%) came into use in 2004, and increased to 11% in 2014 (MDPS 2017). Today, municipal potable water is obtained 99% from desalination and 1% from groundwater (Government of Qatar Web Portal undated). At the same time, the waters of the Gulf are becoming increasingly polluted due to desalination, over-fishing and pollution from hydro-carbon industries and shipping. Researchers believe that the Gulf marine environment has already surpassed its buffering capacity, with very poor circulation of water and extreme pollution (Nadim et al. 2008). In the face of climate change, rising sea levels, food, water, energy insecurity, and loss of biodiversity - citizens, communities, and nations must work on common visions to address these future challenges by working with, and not against nature.

In the words of IUCN President Zhang Xinsheng at the 2018 Global Forum on Urban Resilience and Adaptation 'We must connect cities with nature to build resilience in the face of climate change and natural disasters' (IUCN 2018a, b). He underlined IUCN's role in developing the concept of Nature-based Solutions to bring humans closer to nature and to address complex challenges such as climate change in order to achieve SDG Goal 11: 'make cities and human settlements inclusive, safe, resilient and sustainable' (UN.org). A 'Resilience thinking approach' is vital for city dwellers living with the unpredictable disturbances of climate change effects (Meerow et al. 2016). Nature Based urban design must be accompanied by self-organising tools and measures to face fast changes. This includes retaining knowledge in neighbourhoods and cities, valorization of labour and preparing students to translate their intellectual and aesthetic work into effective contributions to their communities, aware of all the other communities that could be affected by their choices (Lombardi 2017).

One such approach is to consider the city as a regenerative system of organic streams, eg. grey water, sewage, organic waste, urban runoff and storm waters. Capturing the rising amount of rain, storm waters and increasing waste water provides the resources and materials for establishing new layers that are key to the city's sustainability and resiliance. With a combination of modular environmental technology, open form infrastructure and landscape urbanism these material streams will become a steady source of power for the new organic layers of the city composed of parks, trees, water elements, wetlands, green roofs, wild nature, urban farms and collective gardens. These layers reduce the heat-island effect of the city, capture carbon, reduce pollution, produce a comfortable micro-climate, promote urban biodiversity, encourage pedestrian movability, produce food, create an attractive public entourage, amongst other ecosystem services (Roggema et al. 2017). Therefore, aside from governments and professional practices, one level at which this nature based approach needs to be addressed is in academia, and new approaches to design education must include a holistic, systems and nature-based approach to landscape, urban and architectural design.

This chapter presents a selection of design projects developed with students at Qatar University in the Master's program in Urban Planning and Design with a strong focus on ecological and nature based solutions. As part of the pedagogic process, students consulted with experts in Ministries and the private sector, and numerous field trips with experts were conducted to Water Treatment facilities, wildlife reserves, abandoned quarries, etc. to address a wide array of urban and landscape themes that are not necessarily seen as design projects by the authorities or the academic establishment. In this way, the pedagogic approach, and the design studio, addresses contemporary needs, and trains the students to identify important and pressing projects and to bring them to life and to competent authorities. As an example, some of the projects were presented to the President of the Public Works Authority as well as the Minister of Municipality and Urban Planning (Abu Nakhla Constructed Wetland). Other projects were developed in collaboration with Research Centers at Qatar University – The Environmental Research Center (ESC) such as the Project for Turtle Conservation. This bottom-up process is not common in the autocratic culture of the Gulf. Associated with a Nature Based and participatory landscape approach to planning, from the ground up, it is the way forward to achieve more sustainable and resilient urban design in the Region.

9.2 From Green to Blue Design

Green has been the colour of sustainability for many years. But is green really a color of sustainability for a yellow desert and dryland? A beautiful green lawn consumes much water and is maintained by pesticides and herbicides that are harmful to people, animals and birds. In a dry land like Qatar it is necessary to look at the Blue, to consider the water resources, the types of water, water consumption and water management (Grichting 2012), and how we can best recycle water – grey and black water – preferably onsite, using natural and organic systems. It is also important to consider the sweet and salt water sources, the impact desalination has on the ocean, and the relationship between urbanization on land and in or at sea. Blue Design is proposed here as a methodology with three approaches:

1. It focuses on the relationship between oceans and urbanization (Blue Urbanism),
2. It is water-based urban design (Blue Networks)
3. It promotes an active, rather than passive, approach to Design (Green to Blue).

The concept of Blue Design was coined by Saatchi and Saatchi (2008) – and goes beyond the Green to look at systems and designs that are not only carbon neutral but that also give back to the planet. Blue Urbanism – a term used by Timothy Beatley – takes as a starting point the facts that our human fate here on the blue planet is intimately tied to ocean health and that two-thirds of our global population lies within 400 km of a shoreline (Beatley 2014). It seeks a balance between urbanization on the land and its effects on the oceans. Green to Blue was proposed by the author as a new way of approaching urban and landscape design in Qatar and was formulated as a framework for teaching design studios in sustainable urban and landscape design (Grichting 2012).

As mentioned previously, there is very little rainfall in Qatar. However, when rain does fall, there is a lot of flooding, highlighting an absence of consideration for the topography, with the assumption that *Qatar is flat*. For a nature-based approach to urbanism, there is a need to radically overturn the approach to planning, which is top down and which lays buildings and infrastructures on the land as a tabula rasa approach, and to begin with topography, natural systems and water sources. In this way, we design the urban landscape as an infrastructure for the city, and its systems. The landscape is no longer a decoration, a beautification, an amenity or the *icing on the cake*. It becomes the foundation for building the city.

A Nature based approach to urban design also looks at buildings as living systems that are in symbiosis with their environment. Similar to the *blue design* concept, adopted by Saatchi & Saatchi, it is also about designing buildings and landscapes that are productive, that return more to the community and the environment than what they take. With *green design* we talk about carbon neutrality while *blue design* creates places that go beyond carbon neutrality and actually add surpluses to the world. To reverse global warming, we can no longer be passive in our designs. Instead, we need to be active and restore the planet and its resources. For example, recycled water can be re-injected into a buildings' water systems or recycled for landscaping and food production. In the past there was a tendency to think about architecture, landscaping and urban design separately. Additionally, we also tend to think of water, energy, food and waste as being separate concepts At present, we start to think of buildings, landscape and urbanism as an integrated system, with symbiotic relationships. Symbiosis is about interaction between two organisms and typically to the advantage of both. For example, buildings can produce more energy than they consume and the water they use can be recycled for surrounding landscapes including trees and green roofs. The trees can be planted in a way so that they will produce shade to buildings and help cool them, the green roofs also help to reduce the heat island effect and insulate the building, therefore reducing the use of energy for heating and cooling. Furthermore, the buildings and landscapes become habitats for diverse species, promoting urban biodiversity.

This new approach to design represents a fundamental shift in how engineers, designers and architects should view the challenges of local projects and thinking in systems, rather than designing icons. The measures of successful design in future will include the level of giveback the project generates for its occupants as well as to the greater global community. The future of the design and planning sector will demand an intervention and expertise of a wide range of professionals such as economists, biologists, chemists and also a range of social science experts such as demographers, anthropologists and geographers. We will present two groups of projects developed with students at Qatar University. The first group is based on Nature based design in the Coastal Interface, the second group focuses on nature-based design using different types of recycled waters.

9.3 Blue Design in the Coastal Interface

Military conflicts, lack of harmony among Gulf littoral states, and diverse prioritization of coastal issues have impeded the implementation of a sustainable coastal management program in the Persian Gulf region, even though the GDP has increased rapidly in the past few decades (Nadim et al. 2008). With the current Gulf crisis, which includes an embargo/blockade of Qatar since June 2017 (Ulrichen 2018), it is even more important for authorities, planners, designers, environmental scientists, and coastal managers in the Gulf states to focus on smart-planned development, which includes coastal preservation and ecological protection. Nature based urban design must go beyond national and political borders, to embrace ecosystem and watershed based zoning, that takes into account water systems and species habitats. Additionally, zoning and planning should not stop at the water's edge, or at the coastal boundary, but extend into the sea and the waters.

Timothy Beatley positions Blue Urbanism as an emerging set of ideas and perspectives around developing a mutually sustaining relationship between cities and oceans, where city planning carefully evaluates and regulates the effect of urban development on the marine environments (Beatley 2014). According to Beatley, cities should have jurisdiction over near-shore habitats and extend their zones of planning and management to offshore areas. Coastal zones compromise a very complex system and interacted subsystems, yet the planning of these zones is most often ignored or fragmented by policy and decision makers, regulators, scientists, planners and designers, economists, locals and beneficiaries, project managers and engineers, and international and regional agencies, and their complex and dynamic nature makes their management challenging. Qatar's Integrated Coastal Zone Management (ICZM) plan aims to address these issues and to unify objectives and instruments to achieve desired objectives (Afifi et al. 2016).

Several research and design projects being undertaken by the author and students at Qatar University address these coastal interfaces in Qatar. These projects include visions of new ecologies for the Doha Corniche, which create scenarios for mitigating storm surge and sea level rise, minimizing the pollution of the urban waters, increasing urban biodiversity, creating more socially integrated public landscapes, and designing productive landscapes both in the water and on land using emerging technologies and processes such as micro algae and sea farming as well as ecological engineering (Grichting 2016a, b). The Ecological Conservation Master Plan for Al Fuwairit Beach is intended to develop an ecologically protected area for the critically endangered hawksbill turtle by restoring and protecting turtle habitats, as well as the archeological sites adjacent to the turtle beach, and to strategically plan future developments that will preserve the cultural and archeological assets of Al Fuwairit and curate them for ecotourism. Dohasis, a conceptual proposal developed during a workshop on New Directions in Sustainable Urbanism in Qatar, invites free flooding of the sea to become an active part of the cityscape by introducing a series of controlled cracks in the flood defense walls and by designing free water movement within free spaces in the city (Roggema et al. 2017). Using soft engineering as opposed to hard engineering techniques and processes, it prepares the city to deal with the future rise of the sea level and changing weather conditions, where urban water elements also act as stormwater relief channels and contribute to the microclimate of the city. New, mixed fresh and salt water greenery are introduced into the city and coastal areas, developing them into partial urban mangrove ecosystems and inviting back original site-specific forms of flora and fauna. Offshore ecological engineering with oyster reefs will contribute to cleaning the polluted sea waters while creating barriers from storm surges, and will bring back pearl oysters, an important aspect of Qatar's traditional culture. Other projects along Qatar's coastlines include the Mangrove Belt systems and the Gulf EcoGateway between Qatar and Bahrain.

9.4 Blue Belts. Integrated Coastal Zone Management for Rising Seas

This design project addresses rising sea levels and biodiversity protection through ecological infrastructure. It is based on a concept of ecosystem service modeling and the mapping of informed management scenarios to develop an ecosystem based zoning scheme in the coastal areas. It calls for data and information gathering to create a comprehensive map of human activities and coastal and marine ecosystems for Qatar, combined with stakeholder engagement and review. It is an alternative to the sprawling coastal developments that are threatening the habitats of endangered species such as the Hawksbill turtles, who nest in Qatar. It addresses two main threats to Qatar which are also regional and global threats: biodiversity loss and sea level rise, through a multi-stakeholder approach, with new legislations to implement a nature based coastal development.

The project proposes the implementation of a mangrove belt around the coastline especially around existing mangrove ecosystems and nature protected areas. Moreover, it aims to protect endangered marine animals, such as dudongs and Hawksbill turtles by mapping their habitats and migratory paths. It includes the protection of sea grass and planting grass belts in areas of dugongs and Hawksbill turtles existence as they are critical to their survival. The project proposes the creation of a marine buffer zone consisting of different types of beds (oyster, mangrove or grass beds) that will aid in the decrease of rise in sea level impact. This *Blue Belt* or buffer zone should be complementary to the existing ecosystems and preserved areas and benefits both the biodiversity and citizens of Qatar, introducing a new, nature based approach to coastal planning.

Newly drafted urban design regulations include urban lighting regulations for sensitive species such as the Hawksbill turtle as well as offset of urbanism zoning along preserved beaches and sustainable building codes that include structures and facades that enhance biodiversity. With regard to zoning, it will be important to create protected areas that cover both the land and the sea for critically endangers species that appear on the IUCN Red List, such as the Hawksbill turtle. Currently, protected areas are on land only.

The multiple stakeholders for implementation of the design and projects include: coastal and ocean users: fisheries, tourism development and recreation, shipping and port operations and offshore oil operations; Non-governmental organizations (NGOs) which include local and international environmental NGOs such as RAMSAR, IUCN, UNESCO, ROPME, as well as Friends of the Environment Qatar and local youth organizations; Landowners and Business owners; Investors for Public Private Partnerships; Government and Ministries; Universities, Scientific and Research institutions and other education entities.

The project is implemented in the northern and southern areas of Qatar, which are not yet highly urbanized, but face pressure of urbanization on the coastlines. The northern zones are slated for development in the new master plan that looks to develop secondary cities in Qatar. The southern site is the location of Doha's New Industrial Port. The main areas are Al-Ruwais-Al Shamal, Al Fuwairit and surround

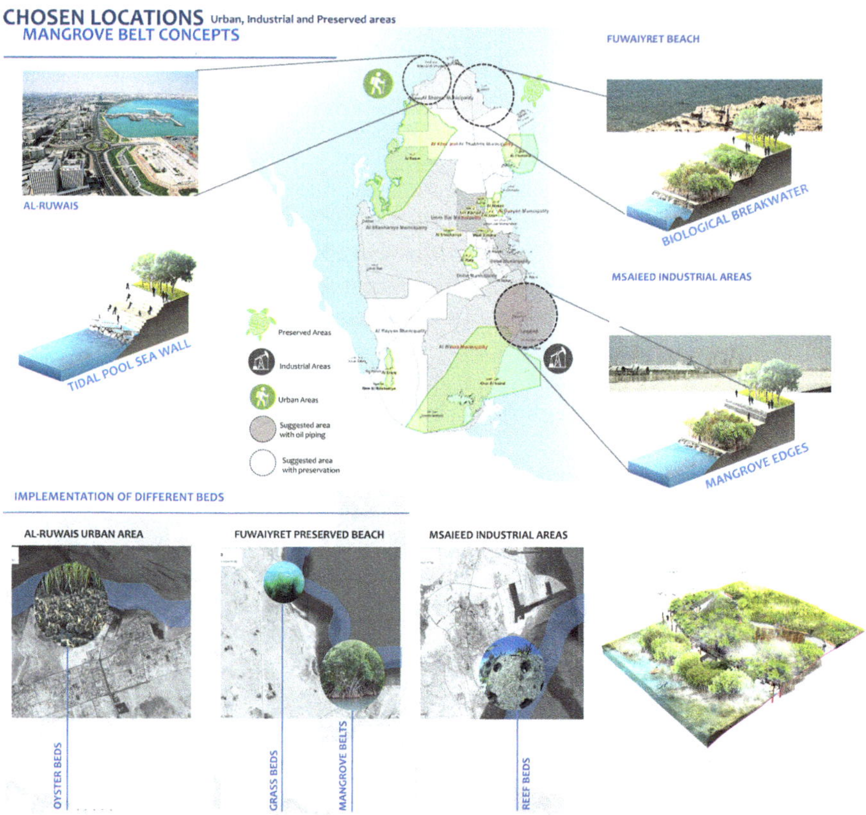

Fig. 9.1 The Mangrove Belt Locations and Systems. (Source: Heba Tannous)

turtle nesting areas, and Al Messaid. In these sites, different typologies of coastal edge landscaping infrastructure, based on Mangrove Belts, are proposed to guide development, mitigate effects of sea level rise and promote and preserve biodiversity (Fig. 9.1). Additionally, the mangrove ecosystem captures carbon and filters and cleans the water. In the Al-Ruwais, where rainfall is the highest in Qatar, the system is based on a constructed Tidal Sea Pool system, associated with Mangrove plantings, which will filter stormwater, provide recreational activities in the developing urban areas, providing car free zones connected to the new transit system being implemented in Qatar. In turn, these will provide new public and civic spaces, connecting people to nature and acting as an ecological infrastructure for water treatment, storm water management, leisure and education. In the Al Fuwairit and turtle nesting coastal zone, the system will act as a biological breakwater, to mitigate storm surges and sea level rise, and to protect the nesting beaches from erosion. Additionally, they will create leisure zones and wildlife reserves for migrating birds and other sensitive species and participate in a careful management and protection of the highly sensitive turtle nesting areas. In the Messaied Industrial coastal area, a system of Mangrove edges will be combined with floating mangrove systems, to reduce the effects of sea-level rise and storm surge and to create an ecological interface between the on-shore and offshore industrial activities.

9.5 Al Thakira Mangroads: Nature Based Tourism

Environmental stress caused by rapid development of the city of Al Thakhira is studied with a focus on mangrove ecology, which is increasingly threatened by rapid urbanization. The Mangroads project proposes to improve the lives of the coastal communities and the quality of the natural environment in the area of Al Thakira on the Western coast of Qatar, using natural systems, ecological engineering and landscape design. The area of Al Thakira is a protected area, with a large potential for nature based tourism which can protect the species and habitat while curating outdoor activities in a natural environment. These activities contribute to the economy of coastal communities and creates an indirect way to change the mindset of the people to create a closer relationship between the communities and the surrounding environment. In the case of Althakhira, the Mangroves are considered an important key to enhance the urban and natural environment. The interactive and education areas are the Mangroves Farms which will be located in the urban and natural areas: near the residential area in Alkhor and close to the existing mangrove areas.

This project builds on a previous project for the Al Thakira protected area, which included a detailed analysis of the sites with the establishment of different zones of intervention, including the core zones of biodiversity, buffer zones and zones for the development of nature based leisure and tourism activities (Fig. 9.2). The Mangroads creates a series of floating mangrove islands including an observation island, an educational center and a recreation center, as well as interactive areas along boardwalks and mangrove nurseries. These mangrove islands also serve as beacons that mark sensitive areas in the gulf waters, acting as a buffer zone to contain movement and tourism and to protect and restore the martime ecosystem.

A sensitivity mapping (Fig. 9.3) was conducted for each of the ecosystem types, with respect to their fragility and their interest for biodiversity (Supreme Council for Environment and Nature Reserves 2007). The types with very high sensitivity are coral formations and dense mangroves, represented by reef inner slopes, seagrass beds and scattered mangroves. Intertidal sand banks, mud flats, and shallow lagoon are of medium sensitivity. Coastal sabkha, beach rocks and subtidal cap rocks deep seagrass and algae beds have been classified as types with a low sensitivity. Pathways and transportation systems are adapted to the sensitivity zones, and modes such as jet-skies and high-speed motor boats will be restricted in these areas. Soft mobility systems that will be used are kayaking, water bicycles, rowboats, waterboards, stand up paddle etc. And for those who are not capable of using them, public boats will be offered in specific timing of the day to commute to and from the islands.

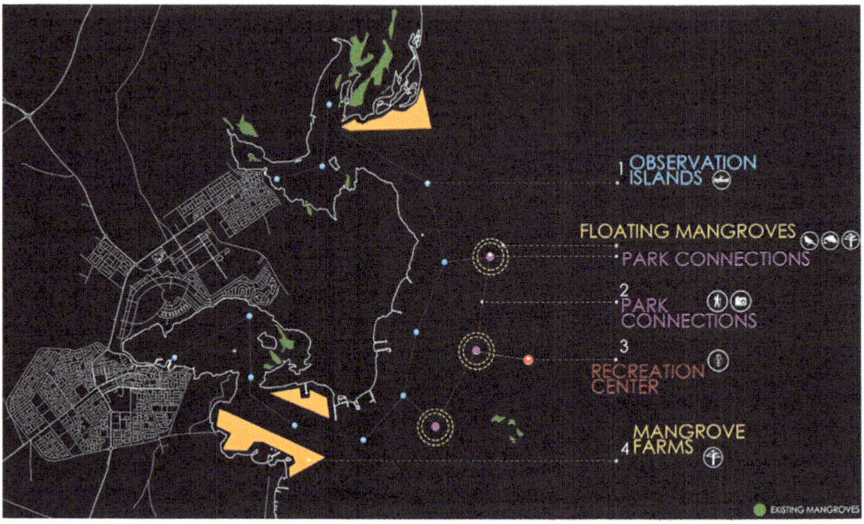

Fig. 9.2 Program and location of the Mangroads Project. (Source: Maryam Al Suwaidi)

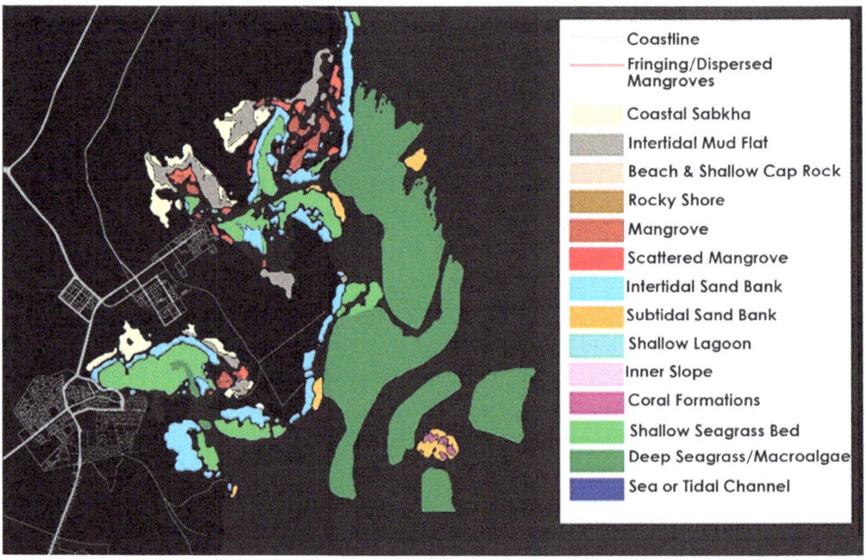

Fig. 9.3 Sensitivity mapping of the Project Area with landscapes and ecosystems. (Source: Maryam Al Suwaidi)

9.6 Blue Line. Biodiversity Conservation for Flagship Species

The Hawksbill turtle nests on Qatar's shores and has been visiting the peninsula for thousands of years. The females migrate thousands of miles and come back to nest on the beaches they were born on. This species is now on the Red List of Endangered Species of IUCN, the International Union for the Conservation of Nature (IUCN Red List 2018b). Of the seven species of sea turtle, it is the most critically endangered. Marine turtles are some of the oldest surviving reptiles on the planet and are irreplaceable ecological resources. They are recognized as playing the role of indicator species for the relative health of its surroundings, both marine and terrestrial. Biodiversity conservation cannot be successful unless local communities both receive their fair share of benefits from local biological resources and assume a greater role in managing them. It is believed that whether sea turtles ultimately vanish from the planet or whether they remain a wild and thriving part of the natural world will reflect the ability of humans to sustainably coexist with the diversity of life on Earth. However, today in Qatar and worldwide, sea turtles face intense threats of extinction directly caused by human acts of consumption and development. Researchers are currently tagging sea turtles with satellite transmitters in order to track these intangible animals and support research in new ecological technologies towards prospective solutions to many global development challenges (Chatting et al. 2018).

Al Fuwairit is an important area of Qatar's coastline, as it represents a popular tourist destination, as well as being a key nesting site for the Hawksbill turtle. Al Fuwairit beach, located on the north-east coast of Qatar, is immediately adjacent to a number of ecologically sensitive habitats including: nearshore coral reefs, foreshore rocky beaches, sand dunes, a brackish lagoon and a mangrove habitat. Al Fuwairit beach serves to protect against coastal erosion and has several exposed geological features of interest. As an entire ecosystem, the Al Fuwairit area is rich in biodiversity, and home to many threatened species of flora and fauna. Habitats surrounding, and including Al Fuwairit Beach, provide foraging grounds for many bird and animal species and are home to diverse plant communities, which together represent a high biodiversity index (Grichting 2016a, b). Overall, the ecosystem of Al Fuwairit Beach can be considered as an ecological 'hotspot' for the rich marine natural history of Qatar. Unfortunately, in recent years, Al Fuwairit beach has been subject to increasing human pressure and is now in urgent need of protection. Threats to the ecological value of the area include tourists visiting and driving across the beach and adjacent sand dunes, as well as leaving behind excessive amounts of rubbish. In particular, sand dune habitat is being compacted and destroyed by vehicle traffic, turtle nests are at risk from poachers, and mangrove habitat and bird nesting areas are being disturbed. All of these activities contribute to the degradation and decline of the ecosystem. The objective of the turtle habitat conservation measures and of the Conservation Master Plan is to strike a balance between controlled access for the public who come to Al Fuwairit Beach for recreational purposes, and the conservation of ecological habitats and biodiversity for the benefit of Qatar. Combined, the project will serve to enhance and protect the ecosystem for current and future generations and will support the Al Fuwairit ecosystem to become an area of international conservation status, with associated green tourism benefits for Qatar.

The project proposes two components. One deals with the more urgent turtle conservation measures that should be implemented with high priority and prevent further population loss in Qatar and the region. This is integrated into the second component: the conservation Master Plan - with phases of interventions based on priority and resources. In terms of the development of the conservation Master Plan (CMP) for Al Fuwairit Beach, and the wider ecosystem, the main aims will be to:

- Develop the CMP for Qatar's first *Eco-Beach and Nature Reserve*;
- Create a safer habitat for endangered turtles by designating and defining the boundaries of the protected area(s);
- Create landscape and ecological designs to restore habitat including: dunes, marine habitats, turtle habitats, etc.;
- Take landscape measures to enhance existing habitats – e.g. turtles light pollution screen, dune creation, tree planting, interpretive signage, bird hides etc.;
- Increase the accessibility of infrastructure – e.g. vehicle access, roads, parking; as well as boardwalks (see below for example), etc.
- Design a permanent Turtle Information Center for the benefit of scientists and visitors;
- Use renewable energy sources e.g. solar power, and reduce waste, noise, air and light pollution in the area;
- Propose "soft activities" for visitor recreation that are compatible with the turtle reserve (no engines, no excessive noise, etc.).

Specific Hawksbill turtle conservation measures under the project may include:

1. The protection of coastal sand dunes. Fencing should be erected around the existing dunes to protect against vehicles driving over the dunes and destroying vegetation. A planting program should be undertaken to restore beach dune vegetation, and to enhance protection against coastal erosion. Ecological engineering shall be implemented to assist in the restoration of the dunes and the protection of the coastline.
2. Sand de-compaction. Vehicles driving over Al Fuwairit Beach and compacting the sand create a major impediment to nesting turtles. As well as restricting vehicle access, a sand harrow could be deployed to de-compact the sand 1 month prior to the turtle nesting season. A sand harrow is a metal linked low impact wire mechanism used to plough compacted shore areas. The harrow allows nesting turtles to dig nests more easily and reduces the stress level of the nesting female.
3. Turtle Hatchery & Information Center. A hatchery should be set up on Fuwairit beach to optimize hatching success and allow for visitors to observe hatching in a sustainable, controlled and educated manner. A permanent Turtle Information Centre at Al Fuwairit Beach would serve to inform the public about the importance of the Hawksbill turtle and its habitat, as well as the value of the ecosystem.
4. An example of habitat conservation for public education. A public walkway (or boardwalk) through the mangrove habitat could be erected to allow controlled public access and increase public awareness and education on the importance of

the mangrove habitat. This would be designed in a way to minimise impact on the mangrove habitat and the bird species through the use of bird hides, interpretive information boards, and ecotourism guides.

This project was developed in collaboration with the Environmental Science Center (ESC) of Qatar University, which has been responsible for conducting Hawksbill turtle conservation on behalf of Qatar Petroleum and the Ministry of Municipality and Environment for many years at Al Fuwairit and other nesting sites in Qatar. The development of a Conservation Master Plan (CMP) for Al Fuwairit beach and its surrounding habitats has been ongoing with students in the Department of Architecture and Urban Planning at Qatar University. This collaborative work with the ESC builds capacity and knowledge in Conservation Master Planning and will also raise awareness amongst the student population and community. (Fig. 9.4).

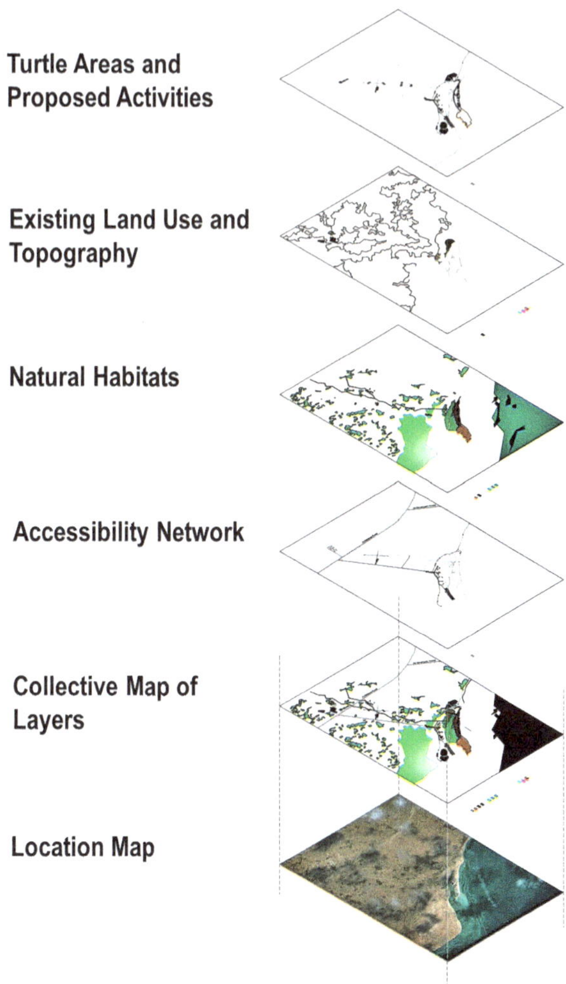

Fig. 9.4 Layers of the Conservation Master Plan for Al Fuwairit. (Source: Anna Grichting and students MUPD Sustainable Urban and Landscape Design Studio 2015. Alifa Muneerudeen, ALMaha Al Malki, Basma Aboukalloub, Deema Al Attar, Fadi Yasin Al Khani, Mooza AL-Mohannadi, Nussyba Eiraibe, Reem Awwaad, Samar Zaina, Sara Nafi', Sara Zaina)

In a bottom-up approach, guided by Qatar's national vision pillars, this project tries to attract decision-makers attention through a *living laboratory* concept Master Plan proposal in the area of Fuwairit beach and archeological site, which will stimulate ecological accountability of municipal bodies and highlight their role undertaking policy formulation, regulation, implementation of networking infrastructure and comprehensive planning projects towards sustainable urban future. The project integrates a system of resilience loops in order to make the project self-sustaining and reduce resource use and pollution. (Fig. 9.5).

Interviewing research members familiar with the site and acknowledging unique heritage and natural habitat within the marine and land interface in Fuwairit resulted in revitalization strategies that focus on habitat, development, social responsibility and resilience, aimed to discover distinctive ways urban master plans can foster collaboration between diverse sectors of community, organizations, stakeholders and inspire holistic vision of the problems and opportunities presented by coastal development. From an organizational standpoint, the elaboration of existing urban heritage, land use and green systems in Fuwairit avoiding any pressures on the natural and ecological systems whilst focusing on disseminating this experience to our communities will introduce ecological practice in planning and development as future development model.

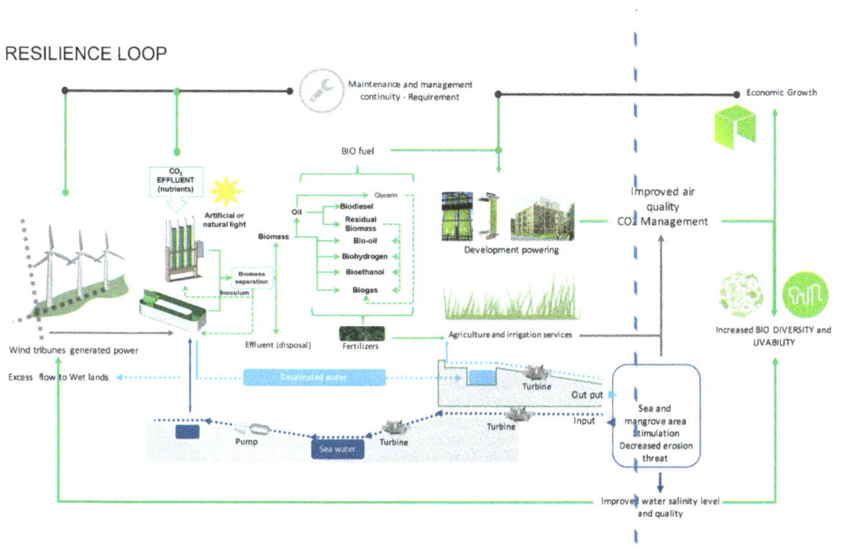

Fig. 9.5 Resiliance loops and systems. (Source: Maryam AlFaraidy & Maryam AlSuwaidi)

9.7 Blue Bridges. The Gulf Eco Gateway. Transboundary Planning for Biodiversity Conservation

Prior to the current Gulf crisis, the concept of the modern state—introduced into the Gulf by the European powers—coupled with the increasing importance of boundaries to define ownership of oil deposits—were at the origin of territorial and boundary disputes in the Gulf region. The contestation between Bahrain and Qatar over the Hawar Islands was settled only in 2001 (Wiegand 2012), after 36 years of dispute, and today they belong to the island of Bahrain. Plans for a friendship bridge to link the peninsula of Qatar to the island of Bahrain, inscribed in a new regional rail network for the GCC countries, have been drafted to increase exchanges and mobility between the two states. However, recent events - the blockade of Qatar in June 2017 - have shattered the Gulf geographies and the Gulf Cooperation Community, and not only interrupted the flow of resources to and from Qatar, but also shut down the air, sea, and land borders between Qatar and the neighbouring states.

The Hawar Islands, situated off the East coast of Qatar, lie on the path of thousands of endangered migrating birds, playing an important role in regional marine ecologies. Khor al Adaid is also a formerly disputed border zone, and it is a unique assemblage of terrestrial and marine environments with a large tidal embayment lying in an area of mobile dunes that straddles the border with Saudi Arabia, recognized as an area of global ecological importance.

Both of these border areas have been nominated as UNESCO World Heritage Site and could one day become zones of cross-border ecological cooperation that could guarantee lasting human and ecological security between Qatar and its neighbors as well as provide well-designed and managed spaces for nature observation. Scientific research, stakeholder engagement and ecological Master Planning should be undertaken as a way to engage cross-border collaboration between scientists and nature conservationists on both sides of the border, with the aid of international NGOs such as UNESCO, International Union for the Conservation of Nature (IUCN), and the Ramsar Convention on Wetlands. The example of the Korean DMZ, the Cyprus Green Line Buffer Zone, the Jordan River Valley, and other transboundary areas in conflict or contested zones can be examples for these projects when scientists, artists, academics, and NGOs on both sides collaborate in common future visions despite the current crisis/embargo.

One research and design project that aligns with these visions, undertaken with students at Qatar University's Master in Urban Planning and Design, is the Gulf Eco Gateway and the Master Planning for the Hawar-Al Reem Transboundary Protected Biosphere Reserve. The First project builds on the proposed Bahrain Friendship bridge, which would connect and link the high-speed rail network that is planned for the GCC. The second is a proposal to link the Al Reem Biosphere reserve, proposed as a UNESCO site, with the Hawar Island Protected Area. The conservation of Nature, species, and ecosystems cannot be undertaken in isolation, and this project is a great opportunity to continue working on collaborative visions despite the current crisis. As mentioned earlier, in these research projects on borders and transboundary cooperation - between scientists, nature conservationists, and artists, - it is imperative to continue the visioning of these projects, and to prepare the ground for when the conditions are ripe for implementation (Grichting and Zebich-Knos 2017). While the current Gulf crisis precludes this type of transboundary planning with its neighbours, ecological and nature based planning, based on species and habitat conservation, must go beyond borders to be effective.

The Gulf-Eco Gateway aims to reweave the land-sea interface between the State of Qatar and the Gulf of Bahrain Hawar Islands. (Fig. 9.6) This region is recognized

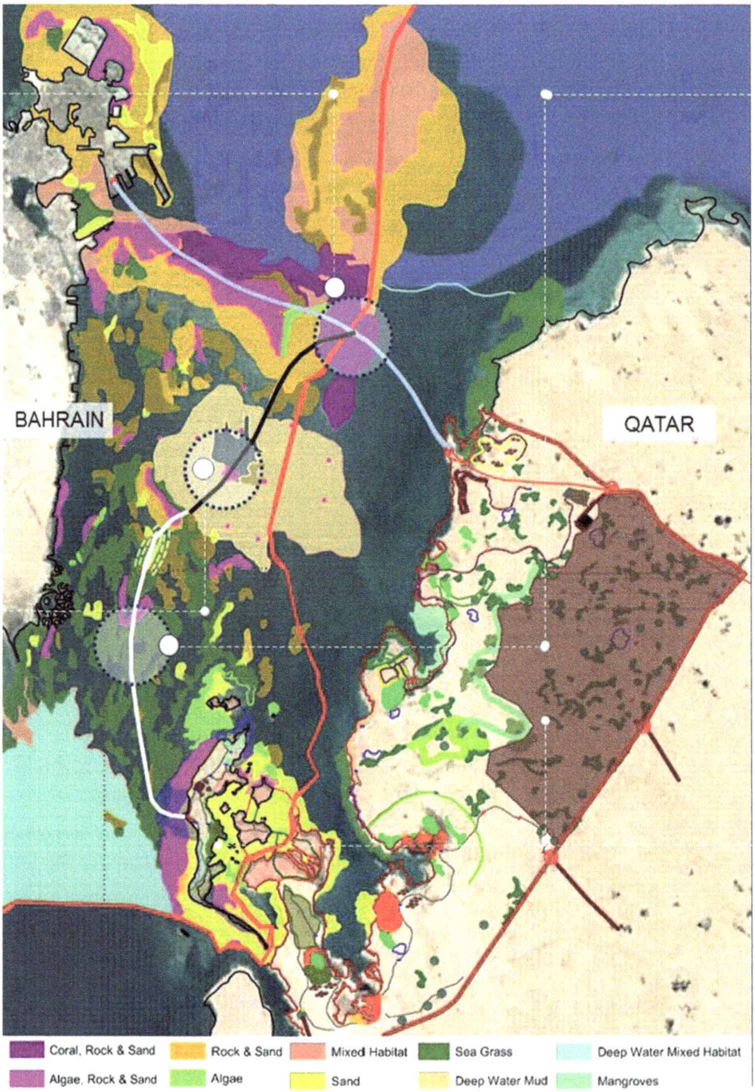

Fig. 9.6 Gulf eco-gateway. (Source: Najeeba Ali)

as hosting on of the Gulf's richest biodiversity hotspots of coral, algae and seagrass ecosystems and includes endangered turtles and dugong species. Based on existing sensitivity mapping and ecosystem studies of flora and fauna (Supreme Council for the Environment and Natural Reserves, SCENR 2008) the project is structured around the proposed Qatar-Bahrain Friendship bridge that was proposed after the resolution of the territorial dispute. While the bridge has been criticized as being unfriendly and a threat to the environment, it is proposed here as a gateway marking a new protected area, and as an ecological infrastructure to contribute to the remediation and regeneration of the threatened ecosystems, flora and fauna. The degradation and threats include oil and gas pollution, fishing nets and debris, coral bleaching due to rising temperatures and lack of currents, increasing salinity from brine from desalination plants, amongst others. The floating bridge will act as a plug-in biome and host the border facilities, as well as bio-reactors, algae open ponds and ecology farms, food farms, a floating island museum, an educational center for endangered species – Dugong and Turtles – and an underwater aquarium. An Eco hub will contain a research center and eco-lodges for tourism. The entire area will be classified as a transboundary biosphere reserve, and host Bird sanctuaries, Turtle Marine reserves, dugong reserves, educational facilities, eco-villages and ecological activities for eco-tourism – kayaking, stand-up paddle, camel safaris, etc.

9.8 Cyan Corniche. The Fusion of Blue and Green Urbanism on Doha's Corniche

9.8.1 The Corniche

As a result of landfill policies over recent decades the Doha Corniche now forms almost a complete circle, and this influences the macroform of Doha city. While old Doha, located on the south bank of the Corniche, originally had an intricate circulation system, the semicircular shape of the coastline gave rise to a new planning approach based on ring roads and radial arteries. The Corniche is perceived as a public space and a marker of Doha's city edge; initiating a dynamic dialogue between blue and urban scapes. One of the challenges of urbanism in Doha is spatial division and social segregation, yet the Corniche remains one of the few places where all communities of the city come together, acting as a social interface that is accessible without cars. Today, reweaving the public realm is one of the major potential legacies to achieve the Qatar National Vision 2030 for a sustainable future, with the potential for Corniche Park and Promenade to become more integrated and accessible, with improved connections to the city and surroundings, as well as increased social and cultural diversity.

Since its construction in 1970s, the Corniche has been the site of numerous urban and landscape projects developed through competitions sponsored by the government. In 1998 the Doha Corniche Project Competition (INNOCENT 2003) involved Jean Nouvel, Zaha Hadid, Martha Schwartz, and other international architects but none of the designs was implemented. Recent landscape and urban design projects envisaged for the Corniche to include mangrove parks, beaches, and biodiversity biomes, following new trends in landscape design that are shifting toward ecological planning, systems thinking, and more holistic and sustainable approaches to designing public spaces and urban environment. Landscape Urbanism concepts involve bio-cultural landscape ecologies, linear urban infrastructures, social and economic functions of landscapes and ecological networks (Grichting 2016a, b). To sum up, the Corniche can be defined as:

- A backbone that ties esplanades, public spaces and parks;
- A landscape system physically connecting city with sea;
- A cultural and social interface for diverse communities;
- A climatic front bringing sea breeze into the city;
- An ecosystem edge bringing marine & terrestrial ecosystems.

9.8.2 Cyan Corniche

Cyan is a color halfway between Blue and Green in the color wheel, representing the fusion between the Blue and Green Urbanism and thus representing the project goal and intention. The Cyan Corniche project is a proposal to implement *Saline Agriculture* in the Corniche of Doha, tackling the issues of urban agriculture, food security, scarcity of ground water as well as sustainable urban realm and landscapes and a productive waterfront for the city. The project aims to create a sustainable agricultural landscape in the waterfront of Doha, which serves functionally as a means to increase food security, as well as providing sustainable public realm that allows interaction and engagement with the environment. Cyan Corniche proposes resilient planning for urban food during a critical time in Qatar's history. The recent embargo on Qatar in June 2017 interrupted food supplies from Saudi Arabia, one of the country's main supplier of food. This project addresses Qatar's food security and integrates local culture and agricultural heritage into modern urban food practices. It creates a multifunctional Green and Blue infrastructure to serve today's and tomorrow's challenges within the dynamic urban area. It enhances Doha's waterfront space, making it into an interactive, productive and ecological space for all communities – human and non-human. It provides a platform for regional and international exchange of innovation and technologies on saline and urban agriculture. The project approaches the Corniche through a site analysis of the flows of food, water, energy and waste. (Fig. 9.7) While it considers all potential and future water sources on the site, it focuses on saline agriculture as a nature-based solution to many of the country's resource and ecological problems.

The project responds to a number of existing problems such as food insecurity, scarcity of groundwater, global warming and climate change, rising sea levels, decrease in biodiversity, unsuitable land for agriculture, lack of sustainable public realm, organic waste management, lack of urban trees, air pollution and CO_2-

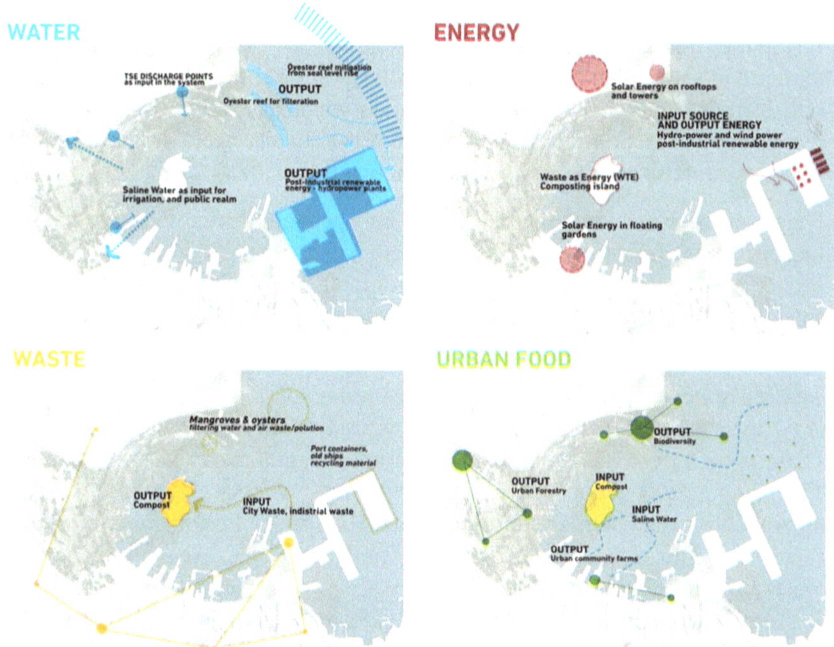

Fig. 9.7 Cyan Corniche. Water, Energy, Wast and Food systems. (Source: Asmaa S Al-Mohannadi, Albandari S Al-Harami, Heba S. Elgahani)

emissions and energy efficiency. The solutions proposed in the project include local urban agriculture, alternative methods of irrigation using Treated Sewage Effluent (TSE), saltwater, produced water from the towers in West Bay, coastal protection infrastructures, species habitats, public spaces for inclusive social interaction, waste recycling, renewable energy sources, innovative technologies and applied research on sea farming.

An innovative strategy for enhancing land and water availability is the use of salted soils and salted water, in a strategy designated as saline agriculture. Saline agriculture can be defined as the profitable and integrated use of genetic resources (plants, animals, fish, insects and microorganisms) and improved agricultural practices to obtain better use from saline land and saline irrigation water on a sustained basis (Saline Agriculture Worldwide 2018). It is a rich collection of possible systems for the use of saline resources. The components of these systems will vary according to the needs of the farmers and the capabilities of the land and water. The saline water that may be used in halophyte crop irrigation can be, for example, seawater, salt-contaminated phreatic sheets, brackish water (from estuaries), drainage water from other plantations, drainage water from human areas, such as sewage, or even water derived from aquaculture waste.

Qatar is a peninsula and an arid region, much of its land is unsuitable for agricultural activities due to salinized soil and the dominance of coastal deserts. The agricultural use of saline water or soils can benefit Qatar, specifically by the integration of salt-tolerant plants that can utilize land and water unsuitable for salt-sensitive crops (glycophytes). Salt-tolerant crops and salt-tolerant vegetable gardening can produce high salt tolerant to low salt tolerant plants.

- High salt-tolerant plants include beets, bell peppers, broccoli, cabbage, kale, loquats, spinach, tomatoes.
- Moderate salt-tolerant plants include carrots, cauliflower, lettuce, peas, potatoes, squash, sweet corn.
- Low salt-tolerant plants include beans, celery, cucumbers, radishes.

The sea water can be mixed with sweet water to create different degrees of salinity, dependent on the type of vegetable desired. Moreover, common halophytes of Qatar were investigated including Aeluropus Lagopoides, Anabasis Setifera, Arthrocnemum Glaucum, Avicennia Marina, Cressa Cretica, Cyperus Conglomeratus, Halocnemum Strobilaceum, Halopeplis Perfoliatam, Limonium Axiliare, Salicornia Europaea, Salsola Marina, Suaeda Aegyptiaca, Suaeda Vermiculata and Zygophyllum Qatarense.

Based on site analysis and several case studies, the comprehensive Master Plan allows free flooding of the sea and recreates a canal that marks the former shoreline of Doha, bringing salt water into Al Bidaa park to create salt water lakes and provide water for the halophytes and productive landscapes. (Fig. 9.8) The design project radiates out from the circle of the Corniche edge, into the city landscapes and out to the sea (Fig. 9.9) The island in the center is used for composting of organic materials and builds up a natural landscape through the accumulated materials. It is extended with breakwaters to protect the corniche edge from storm surges. Along the corniche, towards the historic center to the south, a series of piers jut out into the water, recalling the former jetty's that were used to dock ships, before the modern city port was constructed. These piers host community herb gardens with local species and medicinal and culinary herbs as well as productive gardens with halophytes. In the center of the Corniche, the road is buried underground and the Al Bidda park extends into the sea, creating a more natural edge with the city. Towards the northern edge of the corniche, towards the business district of West Bay, a system of boardwalks follow the curve of the corniche, and are intermingled with urban and floating mangroves, creating a biodiverse ecosystem that attracts birds and fish species. The abandoned towers in the West Bay are restructured to host hydroponic and aquaponic farming, increasing food security in the district. Radiating out to the sea, a series of oyster reefs provide food and protect the land from flooding and coastal erosion. The oysters also filter water and remove nitrogen, which is the cause of algae blooms and dead zones in coastal waters. Further afield, we find floating gardens and fisheries. The Old Doha Port, which is no longer used by commercial boats, aside from cruise boats, is converted with underwater farms, market zones, and a waste collection and transformation center.

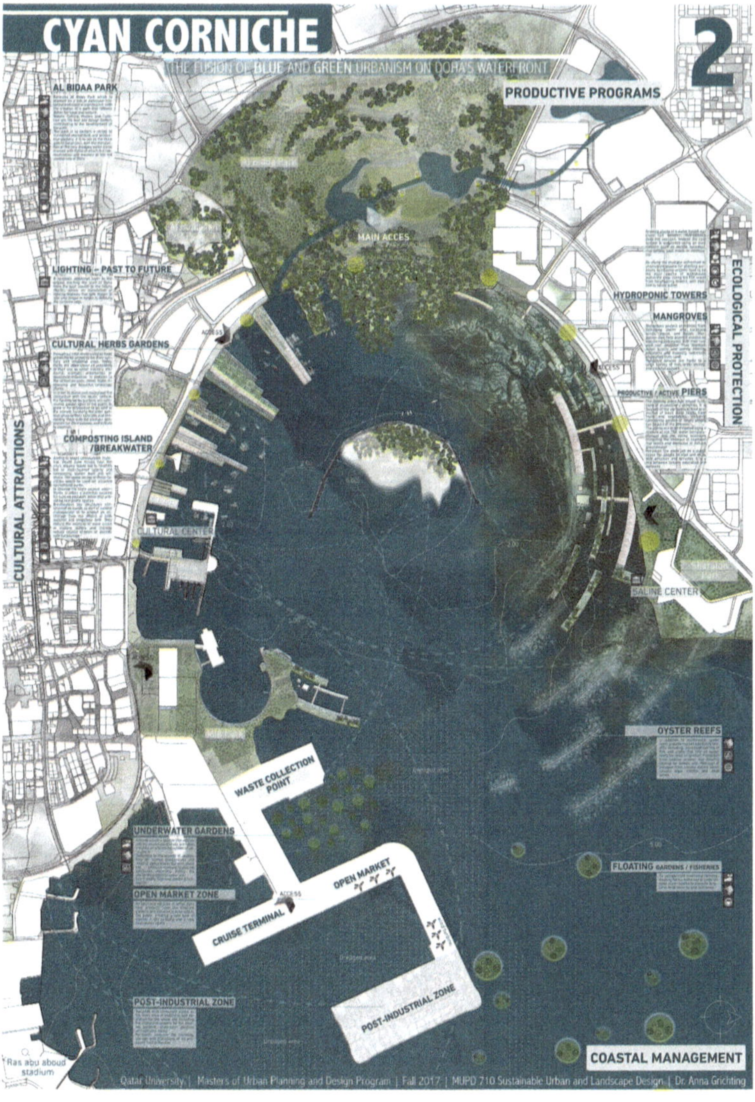

Fig. 9.8 Plan for Cyan Corniche. (Source: Asmaa S Al-Mohannadi, Albandari S Al-Harami, Heba S. Elgahani)

9.9 Blue Green Design with Urban Water Systems

Sustainable development must include water quantity and quality as indicators applied to both human and ecosystem health and this across a full range of land use, including urban agriculture and industry. New approaches to sustainable urban design – more precisely from the Landscape Urbanism field (Bélanger 2010) – propose that water supply and treatment infrastructure must be incorporated into Master

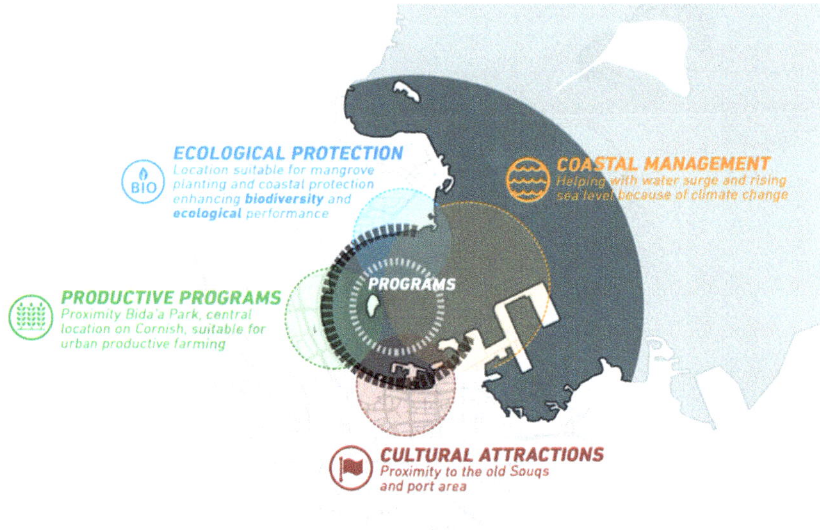

Fig. 9.9 Cyan Corniche. Programs. (Source: Asmaa S Al-Mohannadi, Albandari S Al-Harami, Heba S. Elgahani)

Planning. Retrofitting existing cities and systems is possible and also necessary and requires coordinated efforts from government agencies, designers and citizens as well as the private sector. Increasing land and coastal flooding are bringing this imperative to the forefront of resilient urban design for cities and green and blue infrastructure will become the foundation for sustainable urbanism. The form and hardscapes of the city no longer play the ecological role required of a city today (Austin and Yu 2016).

Because of rapid agricultural, industrial and social developments and the massive increase in Qatar's population, conventional water resources have become seriously depleted and non-conventional alternatives such as desalinated water and, to a lesser extent, treated sewage effluent (TSE), now have an increasing role in the planning and development of an expanded water supply (Al Mohannadi 2010). Additional water resources are also currently being researched, such as saline water and halophytes for food and medicinal plants as well as microalgae, produced water from the oil and gas industries as well as from air conditioning units, and treated sewage effluent for food production. Until now, TSE is only used for landscaping, and this is regulated so as not to include public parks where children play. However, it is proposed in a number of the projects presented here to grow food. The current technology of water treatment is of very high quality, and it is possible to use the water for agriculture. Currently cultural issues and the regulation do not allow the use of TSE for food production that is for human consumption.

In just 35 years, the frequency of disasters worldwide has more than doubled, driven by climate-related and weather-related hazards like flooding, tropical cyclones and droughts. UN Water estimates that 90 per cent of all natural hazards are water-related (United Nations Office for Disaster Risk Reduction UNISDR 2015), and the Intergovernmental Panel on Climate Change (IPCC) predicts even more extreme events going forward (Intergovernmental Panel on Climate Change IPCC 2018). For the year 2017, the Standing Committee of the Ramsar Convention on Wetlands approved Wetlands for Disaster Risk Reduction as the theme for World Wetlands Day, so as to 'raise awareness and to highlight the vital roles of healthy wetlands in reducing the impacts of extreme events such as floods, droughts and cyclones on communities, and in helping to build resilience' (RAMSAR 2017). Wetlands are essentially land areas that are flooded with water, either seasonally or permanently. This way they form a natural buffer against disasters. The Ramsar Convention website sums up much of their significance: 'Along the coastline, wetlands act as a natural protective buffer. For example, they helped avoid more than USD625 million in damages from Hurricane Sandy in 2012. Inland, wetlands act as a natural sponge, absorbing and storing excess rainfall and reducing flooding. During the dry season, they release the stored water, delaying the onset of droughts and reducing water shortages. When well-managed, wetlands can make communities resilient enough to prepare for, cope with and bounce back from disasters even stronger than before' (RAMSAR 2017).

Landscapes and ecosystems everywhere around the world are increasingly threatened by urbanisation. There are many types of wetlands, from coastal to inland, and from freshwater to saltwater. There are also natural as well as constructed wetlands. Qatar has several important coastal wetland areas, mainly mangrove ecosystems, which are important not only for the habitats and biodiversity, but also for their contribution towards building resilience against coastal erosion and acting as an interface between coastal and terrestrial ecosystems (Gulf Times 2017). Qatar also has some very interesting constructed wetlands. These were not initially designed as wetlands but were built as ponds to receive excess TSE (Treated Sewage Effluent). Gradually, as in the case of the Abu Nakhla Wetland, these ponds naturally evolved over time with riparian vegetation, creating habitats for sedentary and migratory birds. These wetlands are important as they create habitats for biodiversity and landing spots for migratory birds, as well as become natural landscapes that the people of Qatar and visitors can enjoy. Students at Qatar University designed landscapes and urban developments for reconstructed wadis (Wadi Jalal) and artificial lagoons (near Al Raqqiyah farms and in urban areas) as well as remediating neighbourhood areas that suffer from flooding and water infiltration to create productive landscapes and leisure parks including butterfly parks. By using TSE and storm water in this manner it can be avoided that huge amounts of water are wasted by pumping it deep into the ground as there is no sensible use for it. In what ways can wetlands help urban planning of a city? If water management is better integrated into urban planning, recycled water, whether it is grey water, TSE, or storm water, can be used to create attractive urban landscapes. An integrated urban water system not only makes better use of all the types of water in a city, including seawater and TSE, but also creates attractive landscapes for inhabitants and a more resilient urban realm that can mitigate the impact of extreme weather, be it heavy rain or sandstorms.

Urban Forestry, an emerging field in urban landscape planning, encourages the planting of trees to provide many urban ecosystem services, which include improving air quality, mitigating the urban heat island effect, producing food, creating shade, capturing CO_2 and creating attractive and healthy landscapes for the city's inhabitants. These urban forests and greenways can be created with the city's waste water and storm water networks, as well as the sea water along the coastal areas.

The following projects use recycled and alternative water sources to create natural systems and nature based approaches to resilient urban design.

9.10 Blue-Green Belts and Networks. Urban Forestry and TSE

This nature-based approach to urban planning and design is based on trees and citizens and contributes to improving air quality in the city, along with numerous other benefits and ecosystem services. The concept behind this is 'better cities through better citizens' (Food and Agriculture Organization, FAO 2016). The idea is that the urban forester becomes a better citizen. Urban forestry is the management of trees for their contribution to the physiological, sociological and economic well-being of urban society (Grey and Deneke 1986). There are multiple benefits from implementing a city-wide Tree Charter supported by local business and academic institutions to spotlight the importance of indigenous trees and plants in modifying microclimate, stabilising soils, beautifying the city and inspiring an urban forestry movement at local and city-wide level. An urban forestry program can also include a community heritage tree component, which would include the propagation of indigenous species from known heritage trees in the city, with certain areas of the city promoted as urban forest reserves (Gibbons 2018).

The rapid urbanization in Qatar leads to environmental challenges. According to the World Health Organization (WHO) Qatar is number 12 among the 20 most polluted cities (WHO 2016). This is primarily because of its dominant fossil fuel industry and consumption. Additionally, due to its geographic and climatic conditions, on a yearly basis Qatar faces several sand storms that affect the health of its citizens. Moreover, the rapid urbanization is not always accompanied by sufficient or suitable green spaces. Qatar National Vision 2030 (QSA 2008) defines the five major challenges in the development of the country and four main pillars, by which the challenges must be addressed. Environmental development and sustainability is one of the four pillars with which Qatar's future development must be aligned. The project on Urban Forestry aims to improve the urban environment of Qatar using minimum resources to create a maximum of ecosystem services. Urban Forestry in Qatar is expected to build more resilient landscapes through better resource management and to respond to major environmental challenges through an Urban Forestry (UF) approach, using excess TSE (Treated Sewage Effluent). It delivers a number of ecosystem services, as well as shade and leisure landscapes using a waste resource, TSE that is currently being discharged into the sea or into the ground as it is in excess quantities due to the increasing desalination and water recycling.

The Urban Forestry project aligns with the projected Green Belt that surrounds the city of Doha, and branches through its major transportation routes towards the heart of the city, creating a system of Greenways. The Green Belt is proposed as a Productive Greenbelt for the growth of food, filtering dust storms, leisure landscapes and biodiversity conservation, and links via the green network to the coastal forestation of the Corniche. A larger Green Climatic Greenbelt is proposed beyond the city limits to filter the frequent dust storms and to mitigate their effect on the city, as well as to capture carbon.

The project builds on the concept of the Greenbelt from the Qatar National Development Framework (MME 2016). The research proposes scenarios and a series of urban interventions for Urban Forestry in Doha. The forestation concept is developed to address a series of urban functions:

1. Coastal protection;
2. Urban connections;
3. Urban protection;
4. Climatic protection.

An analytical study of the concept was taken into consideration starting with climate, accessibility, agriculture and existing functions in proximity to the green belt. Moreover, the various systems of TSE, trunk water and infrastructure were investigated for the implementation of the project and its feasibility. (Fig. 9.10)

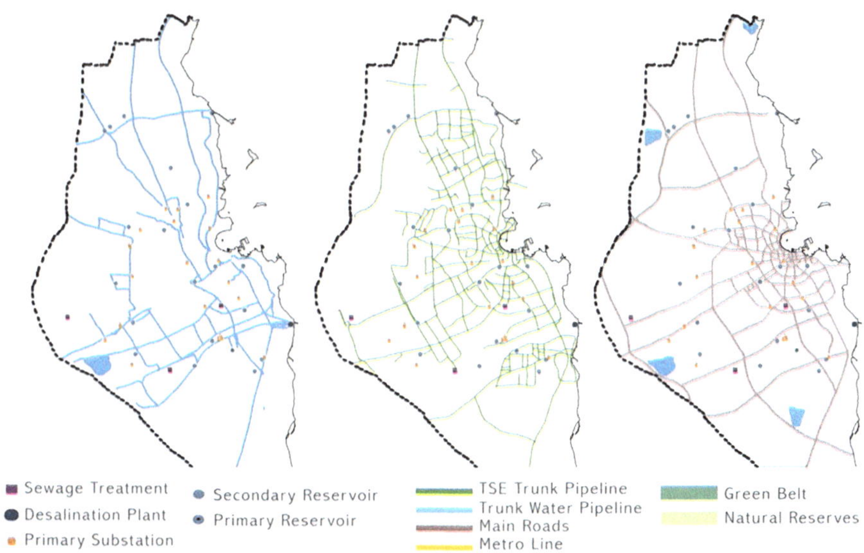

Fig. 9.10 Systems used for the Green Belt Urban Forestry Project. (Source: Asmaa S Al-Mohannadi, Albandari S Al-Harami, Heba S. Elgahani, Maryam Abbara)

Additionally, an analysis taking into consideration the stakeholders and actors, and the policies and regulations was considered that would build up the legislative framework of the project for implementation. Finally, various nodes within the proposed green shield were further developed into more focused design proposals. These design scenarios were presented to professionals from landscape forestry, urban planning and urban legislation, as well as water policy fields to discuss the feasibility and implementation.

The final design proposals focused on different layers: (Fig. 9.11)

Fig. 9.11 Master Plan for the Green Belt Urban Forestry Project. (Source: Asmaa S Al-Mohannadi, Albandari S Al-Harami, Heba S. Elgahani, Maryam Abbara)

1. The Climatic Green Belt focusing on forests and fields;
2. The Urban Green Belt, which is more people and society focused;
3. The Green corridors, connecting the urban and agricultural greenways;
4. The Coastal Green-Blue Belt which interfaces the city and the sea.

As a strategy of the QNDF 2032 (MME 2016), this research contributes to knowledge on green and blue infrastructure, and it is an opportunity to understand the merits of the green belt in Qatar and Drylands, and to start to envision the implementation of such a project at multiple scales.

9.11 Blue Oasis. Constructed Wetlands and Wildlife Conservation

Green and blue infrastructure are well-known natural features in cities. Wetlands are found in virtually all cities and because wetland ecosystems have both terrestrial and aquatic characteristics, they represent both types of infrastructure. And because the colors green and blue together make turquoise, urban wetlands provide *turquoise services*. Urban wetlands designed to treat wastewater effluent are expected to provide nutrients and the uptake of contaminants (Childers et al. 2015). In Qatar, one of the largest inland water features is the Abu Nakhla Wetland, which equals approximately the size of the Doha Corniche Bay.

Constructed wetlands are conceived as productive ecosystems that provide benefits and services to the people and contribute positively to the overall ecosystem. They are defined as 'areas of marsh, fen, peat land, or water, whether natural or artificial, permanent or temporary, with water that is static or flowing, fresh, brackish or salt, including areas of marine waters, the depth of which at low tide does not exceed six meters' (Hollis 1990). Negative factors influencing the wetlands involve adverse climatic effects, environmental changes, non-sustainable alternative uses, and disruption (Babiker and Osman 2012). A few of the wetlands in arid zones are supported by locally generated water supplies. Abu Nakhla Wetlands is a treated wastewater pond that was constructed in 1979 and was established in 1982 (Abdulfatih et al. 2002). The wetland is located approximately 12 km outside the city of Doha along the southern borders of Abu Nakhla village It is a special place for both humans and the wildlife as it is considered as a more moist point than in other regions. The pond is supplied by municipal water (TSE) on a regular basis. Treated municipal water from Doha-West and Doha-South wastewater treatment plants is discharged to the pond. The treated wastewater pond is approximately 37-38 m elevated above sea level while the water depth ranges from one to two meters. The capacity of the lake is *five* million sqm., taking 2 km length by 2 km width. In case of rain, the water level rises without changing the outer boundaries of the lake. Since 2006, the borders have been fixed so that it does not flood and damage the surrounding areas.

Presently, at Abu Nakhla, the treated sewage effluent is used solely for fodder crops for animals and for landscape irrigation. In this project, the optimization of treated wastewater is proposed with systems that enable recycling. Implemented systems emphasize maintaining further treatment to Abu Nakhla treated wastewater through constructive wetlands technologies. The purified water could then be used for irrigation of animal fodder, landscaping, and irrigation of crops. The resulting food production can be sold in the market to serve local demands and support diversification of the economy. Harvest waste can be recycled and composted to become natural fertilizers that can be utilized in the food production process. To overcome cultural sensitivies, the food products can be labelled to indicate the water source. Such a system supports sustainability and contributes to the local economy in an environmentally friendly way.

Students worked on a regeneration plan for Abu Nakhla, to mitigate the negative effects of the pond such as seepage, smell, inaccessibility, danger, etc. and to design it as a biodiversity hotspot and eco-leisure zone. An analysis of the plant and animal species, including sedentary and migratory birds, was previously conducted (Abdulfatih et al. 2002) and used as a basis for the design of the eco-reserve and habitats. Proposals to resolve the seepage through bio-drainage, that is the plantation of deep rooted and fast-growing trees, would contain the waters of the pond, provide shade for public areas as well as habitats for wildlife. A zoning was established with priority and core conservation areas, as well as more active zones for the public, including a botanical garden, food growing zones, community gardens, and a Center for Research on Wetlands in Drylands (Fig. 9.12).

Unfortunately, despite efforts to save Abu Nakhla, it has been decommissioned as a TSE pond and is gradually drying out. On a positive note, the planning authorities and Public Works are constructing a landscape, along the same lines as the proposed project for Abu Nakhla, at Doha North Sewage Treatment Plant. This will become a landscape for nature, biodiversity and leisure and hopefully will lead the way for other similar projects in Qatar. However, this is a constructed landscape being built from scratch, whereas the Abu Nakhla landscape evolved naturally over 40 years. (Fig. 9.13) So there is a great economic cost to creating fast, artificial landscapes as opposed to planning and allowing more natural landscapes to emerge that function with recycled water and also act as ecological infrastructures for water management.

9.12 Blue Ways for West Bay. Transit Oriented Development and Public Space

Located nearby Doha city's coastline, West Bay is the major business district in Qatar. It was built on reclaimed land as part of the Master Plan by Pereira and Associates for the Doha Corniche (Adham 2008, p. 233). It was designed as a

Fig. 9.12 Master Plan for Abu Nakhla Reservoir. (Source: Anna Grichting, Abdullah Al-Qahtani, Amna Al-Jehani, Ayla Shawish, Dalal Harb, Esra Mutlu, Nada Ghanem and Rana Awwad)

typical downtown business center, car oriented with huge spaces dedicated to open parking, and very little public space or connectivity. As part of future plans for a Transit Orient City the district will host two of the major train stations on the Red Line rail route, the West Bay Central Station (WCS) and the Doha Exhibition and Convention Centre (DECC). This design project combines new public spaces generated by the transit stations with landscape connections along blue ways, urban rivers that are supplied with treated waste water, storm water or sea water. (Fig. 9.14) Currently, there is a large outlet for TSE at the Sheraton Hotel, which is discharged directly into the sea. These blue ways will be landscaped with bio-landscapes to operate the bio-remediation of the water before it enters the sea. (Fig. 9.15).

The project begins with an analysis of the topography and natural hydrology of the West Bay, and links it the integrated water system, where waste water from the surrounding buildings will be directed to the blue ways, after treatments. Storm water and other TSE sources can be cleaned through the landscape, so that the water that arrives at the public spaces of the Corniche, and is led into the sea, is cleaned.

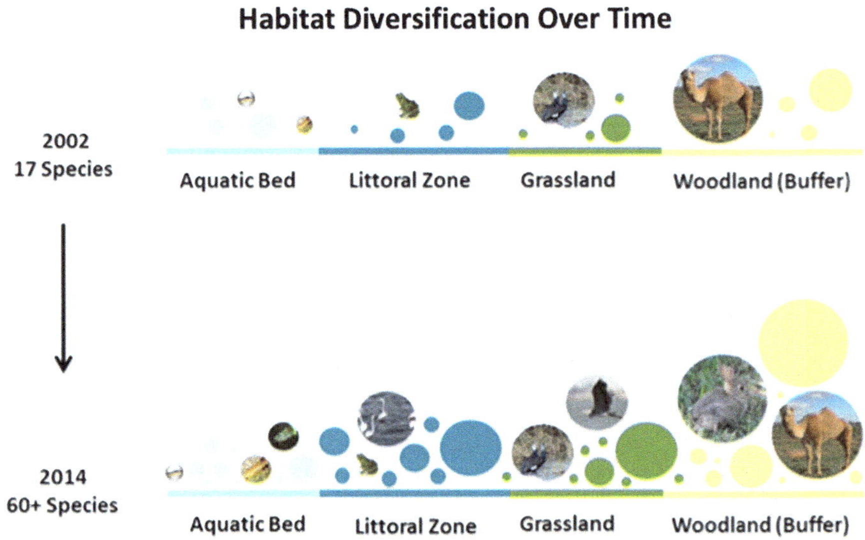

Fig. 9.13 Abu Nakhla Reservoir. Natural Habitat Diversification Over Time. (Source: Anna Grichting, Abdullah Al-Qahtani, Amna Al-Jehani, Ayla Shawish, Dalal Harb, Esra Mutlu, Nada Ghanem and Rana Awwad)

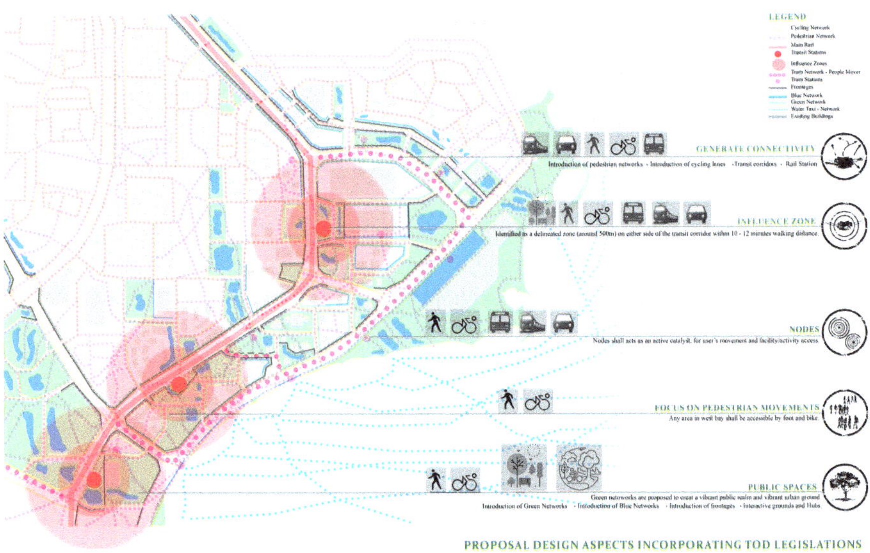

Fig. 9.14 Master Plan for Blue Ways with Transit Oriented Development. (Source: Dina Saleh, Angelica Caccam)

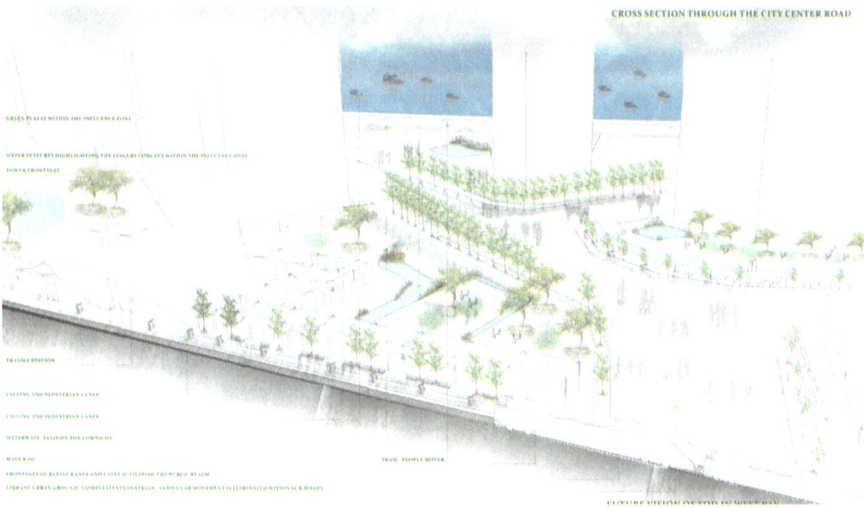

Fig. 9.15 View of Blueways for West Bay (Source: Dina Saleh, Angelica Caccam)

9.13 Conclusions

With the blockade of Qatar in June 2017, including Saudi Arabia through which 80% of its food transited, Qatar had to rapidly resource its imports of food through Iran and Turkey as well all its building- and raw materials. While it has ample reserves of non-renewable energy through its offshore gas fields, it relies on desalination for the majority of its sweet water resources. Additionally, the air quality in the main city of Doha is very poor because of the effects of the fossil fuel industries and automobile reliance. While it may seem contradictory to propose Nature-based urbanism in a desert or dryland, that is in a place that does not naturally have much *green* nature, Nature-based urbanism that works with natural resources and systems, creating regenerative systems, is introduced in this chapter. The city as a regenerative system is like an organism, or ecosystem, that runs on positive and negative feedback loops. Capturing increasing amounts of water, dust and organic waste provides resources and materials for a new urban organic layer, which is key to urban sustainability. Materials brought to the city, whether it is water from the sea or dust from storms, have the potential to transform a desert into an oasis. These material streams become a steady source of power for this layer composed of parks, trees, water elements, wetlands, green roofs, wild nature, urban farms and collective gardens. These layers reduce the heat island effect of the city, produce a comfortable microclimate, encourage pedestrian movability, and create an attractive public entourage and a new urban ecological luxury (Roggema et al. 2017).

A recent conference in Qatar on Future Landscape and Public Realm (Qatar University 2017) brought together the landscape industry, consultants, the public sector, and academia. The future of the landscape industry in relation to the Gulf crisis was discussed. Aside from the opportunities that any crisis offers, with more resource efficiency, self-reliance and resilience, there is also a concern about the status of landscape projects in the hierarchy of economic interests and budgets, as well as in the project management process. With reduced spending, landscapes projects, which are still considered beautifying amenities, are often largely reduced in budget or even eliminated. Landscape departments have been closed in important public works and planning administrations. At the same time, the landscape industry is coming up with innovative and highly cost-reduced solutions such as use of native plants and reducing water consumption, but these projects are not being accepted in the planning and construction authorization permission process.

It is time to elevate landscape to the role of infrastructure—and to position it as a foundation for urban and architectural projects. In the words of the great landscape architect Ian McHarg (1992), who has inspired many landscape urbanists, it is imperative that we *work with, and not against, nature* in shaping our cities and environments. This requires a shift in the conceptualization of urban plans to begin with natural systems and integrate technological solutions, with the aim of giving back to nature and the biosphere (Grichting 2017). Perhaps the Gulf crisis will be an opportunity for landscape urbanism to take a leading role and take its place in the design and development of future Gulf cities and landscapes. Hopefully, designers and scientists will continue to envision common futures across boundaries, looking beyond the borders to create more efficient, resilient and sustainable systems at the local and regional scale. As with the Blue Peace initiative in the Middle East (Waslekar 2011): 'an innovative approach (…) for harnessing and managing collaborative solutions for sustainable regional water management', water can become an agent for future nature-based solutions in the Gulf and in the Region. Additionally, a less anthropocentric, and more biodiverse approach will enable us to re-create habitats for some of the endangered and lost species, and perhaps one day bring the dolphins back in to Doha bay'.

References

Abdulfatih HA, Al-Thani RF, Al-Naimi IS, Swelleh JA, Elhag EA, Kardousha MM (2002) Ecology of wastewater ponds in Qatar. Scientific and Applied Research Centre (SARC), University of Qatar, Doha

Adham K (2008) Rediscovering the island: Doha's urbanity from pearls to spectacle. In: Elsheshtawy Y (ed) The evolving Arab City. Routledge, New York

Afifi A, Al Muqaddam S, Hagan S (2016) Overview of the integrated coastal zone management (ICZM) plan project for the State of Qatar. Ministry of Municipality and Urban Planning/ Wataniya Environmental Services, Doha

Al Jazeera (2018) https://www.aljazeera.com/news/2018/11/raining-qatar-season-181112125930905.html. Accessed 17 Dec 2018

Al Mohannadi, H. (2010) Water management in Qatar. State of Qatar: General Secretariat for Development Planning

Austin G, Yu K (2016) Constructed wetlands and sustainable development. Routledge, Avbingdon

Babiker A, Osman A (2012) Abu Nakhla Pond. Interview by Rana Amawi. Chief operations and maintenance engineer at Ashghal public works authority 28 April 2012

Beatley T (2014) Blue urbanism: exploring connections between cities and oceans. Island Press, Washington, DC

Bélanger P (2010) Redefining infrastructure. In: Mostafavi M, Doherty G (eds) Ecological urbanism. lars Muller Publishers, Baden, pp 341–344

Chatting M et al (2018) Nesting ecology of hawksbill turtles, *Eretmochelys imbricata*, in an extreme environmental setting. PLoS One 13:e0203257. https://doi.org/10.1371/journal.pone.0203257

Childers DL, Bois P, Sanchez CA, Tallman D and N.A. Weller (2015) Turquoise infrastructure in cities: how a constructed treatment wetland provides key urban ecosystem services in Phoenix, AZ. In: Proceedings 100th ESA annual meeting. CITY, 9–14 August 2015. https://eco.confex.com/eco/2015/webprogram/Paper52815.html. Accessed 13 October 2018

Floodlist (2018) http://floodlist.com/asia/qatar-floods-october-2018. Accessed 17 Dec 2018

Food and Agriculture Organization, FAO (2016) Guidelines on urban and Periurban forestry. *R*eport by Fabio Salbitano, Simone Borelli, Michela Conigliaro, Yujuan Chen. http://www.fao.org/3/a-i6210e.pdf. Accessed 17 Dec 2018

Gibbons J (2018) Homegrown urban forestry. In: Bullivant L, Grichting A (eds) Sustainable urbanism, new directions. Qatar University, Doha, pp 11–13. www.sustainableurbanismqatar.org. Accessed 17 Dec 2018

Government of Qatar Web Portal (undated) http://portal.www.gov.qa/wps/portal/topics/Environment+and+Agriculture/waterанddesalination. Accessed 13 Oct 2018

Grey GW, Deneke JD (1986) Urban Forestry. Second Edition. John Wiley and Sons. New York, NY

Grichting A (2012) Is blue the new green? The changing hues and paradigms of sustainability. Invited lecture. Qatar Society of Engineers. Diplomatic Club, Doha, Qatar. October 2012

Grichting A (2016a) Contemporary landscape as Urbanism: emergent ecologies of the Doha corniche. In: Gharipour M (ed) Contemporary urban landscapes of the Middle East. Routledge, New York, pp 207–230

Grichting A (2016b) Seascapes-lifescapes. Al Fuwairit Turtle Beach and ecological reserve. Draft summary of Master Plan elaborated with students in Masters in Urban Planning and Design, Qatar University. (Not published)

Grichting A (2017) Extradisciplinary investigations - Antidisciplinary spaces sustaining future urban and social systems. In: Bullivant L, Grichting A (eds) Sustainable urbanism, new directions. Qatar University, Doha, pp 11–13. www.sustainableurbanismqatar.org. Accessed 13 Oct 2018

Grichting A (2018) Edges, interfaces, and Nexus: new paradigms for blue urban landscapes in the Gulf in International Journal of Middle Eastern Studies, 50(3) (Environment and Society in the Middle East and North Africa), August 2018, pp 580–585

Grichting A, Zebich-Knos M (2017) The social ecologies of border landscapes. Anthem Press, London

Gulf Times (2017) Vetting wetlands. Interview of Dr Anna Grichting by Anand Holla, 2 February 2017. https://www2.gulf-times.com/story/531267/We-have-to-integrate-biodiversity-more-than-ever-i. Accessed 17 Dec 2018

Hollis GE (1990) Environmental impacts of development on wetlands in arid and semi-arid lands. Hydrological Sciences Journal-des Sciences Hydrologique 35(4):411–428

INNOCENT Innovation Center for Design and Technology: Ankara (2003) Design brief for Doha Corniche: Qatar's Centre for Arts & Culture. https://archnet.org/sites/4172/publications/3394

Intergovernmental Panel on Climate Change IPCC (2018) Special report. Global warming of 1.5 °C. https://www.ipcc.ch/sr15/. Accessed 17 Dec 2018

IUCN (2018a) Cities must connect with nature to build resilience. https://www.iucn.org/news/secretariat/201805/cities-must-connect-nature-build-resilience-%E2%80%93-iucn-president. Accessed 13 Oct 2018

IUCN (2018b) The IUCN red list of threatened species. https://www.iucnredlist.org/. Accessed 17 Dec 2018
Lombardi P (2017) Sustainble urbanism. Gaps in current research. In: Bullivant L, Grichting A (eds) Sustainable urbanism, new directions. Qatar University, Doha, pp 143–147
McHarg I (1992) Design with nature. Wiley, Hoboken
MDPS Ministry of Development, Planning and Statistics (2017) Water statistics in the State of Qatar 2015. https://www.mdps.gov.qa/en/statistics/Statistical%20Releases/Environmental/Water/2015/Water-Statistics-2015-En.pdf
Meerow S, Newell JP, Stults M (2016) Defining urban resilience: a review. Landsc Urban Plan 147(2016):38–49
MME Ministry of Municipality and Environment (2016) Qatar national development framework 2032. Qatar National Master Plan, Urban Planning Department
Nadim F, Bagtzoglou AC, Iranmahboob J (2008) Coastal management in the Persian Gulf region within the framework of the ROPME programme of action. Ocean Coast Manag 51:556–565
Qatar University (2017) Future landscape and public Realm. Doha, 30–31 October 2017. http://www.futurelandscapeqatar.com/. Accessed 17 Dec 2018
QSA Qatar Statistics Authority – Ministry of Development Planning and Statistics (2008) Advancing sustainable development – Qatar National Vision 2030
RAMSAR (2017) World Wetlands Day 2017. https://www.ramsar.org/activity/world-wetlands-day-2017. Accessed 17 Dec 2018
Roggema R, Grichting A, Casagrande M (2017) Dohasis: the biourban restoration of Doha. In: Bullivant L, Grichting A (eds) Sustainable urbanism, new directions. Qatar University, Doha, pp 117–122
Saatchi & Saatchi (2008) The birth of blue. http://saatchi.com/en-us/news/the_birth_of_blue. Accessed 17 Dec 2018
Saline Agriculture Worldwide (2018). https://www.salineagricultureworldwide.com/saline-agriculture. Accessed 17 Dec 2018
SCENR (2008) Sensitivity mapping of the eastern coast of Qatar (Phase I). Supreme Council for the Environment and Natural Reserves (SCENR), Doha-Qatar
Supreme Council for Environment and Nature Reserves (2007) Sensistivy Mapping of the Eastern Coast of Qatar
Ulrichen KC (2018) How Qatar weathered the Gulf crisis. In Foreign Affairs, June 11 2018. https://www.foreignaffairs.com/articles/middle-east/2018-06-11/how-qatar-weathered-gulf-crisis. Accessed 21 Dec 2018
UN.org. https://www.un.org/sustainabledevelopment/cities/. Accessed 17 December 2018
United Nations Office for Disaster Risk Reduction UNISDR (2015) The human cost of weather related disasters 1995–2015. https://reliefweb.int/sites/reliefweb.int/files/resources/COP21_WeatherDisastersReport_2015_FINAL.pdf. Accessed 18 Dec 2018
Waslekar, S. (2011) The blue peace. Rethinking middle east water. Mumbai: Strategic Foresight Group. https://www.eda.admin.ch/dam/deza/en/documents/.../198458-the-blue-peace_EN.pdf. Accessed 18 Dec 2018
Wiegand K (2012) Bahrain, Qatar, and the Hawar Islands: resolution of a Gulf territorial dispute. Middle East J 66(1):2012
World Health Organization (2016) WHO global urban Ambient Air pollution database (update 2016). https://www.who.int/phe/health_topics/outdoorair/databases/cities/en/. Accessed 17 Dec 2018

Chapter 10
South Creek in Far Western Sydney: Opportunities for a New Waterway Focused City

Phillip James Birtles

Abstract As Sydney continues to expand into a polycentric city, the far west faces rapid urbanisation around a new airport. This new inland city is not located within striking distance of Sydney's iconic coastline but rather centred around a small ephemeral creek on flat alluvial plains. There is a significant opportunity for this natural environment, though maligned by urbanisation styles of the past, to become a key focus of attention and driver for a new ecology-based urban concept and form. By re-imagining the design and construction of water systems, urban layouts and green infrastructure, a new city vision can provide broad benefits to future populations and may become a template for urban waterway management in other regions.

Keywords Urbanisation · Waterway health · Liveability

10.1 Introduction

Sydney is a coastal city, it's identity and culture focused on its beaches and the harbour. The Central Business District is within a few kilometres of the Pacific and any part of the city inland is generalised into the category of 'Western Sydney'. Limited by protected national parks to the north and south, over the last 200 years a disjointed sub-urbanity has sprawled inland past Parramatta and onwards, 30-40kms to the west, to the vast agricultural Cumberland plains at the foot of the Blue Mountains.

This far western Sydney, has been a blind spot on the collective minds of city planners until very recently. In 2016, a new quasi-independent metropolis planning agency, the Greater Sydney Commission, detailed their three-city vision for Sydney (GSC 2017). Breaking the city up into regional centres they prescribed an entire city population could be imagined around a small ephemeral creek named South Creek. The local industry association of landscape architects advocated to design this regional centre as a park. The NSW charter of the Australian Institute of Landscape Architects wrote a letter to the then planning minister on 29 February 2016 offering

P. J. Birtles (✉)
Sydney Water, Sydney, Australia
e-mail: phillip.birtles@sydneywater.com.au

R. Roggema (ed.), *Nature Driven Urbanism*, Contemporary Urban Design Thinking, https://doi.org/10.1007/978-3-030-26717-9_10

a role "promoting Western Sydney's livability and build a design narrative for its future" (AILA 2018). The new "Parkland City" became an opportunity to re-establish urban form within the landscape. It was never going to be easy.

10.2 Natural Waterways

In the West of Sydney, freshwater creeks generally run in a northern direction before combining into the Hawkesbury Nepean River, which skirts the basin edge between the neighbouring Blue Mountains and empties into the Hawkesbury Estuary to the Pacific (Fig. 10.1). The flat alluvial shale basin provided land for agriculture and was the setting of numerous skirmishes with Aboriginal warriors at the turn of nineteenth Century as sections of the catchment were seized and sold to settlers for western farming practices (Corr 2016). In the Aboriginal Dharug language, the creek is called "Wianamatta", meaning "mother place". The clearing of land for pasture and agriculture was the first wave of dramatic negative changes to the catchment and remains the prominent land use to this day.

The pre-European environment of the Cumberland Plain supported an array of unique life. The major vegetation community 'Cumberland Plains Woodland' (Tozer 2002) is defined by a small range of trees and shrubs with extensive

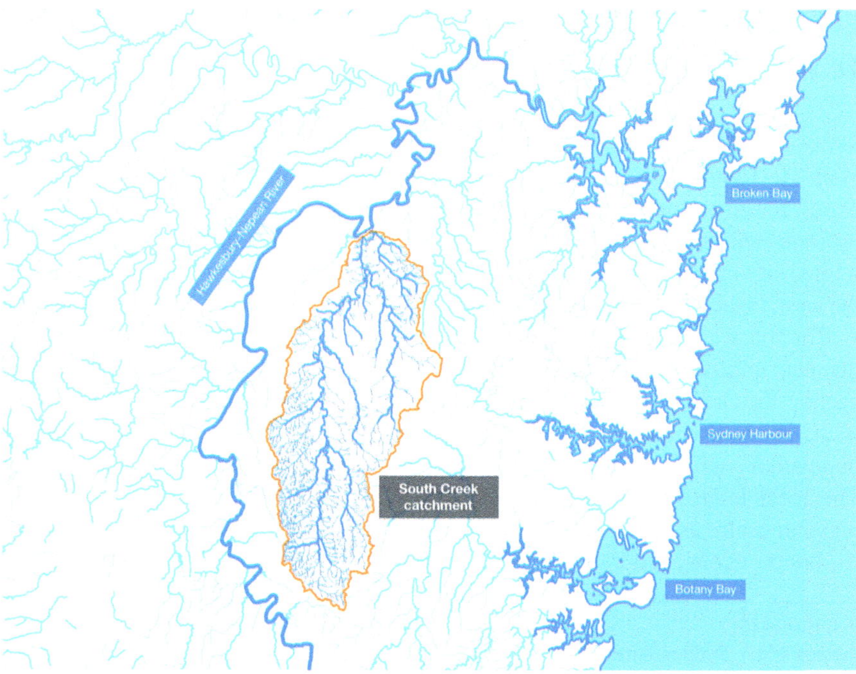

Fig. 10.1 Location of the South Creek Catchment. (Courtesy of M. Dean 2019)

Fig. 10.2 Photograph of an Australian Bass (Macquaria novemaculeata). An iconic freshwater species in Western Sydney, prized by game fishers. (Photo: Gunther Schmida. Courtesy of the NSW Department of Primary Industries Fisheries)

complexity in grass and herb species. This landscape supported a range of large fauna including kangaroos, wallabies, quolls, dingoes, emus, goannas and birds of prey such as the wedge-tailed eagle. Biodiversity is concentrated in the waterways of the South Creek Catchment. Slow moving, shallow creek lines that often presented a series of ponds rather than a continuous channel would have been cluttered with tree branches, dense tickets of reeds and melaleuca trees providing habitat for platypus, dragonflies, water birds and native fish such as the famous Australian Bass (Fig. 10.2).

10.2.1 The Impacts of Suburbanisation on Waterways

The suburbanisation of Sydney has sprawled to the margins of the basin and followed the main transport lines across the catchment (the rail line westward to the Blue Mountains and the interior). The creek lines continue to experience significant modification and regime change (Hoban et al. 2015b). Urbanisation of catchments impacts waterways directly via flow path modification (such as hardening with concrete or piping) and vegetation removal. The greatest impact, however, comes from the reduction of the land's ability to soak up water (Walsh et al. 2004). Hard surfaces replace vegetation and rainfall rapidly washes to drains and pipes delivering vast amounts of polluted water to the creeks. This powerful pulse of flow erodes creek banks and deepens channels, simplifying the complex, meandering, pool and riffle habitats into barrelling channels that blast with water after rain, yet dry up completely shortly after. Adding to this change in the water cycle, wastewater treatment plants discharge treated effluent into waterways, independent of the localised rainfall. This constant flow volume can create erosive results in channels where discharge points are located. The traditional approach to address this instability is to further harden the creek or river channel with boulders or concrete. A well-known, though extreme, example of this is the concrete channel of the Los Angeles River, famously depicted in Hollywood car chase scenes.

The waterways of South Creek are typical of a catchment experiencing the first symptoms of this 'urban stream syndrome' (Kermode et al. 2016). By 2015, approximately 20% of the catchment had been urbanised from its agricultural state (Hoban et al. 2015a) and channel simplification and active erosion is now easily observable in the waterways. Once common native species such as the platypus have not been recorded in the catchment for decades. However, a population of the Australian Bass still exist in the catchment and are regularly reported on by an active recreational angler presence (South Creek Bass Club 2019).

Attempting to mitigate this declining ecosystem trajectory has had limited successes in Western Sydney. This is partly due to the nature and aspirations of urban development in the catchment.

10.3 Urban Development in Far Western Sydney: Re-imagining a Landscape

Urban development of far Western Sydney has taken a series of forms and direction over the last few decades. By the mid 1990's the concept of master planned 'greenfield' estates were being designed and constructed by a variety of major development corporations. These estates curated the look, feel and form of the finalised estate lifestyle narrative as a marketing strategy for lot sales. Large modern two-story houses encouraged clean streets, large garages and little need to interact with the streetscape from outside a vehicle. These car-dominated estates were labelled 'McMansions' for their on-mass construction and their lack of character. Some estates focused on a romanticised Europeanisation of the landscape. A good example being the original Harrington Park estates within Camden Council with exotic deciduous street tree plantings and the transformation of creek lines into duckponds and picnic lawns.

However, toward the mid 2000s Western Sydney estates were delivered with a greater focus on stormwater management. Targets were adopted from a progressive stormwater industry in Victoria. These mostly focussed on reducing dissolved nutrient pollutants in stormwater (Brown and Clarke 2007; DEC 2006) and were mandated into the growth planning from a state government level. Added to this was a stronger direction on riparian land reservation for environmental purposes under the NSW Rivers and Foreshores Improvement Act.

These directions were adopted in varying degrees by the development industry. A pioneer development was 'the Ponds' in Blacktown Council area just outside of the South Creek Catchment. This 3200-lot subdivision included 88 ha of parkland area mostly along the creek lines and was developed from the mid 2000s with design principles that encouraged 'Engaging with Water' and integrated environmental drainage engineering with landscape design (O'Dea and Nakkan 2012). The Ponds was a successful development, with premium house prices being realised. However, it was the exception rather than the rule and did not create a re-configuration of practice towards a more specific natural aesthetic.

Generic 'best practice' frameworks failed to target specific ecological outcomes for the catchment of South Creek. These pragmatic targets were endured by developers as subdivisions continued to roll out that played tribute to, but rarely integrated land- and water-scape forms. The 'environmental infrastructure' often being rendered to a corridor on the edges of new suburbs.

The resulting urban form is not sympathetic to the landscape, the ecology or the water cycle of western Sydney. The full set of benefits available to future residents is compromised by continued cookie cutter housing into bulldozed clay plains.

10.4 Opportunity and Vision

In 2015, the regional water utility, Sydney Water, began planning for their servicing of the far western side of the city. Championing a more holistic approach to water management, Sydney Water planners ran a series of workshops to consider the improvement that alternate waterway outcomes could bring to this part of the city (Hoban et al. 2015a). A multidisciplinary project team consisting of an engineer, urban designer, ecologist, waterway specialist and urban planner guided a selection of staff from the utility through a workshop-process to conceptualise at a high level what the future options for urban outcomes for the development could become (Hoban et al. 2015a).

A simple model was developed of the South Creek Catchment and it's overlay with expected levels of future development (Fig. 10.3). This revealed the clear relationship between the proposed urbanisation and the footprint of the catchment. Urbanisation was expected to increase from the current 20% up to 80% in the next 50 years.

A simple model was developed to estimate the flow impacts on the waterway (Hoban et al. 2015a). An annual water balance for the South Creek catchment was updated (from Singh et al. 2009) with growth projections and likely water demands from Sydney Water. It showed that water demand from future residents was expected to increase to 125GL per year. This creates a considerable challenge in dealing with treated wastewater discharge should it be expected to discharge into the South Creek Catchment.

Further, the change from permeable to impervious surfaces will increase stormwater discharge flow into and through South Creek with a total increase to 234GL/year. It is important to note that this annual stream flow figure is somewhat misleading in the magnitude of potential impact. The likely impacts of urbanisation on this ephemeral waterway include a persistent elevated base flow in the main creek trunk from discharged treated wastewater and extreme high flood volumes during and just after rainfall events (Fig. 10.4).

Whilst there is a lack of detailed study into the geomorphic response of Western Sydney shale plain creeks to urbanisation, South Creek is likely to further erode,

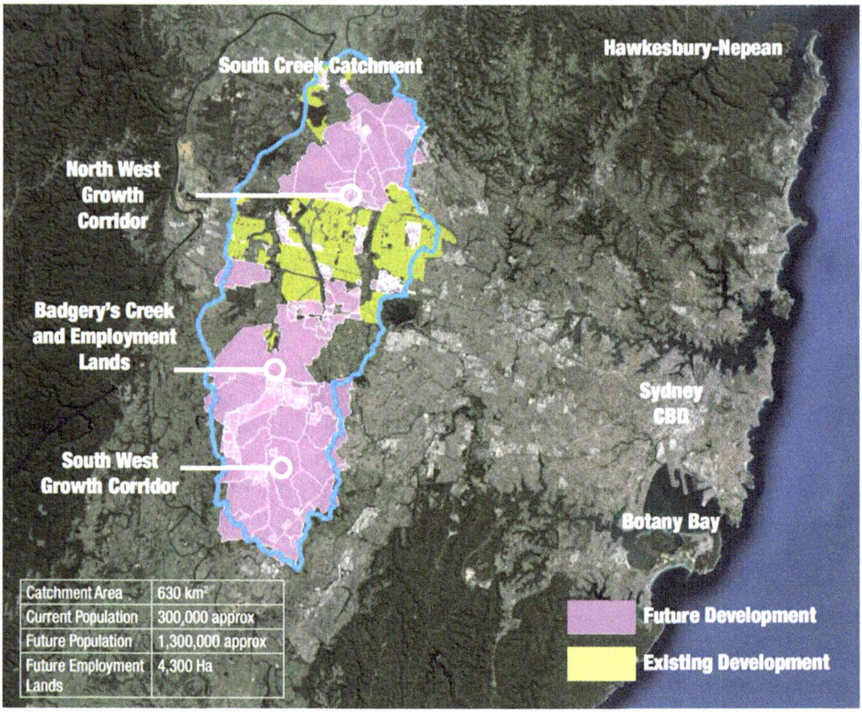

Fig. 10.3 Existing and predicted urbanisation in Western Sydney for the next 50-year period. (From Hoban et al. 2015a)

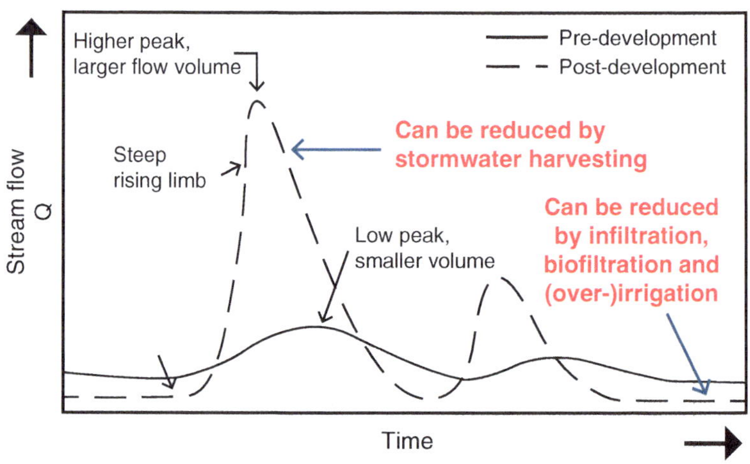

Fig. 10.4 Typical pre and post development hydrographs with management options for urban stream catchments. (From Fletcher et al. 2012)

becoming wide, deep and uniform. Habitat value and biodiversity will be lost as the stream continues to degrade over time. Bankside living for future residents is unlikely to be seen in a positive light.

With this stark information now in focus, the project team then workshopped scenarios moving forward. They focused more specifically on outcomes for the new population of the South Creek Catchment and their potential aspirations.

10.5 Eco-Centric City Planning in Sydney

Water service management within Sydney is being influenced by emerging sustainable management principles such as the Water Sensitive City concept (Brown et al. 2009). Government agencies associated with the water sector had commenced discussions that resulted in principles being articulated and published (Fig. 10.5), Water Sensitive Greater Sydney 2016).

A growing consensus regarding the need for urban communities to connect with, and experience, nature to gain physical and mental health, aesthetic and general wellbeing outcomes continues to gather pace. Birtles et al. (2013) identified that the multiple levels of ecosystem services to generate ideal liveability required such services being sourced from within the context of the city rather than being imported to it. This followed the Water Sensitive City paradigm (Wong and Brown 2009) that affirmed the need for ecosystem services as one of 3 'pillars' to make a city resilient. This frame was further developed by Johnstone et al. (2012) who identified a

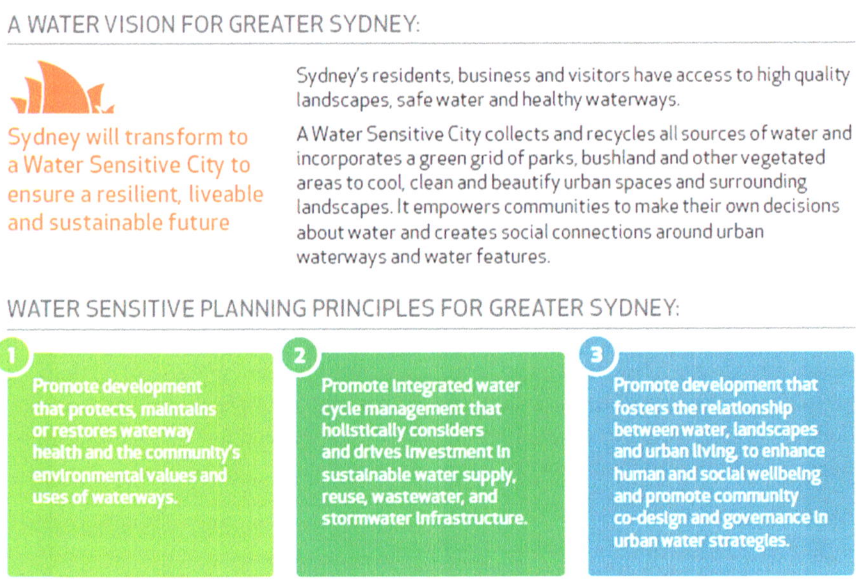

Fig. 10.5 A water vision for Greater Sydney (Water Sensitive Greater Sydney 2016)

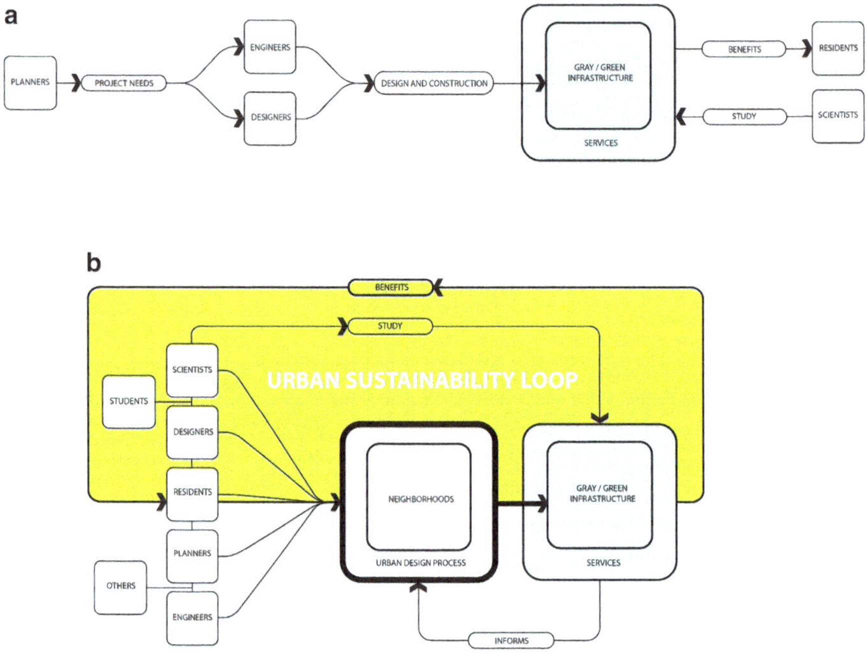

Fig. 10.6 Typical linear delivery of urban infrastructure (**a**) in comparison to an integrated, cyclical, multidisciplinary approach (**b**) that includes communities of neighbourhoods as a central element in the design process (Childers et al. 2015)

range of ways that water in all forms could address societal needs working under the Alderfer E.R.G theory (Alderfer 1969).

Childers et al. (2015) described a model for redesigning the connection between ecology and city outcomes in the diagram below (Fig. 10.6). By combining disciplines, the 'Urban Sustainability Loop' concept is more likely to deal with the complexity of systems thinking and urban planning.

In this concept, residents are active in the design process, influencing the urban form of neighbourhoods and infrastructure and gaining ongoing benefits from the newly informed setup. As a model, this set up challenges traditional urban waterway management as it brings a human centred lens to the framing of ecosystem requirements and urban form. However, it is limited in time. In far Western Sydney, the population who would be best placed to inform the ideal grey/green infrastructure does not reside there yet. Therefore, an approach is needed that allows visioning of what potential residents might need.

10.6 Future Scenarios: Business as Usual and the New Australian Dream

The original work presented possible scenarios for the future of urbanisation of South Creek (Hoban et al. 2015a). The first followed the development outcomes likely in the current context under 'best practice':

10.6.1 Traditional Urbanism

The management objectives for this level of development can basically be described as:

- Clean water supply;
- Robust wastewater service;
- Stormwater treatment before discharge;
- Impeding and regulating storm flows to reduce erosion;
- Flood risk management.

The focus of such an approach is the segmentation of water streams to most efficiently provide services to the future residents and some of the objectives of environmental management. The lack of integration creates a significant change in the total water balance of the system. As identified in Fig. 10.7, a final stream flow of 234GL/yr. in volume is more than twice the predevelopment flow. This flushing will lead to a reduced ecological outcome for the waterway.

10.6.2 The New Australian Dream

The workshop groups came up with a new concept that re-defines the idea of waterway management and urban form. It is in its essence an acceptance that the human population can have significant synergies with the western Sydney landscape and the waterways. Here, the concept of ecosystem services was championed alongside the integration of social and health outcomes by interaction with a somewhat new ecological narrative. The outcomes focussed on:

- Ecosystem Services available to residents;
- Wellbeing of residents;
- Social connection of residents;
- Sense of place;
- Greening of urban form;
- Desirability of lifestyle;

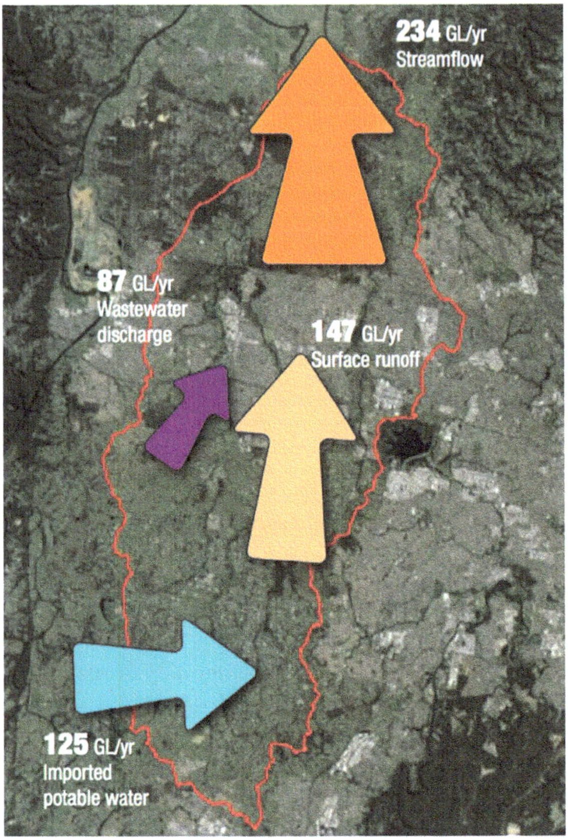

Fig. 10.7 Water balance scenario for traditional urbanism (Hoban et al. 2015a)

The new Australian Dream therefore integrates natural waterways into the urban form and gains the benefits of the ecosystem services provided. The urban form is no longer in the environment but is a part of it. This integration changes the outlook on the environment in terms of water management from being a point to discharge to. Importantly, water cycle management maximises opportunities to recycle waste water and harvest stormwater. This is essential for reducing impact on the receiving waterway of South Creek (Figs. 10.8 and 10.9).

There are several principles worth noting that create an important framework for waterway centric urban planning:

1. **All the water we need is already in the City**
   When considering a city with a broader system-based approach it becomes obvious, somewhat surprisingly, that cities generally provide an overabundance of water. However, due to the traditional approaches of managing these streams of water separately, engineering systems are not optimised to make the most of these resources. Importing drinking water is focussed on securing demand volume and water quality, wastewater systems focus on rapid removal and treatment, stormwater systems are focused on flood risk mitigation.

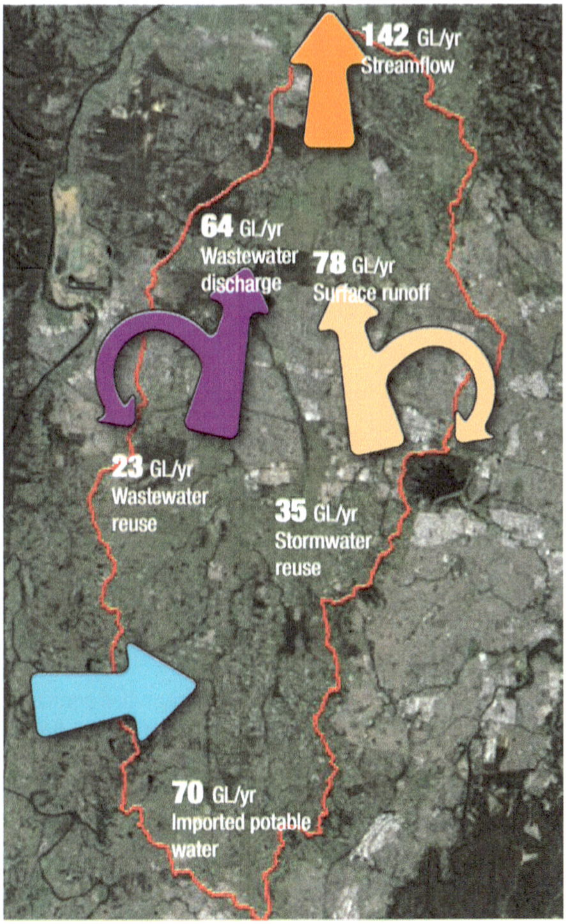

Fig. 10.8 Water balance for the New Australian Dream Scenario (Hoban et al. 2015a)

In their landmark paper of 2009, Wong and Brown, proposed one of three core pillars of a water sensitive city includes the mindset of "*cities as water supply catchments*". Certainly, this creates significant challenges as the "new" water supply will be recycled water, which has public perception challenges, and stormwater requiring large storages to be deeply engineered into the urban form.

However, with Australian coastal cities focussing on larger and larger desalination infrastructure, there is a certain sustainable logic to using the water in the city rather than disposing waste and stormwater to the sea at one point and then harvesting sea water at another to treat and pump back into the city. There are multiple technological options available to close this gap (IWA 2016) and the benefits of doing so considerably assist with the next two principles.

2. **A clean river is a fun river**
Waterway health can be linked to city productivity and liveability. This requires an environmental ethic from city planners that goes beyond meeting standards and rather, working directly with ecologists and engineers to focus on specific

Fig. 10.9 The New Australian Dream as workshopped by a multidisciplinary group in 2015 as a possible scenario for development in the South Creek Catchment (adapted from Hoban et al. 2015a)

environmental outcomes. This somewhat controversial concept goes against the grain for many environmental professionals beholden to an unrealistic philosophy of a pre-European environmental goal. Likewise, it is a hard concept for many planners, who want targets and standards to meet and tick the box for a complying design. What is needed is considered conversations, good science, community data and the courage to, in some ways, curate the likely environmental outcomes we are hoping to achieve.

As indicated earlier, the Australian Bass is a prized fish species for the local fishing clubs of Western Sydney. By investigating what outcome could be achieved with the coming development to optimise the habitat for this species there can be real and tangible options for setting outcomes and goals. For example, to improve the area of the bass by 20% across the catchment. This gives us targets and an adaptive management framework to continuously interrogate and refine. It allows us to consider any range of potential environmental requirements from water temperature to riparian habitat, to flow regime. The benefits of maintaining habitat for the bass will certainly have flow-on effects to other species as ecosystems work with a food web.

The alternative is currently failing – setting generic targets on dissolved pollutants or vague riparian widths is providing little incentive for planners or developers to comply, no public pressure to ensure they take part, nor real understanding from the community to ensure the outcomes are working. The change is to focus on the environmental outcomes from an anthropocentric view. As aghast as this will be to many, this is the reality of driving environmental improvement in an urbanised context.

3. **Healthy communities are connected to water**
   Landscape oriented urbanisation is important not just to achieve waterway health but for improving human health. By creating a multifunctional blue/green corridor each side of South Creek, urban planners are also creating a significant open space asset for the recreational benefit of residents. Walking trails and cycle paths should be integrated into the naturalised elements of the corridor. The goal of balancing the water cycle is significantly assisted by extensive irrigation of recreational sports fields in the corridor.
   By keeping water in the landscape with multiple smaller water bodies and maximising evapotranspiration via irrigated vegetation elements within the urban form, the ambient temperatures can be reduced. This cooling outcome is highly sought after in locations like Western Sydney where summer daytime temperatures regularly top 35 degrees and remain in the mid-20s overnight.
   Finally, the mental health benefits of re-establishing and planning natural environments is becoming an important element of urban place making (West and Jones 2009). Careful water management is essential to ensure such spaces remain green, healthy and wild.

A great challenge to the roll out of such principles in Western Sydney and other urban centres is a coherent and considered conversation between urban planners and water/ecological professionals. Planners lack knowledge of the water cycle options across multiple water technologies. Likewise, water managers and ecologists lack understanding of land use planning and the timing of the land release process to allow integration.

10.7 Current State of Play

In 2017 The Greater Sydney Commission (GSC) began communicating high level urban planning strategic concepts for the city (GSC 2018). This included breaking the metropolis into three major centres labelled as Harbour, River and Parkland Cities. The Western City maintained the concept of South Creek being the 'central element' in the urban design. With Sydney's new airport and associated commercial uplift (labelled the Aerotropolis) also featuring as key anchor for the density proposed. The GSC concept picked up on many of the 'new Australian dream' elements including integrated water cycle management with an irrigated green corridor for cooling, recreation and environmental preservation. Importantly, there was clear indications of integrating the urban form directly with high density buildings to allow for open spaces, stormwater and wastewater capture for localised use, high tree canopy density and soil permeability, rather than expecting all outcomes to be delivered within the corridor (Fig. 10.10).

This landscape-focused concept has continued to be documented into the 'district plans'– the strategic high-level planning documents for the Sydney Metropolitan Area (GSC 2018). Further, the State's major infrastructure co-ordination agency,

Fig. 10.10 South Creek urban design principles were released in 2018 and show that waterway centric urban design is now government policy for South Creek (GSC 2018)

Infrastructure NSW, has been tasked directly with co-ordinating multi-agency input to achieve the South Creek corridor.

At a local government level, the engagement with waterway outcomes is mixed. Some councils are progressing a best practice approach where possible and importantly working to engage their communities to enable healthy waterway discussions. Blacktown Council in particular, has an extensive waterway monitoring program tailored towards a community facing report card that is produced annually (BCC 2018).

10.8 Challenges for the Future

The most pressing challenge for eco-urban outcomes for South Creek does not lie in the technology, political will nor cost. It is time. More specifically, how quickly change can occur within the complexity of government, influence developments and then be implemented on the ground. While multiple studies continue, and the best intentions are being prepared for deployment, development continues via the drivers, market forces and risk perceptions of the past. As one wag put it recently "*the bulldozers are still heading the other way*".

References

AILA (2018) Summary of advocacy for February 2016. Available at www.aila.org.au/iMIS_Prod/AILAWeb/Advocate/Submissions_to_Government/State/NSW/AILAWeb/Chapter/NSW/Advocacy.aspx Sighted 2 October 2018

Alderfer CP (1969) An empirical test of a new theory of human needs. Organ Behav Hum Perform 4(2):142–175

Birtles PJ, Hore J, Dean M, Dahlenburg J, Hamilton R, Moore J, Bailey M (2013) Creating a Liveable City – the role of ecosystem services. State of Australian cities conference 2013, Sydney

Blacktown City Council (2018) Waterway Heath Report Card 2017–2018. Available at https://www.blacktown.nsw.gov.au/Community/Our-environment/Waterways

Brown RR, Clarke JM (2007) Transition to Water Sensitive Urban Design: The story of Melbourne, Australia, Report no. 07/1

Brown RR, Keath N, Wong THF (2009) Urban water management in cities: historical, current and future regimes. Water Sci Technol 59(5):847–855

Childers DL, Cadenasso ML, Grove JM, Marshall V, McGrath B, Pickett STA (2015) An ecology for cities: a transformational Nexus of design and ecology to advance climate change resilience and urban sustainability. Sustainability 7:3774–3791

Corr B (2016) Pondering the abyss: the language of settlement on the Hawkesbury. Available at http://www.nangarra.com.au/documents.html

Dean M (2019) South Creek Catchment. Image

Department of Environment and Conservation (2006) Western Sydney growth centres stormwater guidance for precinct planning

Fletcher T, Walsh C, Vietz G, Burns M, Hamel P, Poelsma P (2012) Stormwater management principles & technologies to restore stream ecosystem function. Blueprint 2012 workshop for the CRC for water sensitive cities, Sydney

Greater Sydney Commission (2017) A metropolis of three cities: the Greater Sydney region plan
Greater Sydney Commission (2018) Western City district plan. March 2018
Hoban A, Shoo B, Tippler C, Davies P, Wright I (2015a) South creek catchment, liveability and waterway health. Bligh Tanner report prepared for Sydney Water. March 2015
Hoban A, Vietz G, Walsh C, Fletcher T, Mills K (2015b) Protecting Little Creek: Marsden Park industrial precinct Little Creek catchment alternate Stormwater management strategy. Bligh Tanner Report prepared for Blacktown City Council. June 2015
International Water Association (2016) Water utility pathways in a circular economy, London
Johnstone P, Adamowicz R, de Haan FJ, Ferguson B, Wong T (2012) Liveability and the water sensitive city. Science policy partnership for water sensitive cities. Cooperative Research Centre for Water Sensitive Cities, Melbourne. ISBN 978--1--921912--17--7, August 2012
Kermode S, Birtles P, Vietz G, Lynch S, Dixon J, Tippler, Dean M (2016) The expanding role of urban fluvial geomorphology: South Creek. Proceedings of the 8th Australian Stream Management Conference. Leura, NSW
O'Dea M, Nakkan K (2012) Putting the D back in WSUD: invisible engineering: a collaborative design approach to integrating water management with liveable public open space at "The Ponds". In: Proceedings Stormwater '12: National Stormwater Conference, Melbourne
Singh R, Nawarathna B, Simmons B, Maheshwari B, Malano HM (2009) Understanding the water cycle of the South Creek catchment in Western Sydney part I: catchment description and preliminary water balance analysis. CRC for Irrigation Futures Technical Report 05/09
South Creek Bass Club (2019) South Creek Bass Club Facebook Page available at www.facebook.com/SouthCreekBassClub/. Sighted 14 February 2019
Tozer M (2002) The native vegetation of the Cumberland plain, western Sydney: systematic classification and field identification of communities. Cunninghamia 8(1):75
Walsh CJ, Leonard AW, Ladson AR, Fletcher TD (2004) Urban stormwater and the ecology of streams. Cooperative Research Centre for Freshwater Ecology and Cooperative Research Centre for Catchment Hydrology, Canberra
Water Sensitive Greater Sydney (2016) Opportunities for a water sensitive greater Sydney. Available at https://watersensitivecities.org.au/content/opportunities-water-sensitive-greater-sydney/
West S, Jones C (2009) The contribution of public land to Melbourne's liveability – final report. McCaughey Centre and University of Melbourne for the Victorian Environmental Assessment Council
Wong TH, Brown RR (2009) The water sensitive city: principles for practice. Water Sci Technol 60(3):673–682

Chapter 11
Nature-Inclusive Cities: Concepts and Considerations

Stewart Monti

Abstract The world is currently experiencing its sixth great extinction event (Ceballos et al., Sci Adv 1(5):1, 2015). The *Holocene extinction*, also known as the *Anthropocene extinction*, is affecting nearly all of the planet's flora and fauna species, with the current rate of extinction, estimated at between 100 (Ceballos and Ehrlich, Science 360(6393):1080, 2018) and 1000 (Pimm et al., Science 344(6187), 2014) times higher than natural background rates. The loss of species from ecological communities, *defaunation*, is primarily driven by human activity (Dirzo et al., Science 345(6195):401–406, 2014). It is estimated that more than 60% of all wildlife has been lost in the last 40 years, and that by 2020 68% will have been lost (WWF, Living planet report – 2018: aiming higher. World Wildlife Fund, Gland, 2018). There are many drivers of this defaunation; from overexploitation to invasive species and pollution (Hoffmann et al., Science 330(6010):1503–1509, 2010). However, by far the greatest cause is habitat destruction and fragmentation (IUCN, The IUCN Red List of Threatened Species. *Version 2018-2*, International Union for Conservation of nature, viewed 15 February 2019, http://www.iucnredlist.org, 2018). Changes in land use from those that naturally formed in response to local geography, geology and climate to those which serve human populations is resulting in a global decrease in biodiversity. Expanding urban areas consequently leads to increased agriculture, rangelands, forestry and mining to service the needs of the growing population (Ripple et al., Bioscience 67(12):1026–1028, 2017). This inherently means that built environment design professionals are directly complicit in the greatest extinction in human history.

Keywords Biodiversity · Climate change · Resilience · Ecosystem services · Urban ecology · Urban design

S. Monti (✉)
Urban Refugia, Sydney, Australia
e-mail: sm@urbanrefugia.com

R. Roggema (ed.), *Nature Driven Urbanism*, Contemporary Urban Design Thinking, https://doi.org/10.1007/978-3-030-26717-9_11

11.1 Introduction

The world is currently experiencing its sixth great extinction event (Ceballos et al. 2015). The *Holocene extinction*, also known as the *Anthropocene extinction*, is affecting nearly all of the planet's flora and fauna species, with the current rate of extinction, estimated at between 100 (Ceballos and Ehrlich 2018) and 1000 (Pimm et al. 2014) times higher than natural background rates. The loss of species from ecological communities, *defaunation*, is primarily driven by human activity (Dirzo et al. 2014). It is estimated that more than 60% of all wildlife has been lost in the last 40 years, and that by 2020 68% will have been lost (WWF 2018). There are many drivers of this defaunation; from overexploitation to invasive species and pollution (Hoffmann et al. 2010). However, by far the greatest cause is habitat destruction and fragmentation (IUCN 2018). Changes in land use from those that naturally formed in response to local geography, geology and climate to those which serve human populations is resulting in a global decrease in biodiversity. Expanding urban areas consequently leads to increased agriculture, rangelands, forestry and mining to service the needs of the growing population (Ripple et al. 2017). This inherently means that built environment design professionals are directly complicit in the greatest extinction in human history.

Meanwhile, humans ourselves are grappling with the greatest threat to our own future existence. Human-induced climate change characterised by the increasing frequency and severity of acute shocks and chronic stresses is the defining challenge of our history (IPCC 2007). The necessary mitigation and adaption to survive unprecedented and unpredictable environmental changes has led to a global rethink of how we inhabit the planet. The consequences for flora and fauna further exacerbate the already dire circumstances caused by human overpopulation (and continued population growth) and profligate consumption. Particularly when considering their inability to adapt fast enough or their capacity to rely on mechanical systems in the same way as urban inhabitants.

The city has the ability to act as refuge not just for humans but all flora and fauna alike (Vink et al. 2017). In many cases our current urban design and planning practices are actually leading to cities that are detrimental to the health of residents. In fact, most of the top ten causes of death (2015) are directly or indirectly influenced by faulty urban design and planning policies (WHO 2015). Fossil fuel-powered car-centric suburbs has led to decreases in air quality (Ayres et al. 1999). While the roll-out of impervious roads intensifies already rising temperatures causing urban heat island (UHI) effects (Mohajerani et al. 2017). The warmer more polluted waters which flow from these cities decreases water quality and puts stress on surrounding ecosystems as well as the traditional grey infrastructure we have constructed to deal with it (Chadwick et al. 2006).

There are many lessons to be learned from nature and the complex natural systems it employs to accomplish many of the things we aspire to in our cities and for which we have constructed expensive infrastructure to accomplish. In many cases the cost of environmental restoration and protection is low (Hamilton 2011) and the re-inclusion of natural systems in the urban built environment inherently results in reduced dependence on mechanical systems and requires less expensive maintenance (Wells and Yeang 2010, p. 130), while providing innumerable supplementary benefits to inhabitants' quality of life. Green infrastructure also invariably has the ability to increase biodiversity in our cities and provide sanctuary for under threat species whose future depends on our ability to adapt, and vice versa (Lisle 2010).

Herewith is a brief exploration of some of the many benefits to be gained from nature-inclusive cities which prioritise the inclusion of biodiversity; including ecosystem services and climate resilience. As well as a short examination of the elementary scales of consideration. The concepts will be familiar to experienced urban designers albeit from a different contextual background. In many cases they have been founded on ecological principles only to become anthropocentric derivatives.

11.2 Advantages

Despite the inherent right to live for all non-human biotic entities there are a number of anthropocentric concepts which explore the benefits to humans of nature-inclusive cities. It is generally accepted that increased biodiversity in our cities provides a number of direct and indirect benefits to humans (Wentworth 2006). These concepts are not new, but in the scale of the history of cities they are fairly recent considerations. Due to the increasing density and complexity of our current and predicted future cities they are beginning to receive significantly more attention in both academic and professional circles.

The first most commonly accepted concept is *ecosystem services*. Championed currently by the Millennium Ecosystem Assessment (MEA 2005) *ecosystem services*, their categories and subcategories are commonly integrated by governments of all levels in urban planning considerations. Also, with adaptation to climate change increasingly directing urban design and planning a series of resilience related benefits are beginning to receive increased attention. Here we will examine benefits received from a nature-inclusive design and specifically how they can be gained from the conscious inclusion of biodiversity during the design process.

11.3 Ecosystem Services

Ecosystem services (Fig. 11.1) are the 'components of nature, directly enjoyed, consumed, or used to yield human well-being' (Boyd and Banzhaf 2006). Popularised by the Millennium Ecosystem Assessment (2003) they are organised into four categories – *provisioning services, regulating services, cultural services,* and *supporting services* (MEA 2005). Despite the human species' increasing congregation in cities which functionally eliminate most forms of nature we are still 'fundamentally dependant on the flow of ecosystem services' (MEA 2005). While parks, gardens

Fig. 11.1 Ecosystem services. (Source: The Economics of Ecosystems and Biodiversity (TEEB))

and other outdoor recreation spaces form a part of our cities their often contrived, selective design excludes many valuable species and thus the consequent benefits of a truly biodiverse ecosystem (Mooney 2010).

11.3.1 Provisioning Services

Provisioning services are the *products* obtained from ecosystems – they are physical and tangible. The most obvious products humans receive are food (agriculture, aquaculture, livestock), fibre (cotton, silk, linen), and fuel (wood). While these tend to come from rural areas one of the most important *provisioning services* in cities is providing sufficient quantity of drinking water (McDonald 2015). If we are to assume a nature-inclusive city is generally more biodiverse in both flora and fauna, there are also a number of additional benefits to be gained by city dwellers.

Biochemical, natural medicines and pharmaceuticals (MEA 2005) are derived from ecosystems. While not at the scale of commercial pharmaceuticals *traditional medicine* is recognised by the World Health Organisation as valuable in the prevention and treatment of illness (WHO 2013). Many traditional medicines have been proven to provide direct medicinal benefits (Fig. 11.2) and are also exploited by pharmaceutical companies. City dwellers with open access to biodiverse natural areas could exploit their use in something as simple as a cup of tea.

From an Australian perspective tea-tree oil (*Melaleuca alternifolia*) has antiseptic properties which kill harmful bacteria (Carson et al. 2006) and applied topically has proved extremely useful for the treatment of ulcers, boils, tinea, insect bites, sunburn and head lice. Eucalyptus oil (*Eucalyptus sp.*) can be inhaled for respiratory problems, throat and viral infections as well as for the common cold (Lu et al. 2004), and is used commercially in mouthwash, throat lozenges and cough suppressants. Lemon myrtle (*Backhousia citriodora*) has the highest concentration of citral in the world and distinctive antibacterial, antiviral and antifungal properties (Wilkinson et al. 2003). Traditionally drunk in a tea it is also good for digestive problems and nausea (Hayes and Markovic 2002).

Provisioning services also include *ornamental resources*, or animal products such as skins and shells, and flowers commonly used as ornaments (Fig. 11.3). The use of natural ornaments may be embedded in traditional First Nations cultures, but their use continues today with many contemporary western versions of ceremonial items reinterpreted for modern ceremonies – the wedding being the best example. Although the value of these resources is culturally determined this is also an example of linkages between the categories of *ecosystem services*.

Fig. 11.2 Australian native plants with medicinal benefits (L-R): *Melaleuca alternifolia*, Snow-in-Summer. (Source: Melaleuca – 'Snow in Summer' by Tatters ✾/Flickr CC BY 2.0); *Eucalyptus kruseana*, 'Bookleaf Mallee' (Source: Eucalyptus kruseana (bookleaf mallee) specimen 118 at the Waite Arboretum by Rhys Moult/Flickr CC BY 2.0); *Backhousia citriodora*, Lemon-scented myrtle. (Source: Lemon myrtle (Backhousia citriodora) by Tatters ✾/Flickr CC BY-SA 2.0)

Fig. 11.3 Ornamental resources (L-R): Papua New Guinea man wearing ceremonial headdress. (Source: Adli Wahid on Unsplash); Native American Man, Pow Wow Regalia. (Source: Andrew James on Unsplash); woman with orange and white floral headdress. (Source: Beto Silvestre on Unsplash); man in blue suit jacket with yellow clipped flower. (Source: Tom Kulczycki on Unsplash)

11.3.2 Regulating Services

Regulating services (Fig. 11.4) are the benefits obtained from the *regulation* of ecosystem processes (MEA 2005). These include *air quality maintenance* – the extraction of chemicals from the atmosphere – particularly important considering the general global trend towards decreasing air quality in densifying urban centres (Ayres et al. 2006). *Climate regulation* and effects on both temperature and precipitation – particularly useful in retaining surface water resources (McDonald 2015) and alleviating the urban heat island (UHI) effect (EPA 2008). *Water regulation* affects the timing and magnitude of runoff reducing the threat of flash flooding and pressure on traditional grey stormwater infrastructure (EPA 2011). *Storm protection* can also be provided by the presence of coastal ecosystem such as mangroves and coral reefs and dramatically reduce the damage caused by storm surge. These are just a handful of *regulating services* directly attributed to the presence of increased biodiversity which benefits urban city dwellers.

11.3.3 Cultural Services

Cultural services are the *non-material* benefits people obtain from ecosystems through spiritual enrichment, cognitive development, reflection, recreation and aesthetic experiences (MEA 2003). Australia is home to a great many distinct endemic animal species that are well recognised internationally and contribute to the nations *sense of place*. The Australian government's Commonwealth Coat of Arms features a shield held up most obviously a kangaroo and emu 'which were chosen to symbolise a nation moving forward, based on the fact that neither animal can move

Fig. 11.4 Regulating services (L-R): One Central Park, Sydney clad in plants which capture airborne particulates. (Source: sun scoop + greenwall by Rob Deutscher/Flickr CC BY 2.0); Canopy from a large street tree provided a cool shaded spot to sit. (Source: Oscar Nord on Unsplash); raingarden filters sediment out of before reaching waterways. (Source: Rain Garden by Roger Soh/ Flickr CC BY-SA 2.0); mangrove boardwalk. (Source: Boardwalk Through the Mangroves by Michael Coghlan/Flickr CC BY 2.0)

Fig. 11.5 Examples of iconic native Australian animals used in emblems (L-R): Commonwealth Coat of Arms. (Source: Australian Government); logo of the Australian national rugby league sports team. (Source: National Rugby League); logo of the Australian national rugby union team. (Source: Australian Rugby Union); emblem of the Collingwood Australian Football Club featuring the Magpie. (Source: Collingwood Football Club)

Fig. 11.6 Tourism Australia advertisements. (Source: Tourism Australia)

backwards easily' (Australian Government 2019). Apart from making up the Commonwealth government's coat of arms many Australian national sports teams have adopted the names of native animals, or a portmanteau, as their nickname (Fig. 11.5): Kangaroos/Jillaroos (Rugby league), Wallabies/Wallaroos (Rugby union), Socceroos (Soccer/Football), Boomers (Basketball), Kookaburras/Hockeyroos (Field hockey), Cockatoos (Tennis) – and can be seen as an example of *cultural heritage values*.

Recreation and ecotourism are also measurably valuable *cultural services* which effect where people decide to spend their leisure time. Tourism Australia, the Australian Government agency responsible for attracting international visitors to Australia, both for leisure and business events, lists *nature and wildlife* (Fig. 11.6) as one of their leading pillars of their marketing focus (Tourism Australia 2019). The unfortunate irony is that almost none of Australia's iconic animal species which the population prides itself on and for which people travel across the world see are no longer present within the urban environment (except for zoos and wildlife parks) and in many cases are struggling to survive in the wild.

11.3.4 Supporting Services

Supporting services are those that are necessary for the *production* of all other *ecosystem services*. They differ from *provisioning*, *regulating*, and *cultural services* in that their impacts on people are either indirect or occur over a very long time, whereas changes in the other categories have relatively direct and short-term impacts on people (MEA 2003). While generally considered beyond our control the fact that we have now entered a new geological epoch defined by humans' intervention in the environment (the Anthropocene) it should be clear that the decisions we make, particularly at an urban scale, will inevitably have flow on effects. Even though the ramifications may be impossible to foresee at such a time scale we should always endeavour to reduce our impact as much as possible and live on the earth with a certain lightness. With two billion more people (68 per cent of the world's population) predicted to live in evermore interconnected cities by 2050 (UN DESA 2018) a continuation of our current business as usual approach at the predicted magnitude of our population will have disastrous effects in the ability of the planet to continue providing supporting services for future generations.

11.4 Resilience

With climate change now the defining force of our collective planetary society going forward, our cities' ability to adapt and respond to the subsequent increasing acute shocks and chronic stresses is of paramount importance. A nature-inclusive city, rich in a diverse variety of native flora and fauna is inherently suited to withstand extreme events as it has evolved over millennia to do so in its current location (Beatley 2016). Natural systems and their consequent *ecosystem services* coupled with traditional grey infrastructure in many cases duplicate services and therefore create an element of redundancy. Simultaneously, natural systems are made up of many diverse constituents all performing a small part of a larger process. This complexity means that when one element fails the system continues to function, albeit at a slightly decreased scale (Wells and Yeang 2010).

11.5 Considerations

It should be clear that biodiverse nature-inclusive cities not only provide a huge amount of tangible, measurable and humanising benefits, but will also be necessary for the survival of our species (and all others) moving forward. Designing these cities is however a large and unfamiliar task for all involved. While specialists in the fields of ecology and biodiversity will bring a huge amount of specialised knowledge to the table it is important for other disciplines (particularly design) to have a grasp of the most basic conceptual considerations when designing for the integration of complex natural systems.

Nature and cities both operate at a number of interlocked (Noss 1992) and overlapping spatial and temporal scales (Fernández-Juricic and Jokimäki 2001). From the inhabitants and their communities, to their habitats and surroundings, to the natural cycles of time, nature-inclusive cities incorporate a myriad of additional layers of complexity. Designers of the built environment are experienced at operating in a multi-scalar way and many of the concepts will be familiar. Following is a brief exploration of those considerations and how they relate to humans in the urban environment.

11.5.1 Spatial Scales (Ecological Organisation)

Ecological organisation is a notion that urban designers will undoubtedly already be considering during their process, albeit from a human-centric perspective. It is the hierarchy of complex biological structures and systems that define life using a reductionistic approach (Solomon et al. 2002). Each level in the hierarchy represents an increase in organisation complexity, with each "object" being primarily composed of the previous levels' basic unit (Solomon et al. 2002).

11.5.1.1 Species/Organism

The first scale of consideration is that of the individual. An organism is an example of what scientists call an open system, an entity that exchanges materials and energy with its surroundings (Campbell et al. 1999) and consists hierarchically of atoms, cells, organs, etc. From the perspective of built environment design many determining factors are decided at the scale of an average person. Most, if not all, developed countries have a construction code which dictates minimum requirements for buildings – ceiling heights, corridor widths, access ramp angles – these in turn have subsequent flow-on effects which go on to shape each expanding level of our city. While the minimum requirements for a comfortable life for the typical human are much researched, regulated and understood by those designing our built environment, the same cannot be said of the other 99% of species. Not least because the functional requirements for a comfortable life in the city differ immensely between a moth and a possum 100-times its size. Putting aside the possible cultural stigma surrounding sharing a home with wild animals (Fig. 11.7), or worse an insect (!), our current profit-driven private-developer-centric cities, with an emphasis on floor space ratio, make accommodating anything else near impossible. Given the opportunity however, those in control of shaping cities would find it much more preferable to allow the inclusion of *a* moth rather than a possum.

Fig. 11.7 Typical Australian urban species (L-R): Green Grocer Cicada (*Cyclochila australasiae*) (Source: Cicadas Green Grocer by babbagecabbage/Flickr CC BY-SA 2.0); lace monitor, or tree goanna (*Varanus varius*) (Source: Lace Monitor.Varanus varius by gailhampshire/Flickr CC BY 2.0); grey-headed flying-fox (*Pteropus poliocephalus*) (Source: Grey headed Flying fox by Duncan PJ/Flickr CC BY-SA 2.0); Long-nosed Bandicoot (*Perameles nasuta*) (Source: 2011-01-17 14:40:47 by Aston Clulow/Flickr CC BY-SA 2.0)

11.5.1.2 Population

The following level of hierarchy is *population*, which from a biological perspective consists of groups of organisms of the same species (Campbell et al. 1999) – a pride of lions, a swarm of bees, a murder of crows. Many studies have demonstrated that there often exists a gradient of change in species diversity and abundance along an urban-suburban-rural gradient, with peaks in suburbia, and that the social structure of many higher animals, particularly mammals and birds, may differ in urban compared to non-urban areas (Jarvis 2015). From the human perspective this could be seen as either the family unit or household. This is one point of difference between animals and humans. While each animal species has a predictable population group, shaped by interactions between individuals and their environment – monogamous pairs, extended families, packs led by an alpha – the social complexity of human population is much higher. In response, humans have designed a variety of housing typologies for the different groups – from the studio to detached single family home to multistorey apartment building. Architects and urban designers' experience incorporating the diversity of housing typologies puts them in good stead when it comes to integrating the predictability of many other species. In some cases, urban habitat typologies for animals already exist – the birdhouse, bat access tiles, beehive, and possum boxes (Fig. 11.8).

11.5.1.3 Community

After *population* comes *community* and the beginning of what could be called biodiversity. This level of *ecological organisation* includes other individuals in a population and populations of other species living in the same area, close enough for interaction (Campbell et al. 1999). It's at this point that the complex relations between animals, and between humans and animals (Fig. 11.9) begin to inform behaviour. Typical biological interactions at this level include *predation*, *symbiosis*, *parasitism* and *competition*.

Fig. 11.8 Urban animal habitat typological examples (L-R): Bird nest box. (Source: Birds in Backyards); microbat roost box. (Source: City of Sydney); beehives. (Source: The Urban Beehive); possum box. (Source: Peter the Possum and Bird Man)

Fig. 11.9 Typical human-animal biological interactions (L-R); *prey* (livestock) relegated to peri-urban/rural areas. (Source: Cows on Carmel Head by Reading Tom/Flickr CC BY 2.0); *symbiote* (dog) 'man's best friend' shares a habitat. (Source: Dog by Susanne Nilsson/Flickr CC BY-SA 2.0); *parasitic* German cockroach (*Blattella germanica*) limited to urban areas. (Source: German cockroach (Blattella germanica) probably by gailhampshire/Flickr CC BY 2.0); humans and kangaroos *compete* for grassland – albeit for different reasons. (Source: @CapitalFootball/Twitter)

While the typical city dweller may not think of themselves as engaged in animal relationships, everyone is daily influencing the animals around them. No longer prey (at least in cities) humans are by far the greatest *predators* globally. Simply controlling our prey, shifting it out of the city and renaming it livestock. The consequences of which are a greater detachment from the realities of our existence and the natural resources which support it. We have *symbiotic* relationships with many animals we have domesticated (dogs, cats, fish, birds, even reptiles). Which, while promoting interspecies interaction, are undertaken entirely under the control of humans and completely neglect natural order. We expend great amounts of money and effort attempting to exterminate some *parasitic* species (rats, cockroaches, ants, spiders) and accept others as equal inhabitants (pigeons, seagulls). Meanwhile we are the greatest source of *competition* for virtually every species on the planet. Consuming almost everything the planet has to offer. Whether it be natural resources, habitat or our own agricultural products.

It's important for us to re-insert ourselves into the bigger picture and appreciate that we are but a single constituent of a series of complex interlocked systems. While our own influence is undoubtedly predominant, innumerable interactions are continually occurring between the other species with which we share our space. Understanding these interactions is key to a nature-inclusive city. We largely appreciate the presence of birds in our urban environment. In fact, *birding*, also known *twitching*, is one of the largest and most popular recreational activities. The total number of *birders* in the United States is estimated at approximately 47 million (U.S. Fish and Wildlife Service 2013) and the industry worth more than $20 billion per year (Kerlinger 1993). However, many of our most common bird species are insect eating (insectivorous) and struggle to find prey in our homogenous urban environments whose design tends to favour invasive species (Douglas 2015). The natural world is also full of many interspecies *symbiotic* relationships. The eastern screech owl, for example, has a novel relationship with the live blind snake, whereby the former captures the latter and brings it back to their nest (Gehlbach and Baldridge 1987). The live blind snake in turn cohabits with the eastern screech owl consuming soft bodied insect larvae within the nest thereby protecting the nestlings. Even if our urban environment provides the necessary habitat for the eastern screech owl to survive it is unlikely it would provide the same for the snake. Which would most likely be viewed as a threatening pest and be removed. Parasites are for the most part viewed as harmful intruders. Humans, and plants for that matter, have constantly waged biological war against aphids. They are however a major food source for ladybugs who typically live on plants, feeding on them in a symbiotic way. By spraying poison we are inadvertently reducing the population of ladybugs, and other insect species, who then go on to become prey for larger animal species (Hallmann et al. 2014). Our narrowminded pursuit of monocultural perfection therefore has far reaching flow on effects. Finally, a fruiting tree in the urban environment can provide food for large numbers of terrestrial and airborne animals, from fruit bats and birds to possums and native rats. These species are all in competition with one another for the same food source. If an area only has a small number of appropriate trees, which in many cases are also protected by humans, then competition will be tough and population numbers small. By integrating a greater quantity and diversity of the right type of tree we can assist with the growth of populations.

11.5.1.4 Ecosystem

Following *community* the next level of *ecological organisation* is the *ecosystem* – groups of organisms from all biological domains in conjunction with the physical (abiotic) environment (Campbell, Reece & Mitchell 1999). In biology, abiotic factors can include water, light, radiation, temperature, humidity, and atmosphere. From an urban design perspective this would be seen at the scale of the entire city and its direct surroundings. We have already seen the potential benefits to be gained from the interactions of these components, but unfortunately our current urban environments tend to neglect the benefits of natural systems, instead replacing them with artificial mechanical means.

Increasingly our cities are paved with impervious surfaces which repel water, forcing it underground into stormwater systems or increasingly segregated engineered surface infrastructure. All of which is costly to construct and maintain (Wells and Yeang 2010), and in many cases unsightly. Our increasingly cost-driven development environment coupled with a desire for consistency in our internal habitable spaces leads to neglecting natural light in favour of uniform lighting systems. This in turn places greater pressure on our building's mechanical systems with flow-on effects to increased energy usage. In fact, energy consumption for interior lighting is rapidly increasing and takes up 17.5% of the total global electricity consumption (Ticleanu 2015). Ultraviolet radiation is an issue of particular concern from an Australian perspective due to a number of both cultural and geographic reasons. The large number of rain-free days and a culture of enjoying the outdoors, coupled with the vast number of stunning beaches and reduced ozone protection present serious issues for skin cancer (Cancer Council 2019). Something for which we have created another synthetic solution – sunscreen. Related to this, and again of particular importance from an Australian perspective, is temperature which is predicted to become far more intense and unpredictable with climate change (CSIRO and Bureau of Meteorology 2015). Increasingly, our human tendency is to retreat indoors and rely on mechanical systems like air conditioning to insulate us, further exacerbating energy usage. While atmospherically the air quality in our cities continues to decrease so much that on many occasions official advice from authorities is to remain indoors with windows and doors closed (Climate Council 2019; Clun 2018).

Abiotic environmental factors undoubtedly are an important factor of contemporary urban design, hence the abundance of constructed systems designed to ameliorate their effects. Increasingly however multidisciplinary urban design teams are exploring natural substitutes not only for their inherent *ecosystem services* but also reduced construction and maintenance costs (Wells and Yeang 2010). As we have already seen water runoff can be drastically reduced and pressure on stormwater systems diminished through a combination of natural systems. Common water-sensitive urban design practices (Fig. 11.10) such as green roofs, rain gardens, bioswales and retention basins all use vegetation to control water while providing potential habitat (McDonald 2015). Access to natural light has many of the same health and well-being benefits as access to greenery (Aranow 2011; Aries et al. 2010; Baggerly et al. 2015) and there are a number of natural strategies for controlling it, the simplest of which is the tried and true method of utilising deciduous trees. While the importance of applying sunscreen regularly when in direct sunlight cannot be overstated, designers should endeavour to provide inhabitants with the ability to remove themselves from direct sun contact as much as possible. Unfortunately, in many circumstances people have no option but to expose themselves to harmful ultraviolet radiation simply going about their daily lives.

Fig. 11.10 Common water-sensitive urban design practices (L-R); green roof at MEC Building, Toronto, Canada. (Source: MEC green roof 2 by Padraic/Flickr CC BY-SA 2.0), rain gardens, Singapore. (Source: Rain Garden after the rain by Roger Soh/Flickr CC BY-SA 2.0), bioswale in median of Grange Avenue in Greendale, WI, USA. (Source: Greendale_GrangeAve_2010_07_15 by Aaron Volkening/Flickr CC BY 2.0); bioretention basin at Clinton Elementary, MA, USA. (Source: Clinton Elementary Bioretention Basins by Christopher B. Hoffman Landscape Architect/ Flickr CC BY 2.0)

Something which in many cases could be eliminated with the incorporation of a natural tree canopy. By using the natural benefits of vegetation as solutions to all of the aforementioned issues designers can consequently mitigate the problems of temperature, humidity and air quality. Exploiting trees' natural ability to reduce the urban heat island effect, lower humidity and absorb air pollutants (McDonald 2015).

11.5.1.5 Biome

Beyond *ecosystem* is the level of *biome* – a distinct biological community that has formed in response to a shared physical climate, and which can comprise a variety of habitats (Campbell, Reece & Mitchell 1999). Here we begin to reach the limits of traditional urban design practice as biomes transcend city, state and national boundaries. Though it is beyond the scale at which metropolitan planning may operate we will see how humans have unwittingly over time led to an alternate view of traditional terrestrial biomes. The generally accepted system of clarifying biomes at present was devised by biologists Olson and Dinerstein (1998) after being convened by the World Wildlife Fund (WWF). It classifies 14 terrestrial biomes ranging from tropical and subtropical forests, grasslands and savannahs, to tundra, desert and mangroves, as well as several other freshwater and marine types (Fig. 11.11).

Our cities, at present at least, are wholly contained within and surrounded by one of these environments defined by both biotic and abiotic characteristics. This means that there are a number of correlative practices appropriate between corresponding biomes on different continents than those directly adjacent. This means that the eastern part of China has more in common ecologically with Florida in the United States as they both inhabit a tropical rainforest biome, than the alpine tundra in the western part of the country. Although obviously wildly different in terms of terrain, population and species type, not to mention the social, cultural and political differences, it is nonetheless interesting to note there would be similarities in abiotic factors. While our society approaches a time when governments may indeed begin the conscious planning and design of our planet at the scale of the biome we have already begun to alter patterns globally simply through accumulation over time (Ellis and Ramankutty 2008). Anthropogenic biomes recognise the irreversible coupling of human and ecological systems at global scales based on sustained direct human contact with ecosystems. Vegetated forms predicted by conventional biome systems can no longer be observed across much of the Earth's land surface (Foley et al. 2005). These Anthropogenic biomes include agriculture, human settlements, forestry and rangelands (Fig. 11.12).

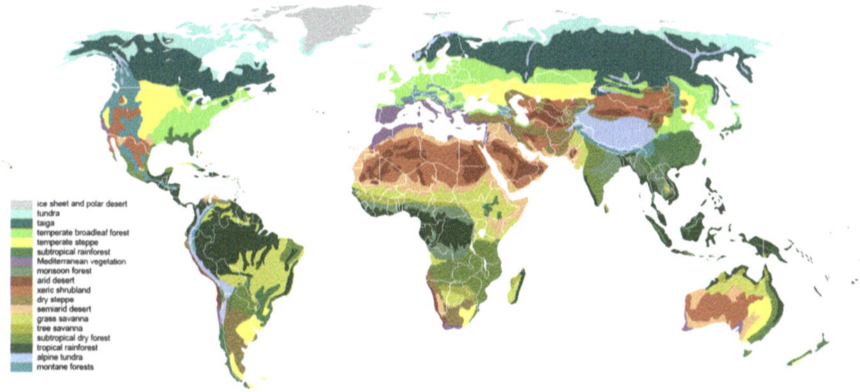

Fig. 11.11 Terrestrial biomes classified by vegetation. (Source: One way of mapping terrestrial biomes around the world by Vzb83/Wikipedia CC BY-SA 3.0)

Fig. 11.12 Anthropogenic biomes (L-R): Wheat farm, Colorado, USA. (Source: 501795 by Mojo Yugen/Flickr CC BY-SA 2.0); urban settlement, Sydney, Australia. (Source: IMG_20160905_160856 by Andrew Harvey/Flickr CC BY 2.0); forestry plantation, Hawaii, USA. (Source: Hawaii Plantations by benmacaskill/Flickr CC BY 2.0); rangelands, Clarke ranch southwest of Big Timber, MT, USA. (Source: Livestock81.tif by USDA NRCS Montana/Flickr)

11.5.1.6 Biosphere

The final and largest level of *ecological organisation* is the *biosphere* (Fig. 11.13). The *biosphere* is the worldwide sum of all ecosystems and their relationships, including their interactions with the elements of the lithosphere, geosphere, hydrosphere, and atmosphere (Campbell, Reece & Mitchell 1999). Far beyond the comprehension of contemporary urban design practice the concept of the *biosphere* still cannot be entirely discounted in the planning of future cities. The actual physical ground cover of human settlements makes up only a miniscule fraction of the overall *biosphere* – estimates range from less than 0.5% (Angel et al. 2011) to 3% (Center for International Earth Science Information Network et al. 2011). However, the additional lands required to support them are staggering. Considering a single human settlement can be recognised as an Anthropogenic *biome* in and of itself, and also requires a certain amount of agriculture, rangelands, forestry, mining, etc. to support it, the influence of a single metropolitan area can begin to have global consequences. These consequences have been speculated for decades, but we are now

Fig. 11.13 Biosphere – Earth (Credit: Image created by Reto Stockli with the help of Alan Nelson, under the leadership of Fritz Hasler). (Source: NASA)

beginning to experience the physical ramifications directly. Climate change is now the defining force of all urban development going forward. The necessity for both climate mitigation and adaptation means that the business as usual approach is no longer feasible and requires entirely new ways of thinking about how we as a species live on this planet (Fig. 11.13).

11.5.2 Temporal Scales

Temporality as with varying spatial scales is by no means a new concept or consideration of contemporary urban designers. The development of our cities has always been characterised by a certain variability and adaptability that has allowed us to live consistently regardless of changes. Taking control of our environment is an innate human characteristic, one that will have to be rethought in a nature-inclusive city.

11.5.2.1 Diurnal

The first and most immediate temporal scale at which we must consider the city is that of the diurnal cycle – one full rotation of the Earth around its own axis. A truly inclusive global city serves its community by operating 24 hours a day, a biodiverse nature-inclusive city does the same for all non-human inhabitants. The advent of the electric light has allowed humans to transform night into day and almost any task is now able to be performed regardless of the time of day or night. The scale of transformation can best be seen in the example of sports stadiums (Fig. 11.14). Mechanical lighting systems have become so large and powerful that we are able to illuminate sports pitches upwards of five acres in area. Irradiated to such a degree that highly skilled and detailed team sports can not only be played in situ but be viewed by tens-of-thousands of spectators and broadcast to millions, sometimes billions in high definition clarity.

Humans however are not the only species to have evolved to conquer the night. Many plant and animal species have evolved to operate nocturnally. These nocturnal creatures have evolved biologically to have highly developed senses of hearing, smell and specialty adapted eyesight. There are a number of reasons animals may have adapted to become nocturnal. Being active at night reduces the potential for resource competition between two species inhabiting the same environment. A hawk and an owl can hunt the same field or meadow for the same rodents without conflict because one is diurnal and the other nocturnal. Predator-prey relationships have also affected the adaptation of nocturnal species. Many predators hunt at night when their prey is at rest, while conversely many smaller animals are active at night to avoid their diurnal predators. Conserving water during the heat of the day is another reason for animals to operate nocturnally, particularly in arid biomes like

Fig. 11.14 Humans conquering the night. (Source: Mario Klassen on Unsplash)

deserts. This is a behaviour mimicked by many plants species also who have adapted to be pollinated by nocturnal creatures like bats.

Humans as we have seen however have begun to intrude on the night at such a scale that it is beginning to have detrimental effects in the lives of nocturnal animals. Light pollution, or *photopollution* (Verheijen 1985), exacerbated by excessive, misdirected, obtrusive uses of light is blamed for compromising health, disrupting competition and spoiling aesthetic environments (Longcore and Rich 2004). Insects are the most obvious example of animals disoriented by artificial lighting. Attracted to lights they are usually killed by heat or current and another example of humans (unwittingly) out-competing other species for the same resources. Nocturnal migratory birds can also become disoriented and lost, leaving them vulnerable to predators. In general, species with specially adapted eyesight to low light are significantly disadvantaged. There can also be detrimental effects on reproductive cycles of nocturnal species who may struggle to find a mate or suitable nest. Nocturnal predators are also at a disadvantage in areas affected by light pollution as their prey is better able to avoid them.

11.5.2.2 Seasonal

Beyond the daily is the annual *seasonal* cycle – marked by the Earth's orbit around the sun and axial tilt relative to the ecliptic plane. This results in the four absolute meteorological seasons with which we are all familiar – summer, autumn, winter, and spring. Again, this is a natural phenomenon very familiar to urban designers and architects, which has played a significant influence in the design of our habitat. During the summer months we open our homes to embrace the added light and warmth the proximity to the sun brings. While in winter we shutter our homes and retreat indoors to insulate ourselves from the frigid conditions. The intermediate spring and autumn seasons bring refreshing transitions and respite from the extremes of the other two.

Biologically, seasonal shifts in climate have a dramatic effect of the behaviour of flora and fauna. Long days and warm weather allow plants and animals to breed and reproduce. They take the opportunity to eat as much as possible and store the necessary energy to survive the cold months ahead. Summer is loud and frenetic. In Australia, summer is filled with the deafening drone of cicadas right throughout suburbia as well as the surrounding wilderness. They have become uniquely adapted to urban conditions which gives Australian cities a distinct summertime buzz. In fact, cicadas are the most efficient and loudest sound-producing insects in existence, able to produce in excess of 120 dB at close range (this is approaching the pain threshold of the human ear). As cicadas converge in their thousands this deafening cacophony can be a source of friction with humans. Conversely however their large numbers and disinterest in concealment makes them easy pickings for carnivorous birds and mammals, and an important source of nourishment to be exploited. Autumn acts as the bridge between summer and winter with the shortening days and reduced temperatures an obvious indicator that it is time to begin serious preparations for winter.

Principally, species will make a decision to either remain where they are and ride out the storm (both figuratively and in many cases literally) or they will leave, migrating to a warmer climate. Those that remain will have gained the necessary additional body weight to persevere the lean months ahead and now begin focusing on appropriate shelter. Burrows are dug, hollows insulated, nests feathered, and, in many cases, refuge sought in existing structures. Once again this is where humans and animals can come into conflict as we are unlikely to be happy about sharing our homes. In many cases they are pests, such as mice or rats, and viewed as intruders. Due to Australia's unique geography and climate it is also home to many native rodent species, the ubiquitous possum and a variety of other small marsupials that are integral to its distinct biodiversity. They seek shelter beneath homes, in roofs and even wall cavities, as well as sheds and out-buildings. Non-migratory birds also tend to nest in eaves and soffits seeking shelter from wind and rain. In many cases conflict can be avoided by simply accepting our non-human cohabiters and allowing for them in the design of our urban environments.

While Australia is known for its temperate, if not altogether hot, climate the long nights and short days still results in a quieter more passive ecological environment. While not home to the most notable hibernating animal the bear there are many Australian native species who enter torpor – a similar short-term energy saving measure. There are several species of pygmy possum who inhabit alpine areas that reside underground, which becomes covered in snow during winter. In many cases with oblivious humans skiing overhead; much to their discomfort. Echidnas in alpine areas have also been found to enter multiday periods of torpor punctuated by periodic arousals to warm up. Several bird species like the tawny frogmouth enter this energy saving state at night or in the early morning between which it feeds and functions as usual. Insectivorous bats often wake up from hibernation to forage on warmer winter days before entering another period of multiday torpor. Reptiles also necessarily enter a state of torpor during winter as a result of falling ambient temperatures. As ectothermic or 'cold blooded' animals, they have limited control over their temperature and metabolic rate therefore they are significantly less active. Many insects also burrow underground during winter, like the aforementioned cicada, or lay their eggs there during the autumn. So, while life rolls on almost unchanged for us humans, winter is the time when we could be doing untold damage to species who are unable to escape. Felling a tree or digging soil which houses hibernating animals can have disastrous effects come spring time, with the loss of valuable pollinators or their prey.

With spring comes a great thawing and revival of plants and animals alike. New grass, soft leaves and vivid flowers signal revival and rejuvenation. Animals begin to slowly replenish the stores that depleted over winter and begin to prepare for mating. There is a delicate balance between ensuring young have as much time as possible to mature before the next winter and ensuring there are enough food stocks available to support that growth. In many parts of Australia spring is also known as 'swooping season'. The Australian magpie – a medium-sized black and white passerine native bird – is ubiquitous in urban areas all over Australia. It is one of Australia's most highly regarded songbirds, with a wide variety of calls, many of which are complex. During breeding season however many become highly aggressive swooping and attacking passers-by. Attacks begin as the eggs hatch, increase in frequency and severity as the chicks grow, and trail off as the chicks leave the nest. So entrenched is swooping season that many councils erect warning signs in parks known as being nesting sites. It has also spawned *Magpie Alert!* a social website to track aggressive swooping magpies, and *Magpie Attack Radar* a smartphone app mapping recorded attacks around Australia. Humans have adopted several novel approaches to discourage attacks. Most notably cyclists, who appear to be one of the main victims, have taken to attaching fake eyes to the back of their helmets or cable ties which acts as deterrent spikes. Despite this, the magpie was voted bird of the year in BirdLife Australia's 2017 annual poll. The acceptance of magpies and their territorial behaviour is one of the few examples of urban residents respecting the natural seasonal behaviour of cohabiting animals. Residents recognise that having to adjust their own behaviour for a few weeks during spring is far outweighed by the *ecosystem services* the emblematic bird provides throughout the rest of the year.

11.6 Conclusion

It should be clear that the benefits of a nature-inclusive city are by no means entirely altruistic. A significant amount of research has been conducted into the direct and indirect economic benefits of *ecosystem services*. The increase in extreme unpredictable weather brought about by climate change also means resilient design strategies will play an increasingly important role in the survival of future cities. The adaptability required to bounce back from acute shocks and chronic stresses can best be achieved by the conscious incorporation of natural systems. While it is easy to examine each spatial and temporal scale individually to determine appropriate ways to achieve this synergy it is the interlocking and overlapping of complex systems that achieves the most robust outcome. Herein has only been an introductory exploration of the elemental principles of intricate nature-inclusive design, and what should be stressed is the importance of incorporating specialists in the fields of biology at the inception of design right through the process to post-construction management. From biogeography to ecology to zoology the future of all life on Earth depends on the cooperation and collaboration of all fields of both the natural and built environments.

The current state of cities worldwide is proof positive that planners and urban designers understand very little about the true benefits to be gained from nature in the city. Transport oriented infrastructure-led urban plans show how little the design of our cities focuses on long-term quality of life. While significant literature exists exploring the myriad economic, physiological and psychological advantages access to earth's systems of natural resources there is very little in the way of how it can be designed. As the profession of urban design becomes multi-disciplinary so to must the institutions educating them. Urban designers have a gift for manifesting spatial outcomes but they must be provided with the right information to understand the full story. Beyond the typical anthropocentric programmatic demands, project briefs should be developed which also include the necessary framework for a thriving community of non-human inhabitants.

References

Angel S, Parent J, Civco DL, Blei A, Potere D (2011) The dimensions of global urban expansion: estimates and projections for all countries, 2000–2050. Prog Plan 75(2):53–107

Aranow C (2011) Vitamin D and the immune system. J Invest Med 59(6):881

Aries MBC, Veitch JA, Newsham GR (2010) Windows, view, and office characteristics predict physical and psychological discomfort. J Environ Psychol 30(4):533–541

Australian Government (2019) Commonwealth Coat of Arms, Department of the Prime Minister and Cabinet, Canberra, viewed 14 March 2019. https://www.pmc.gov.au/government/commonwealth-coat-arms

Ayres J, Maynard RL, Richards RJ (1999) Air pollution and health. Imperial College Press, Singapore

Ayres J, Maynard RL, Richards RJ (2006) Air pollution and health, vol 3. Imperial College Press, London

Baggerly CA, Cuomo RE, French CB, Garland CF, Gorham ED, Grant WB, Heaney RP, Holick MF, Hollis BW, McDonnell SL, Pittaway M, Seaton P, Wagner CL, Wunsch A (2015) Sunlight and vitamin D: necessary for public health. J Am Coll Nutr 34(4):359–365

Beatley T (2016) City as habitat: planning the nature-Connected City. Planning 82(9):50–51

Boyd J, Banzhaf HS (2006) What are ecosystem services? The need for standardized environmental accounting units. Resources for the Future, Washington, DC

Campbell NA, Reece JB, Mitchell LG (1999) Biology. Benjamin/Cummings, Redwood City

Cancer Council (2019) Why does Australia have so much skin cancer? (Hint: it's not because of an ozone hole), Cancer Council Western Australia, viewed 21 March 2019. https://www.cancerwa.asn.au/articles/news-2018/why-does-australia-have-so-much-skin-cancer-hint-i/

Carson CF, Hammer KA, Riley TV (2006) *Melaleuca alternifolia* (Tea Tree) oil: a review of antimicrobial and other medicinal properties. Clin Microbiol Rev 19(1):50

Ceballos G, Ehrlich PR (2018) The misunderstood sixth mass extinction. Science 360(6393):1080

Ceballos G, Ehrlich PR, Barnosky AD, García A, Pringle RM, Palmer TM (2015) Accelerated modern human-induced species losses: entering the sixth mass extinction. Sci Adv 1(5):1

Center for International Earth Science Information Network, C.C.U., International Food Policy Research Institute, I., The World, B. & Centro Internacional de Agricultura Tropical, C (2011) Global rural-urban mapping project, version 1 (GRUMPv1): urban extents Grid. NASA socioeconomic data and applications center (SEDAC), palisades, NY, viewed 21 March 2019. https://doi.org/10.7927/H4GH9FVG

Chadwick MA, Dobberfuhl DR, Benke AC, Huryn AD, Suberkropp K, Thiele JE (2006) Urbanization affects stream ecosystem function by altering hydrology, chemistry, and biotic richness. Ecol Appl 16(5):1796–1807

Climate Council (2019) Sydney heatwave affects air quality, climate council, Sydney, viewed 21 March 2019. https://www.climatecouncil.org.au/sydney-heatwave-affects-air-quality/

Clun R (2018) People warned to stay inside as dust storm engulfs Sydney, the Sydney morning herald, Sydney, viewed 21 March 2019. https://www.smh.com.au/environment/weather/people-warned-to-stay-inside-as-dust-storm-engulfs-sydney-20181122-p50hjv.html

CSIRO and Bureau of Meteorology (2015) Climate change in Australia information for Australia's natural resource management regions: technical report. CSIRO and Bureau of Meteorology, Australia

Dirzo R, Young HS, Galetti M, Ceballos G, Isaac NJB, Collen B (2014) Defaunation in the Anthropocene. Science 345(6195):401–406

Douglas I (2015) Invasive species and their response to climate change. In: Douglas I, Goode D, Houck MC, Wang R (eds) The Routledge handbook of urban ecology. Routledge, London, pp 488–502

Ellis EC, Ramankutty N (2008) Putting people in the map: anthropogenic biomes of the world. Front Ecol Environ 6(8):439–447

EPA (2008) Urban heat Island basics. Reducing Urban Heat Islands: compendium of strategies. U.S. Environmental Protection Agency, Washington, DC

EPA (2011) Green long-term control plan-EZ template: a planning tool for combined sewer overflow control in small communities. U.S. Environmental Protection Agency, Washington, DC

Fernández-Juricic E, Jokimäki J (2001) A habitat island approach to conserving birds in urban landscapes: case studies from southern and northern Europe. Biodivers Conserv 10:2023–2043

Foley JA, DeFries R, Asner GP, Barford C et al (2005) Global consequences of land use. Science 309(5734):570–574

Gehlbach FR, Baldridge RS (1987) Live blind snakes (*Leptotyphlops dulcis*) in eastern screech owl (*Otus asio*) nests: a novel commensalism. Oecologia 71(4):560–563

Hallmann CA, Foppen RPB, van Turnhout CAM, de Kroon H, Jongejans E (2014) Declines in insectivorous birds are associated with high neonicotinoid concentrations. Nature 511(7509):341–343

Hamilton C (2011) Growth Fetish. Pluto Press, London/Sterling

Hayes AJ, Markovic B (2002) Toxicity of Australian essential oil *Backhousia citriodora* (lemon myrtle). Part 1. Antimicrobial activity and in vitro cytotoxicity. Food Chem Toxicol 40(4):535–543

Hoffmann M, Hilton-Taylor C, Angulo A, Böhm M, Brooks TM, Butchart SHM, Carpenter KE, Chanson J, Collen B, Cox NA, Darwall WRT, Dulvy NK, Harrison LR, Pollock CM, Quader S, Richman NI, Rodrigues ASL, Tognelli MF, Vié J-C, Aguiar JM (2010) The impact of conservation on the status of the world's vertebrates. Science 330(6010):1503–1509

IPCC (2007) Climate change 2007: synthesis report. Contribution of working groups I, II and III to the fourth assessment report of the intergovernmental panel on climate change. In: Pachauri RK, Reisinger A (eds) Fourth assessment report, Intergovernmental Panel on Climate Change. Switzerland, Geneva

IUCN (2018) The IUCN Red List of Threatened Species. *Version 2018-2*, International Union for Conservation of nature, viewed 15 February 2019, http://www.iucnredlist.org

Jarvis PJ (2015) Urban animal ecology. In: Douglas I, Goode D, Houck MC, Wang R (eds) The Routledge handbook of urban ecology. Routledge, London, pp 352–360

Kerlinger P (1993) Birding economics and birder demographics studies as conservation tools. In: Finch DM, Stangel PW (eds) Status and management of neotropical migratory birds: September 21-25, 1992, Estes Park, Colorado. Gen. Tech. Rep. RM-229. Rocky Mountain Forest and range Experiment Station, U.S. Dept. of agriculture, Forest Service, Fort Collins, CO

Lisle J (2010) Climate change adaptation: the role of biodiversity in urban open space. Australian Planner 47(2):113–114

Longcore T, Rich C (2004) Ecological light pollution. Front Ecol Environ 2(4):191–198

Lu XQ, Tang FD, Wang Y, Zhao T, Bian RL (2004) Effect of *Eucalyptus globulus* oil on lipopolysaccharide-induced chronic bronchitis and mucin hypersecretion in rats. Zhongguo Zhongyao Zazhi 29(2):171

McDonald RI (2015) Conservation for cities: how to plan and build natural infrastructure. Island Press, Washington, DC

MEA (2003) Ecosystems and human Well-being: a framework for assessment. Millenium Ecosystems Assessment, Washington, DC

MEA (2005) Ecosystems and human Well-being: synthesis. Millenium Ecosystems Assessment, Washington, DC

Mohajerani A, Bakaric J, Jeffrey-Bailey T (2017) The urban heat island effect, its causes, and mitigation, with reference to the thermal properties of asphalt concrete. J Environ Manag 197:522–538
Mooney HA (2010) The costs of losing and of restoring ecosystem services. In: Lindenmayer D, Hobbs RJ (eds) Managing and designing landscapes for conservation: moving from perspectives to principles. Zoological Society of London, London/Malden/Oxford
Noss R (1992) Indicators for monitoring biodiversity: a hierarchical approach. Conserv Biol 4:355–364
Olson DM, Dinerstein E (1998) The global 200: a representation approach to conserving the Earth's most biologically valuable ecoregions. Conserv Biol 12(3):502–515
Pimm SL, Jenkins CN, Abell R, Brooks TM, Gittleman JL, Joppa LN, Raven PH, Roberts CM, Sexton JO (2014) The biodiversity of species and their rates of extinction, distribution, and protection. Science 344(6187)
Ripple WJ, Wolf C, Newsome TM, Galetti M, Alamgir M, Crist E, Mahmoud MI, Laurance WF (2017) World scientists' warning to humanity: a second notice. Bioscience 67(12):1026–1028
Solomon EP, Berg LR, Martin DW, Morales E, Graphics R (2002) Biology. Brooks/Cole, Thomson Learning, South Melbourne
Ticleanu CM (2015) Aiming to achieve net zero energy lighting in buildings ∗. Revista Romana de Inginerie Civila 6(1):63–78
Tourism Australia (2019) Our campaigns, Australian government, viewed 14 March 2019. http://www.tourism.australia.com/en/about/our-campaigns.html
U.S. Fish and Wildlife Service (2013) Birding in the United States: a demographic and economic analysis, Arlington, VA, 2011-1
UN DESA (2018) World urbanization prospects: the 2018 revision. United Nations Department of economic and social Affairs, New York City
Verheijen FJ (1985) Photopollution: artificial light optic spatial control systems fail to cope with. Incidents, causation, remedies. Exp Biol 44(1):1–18
Vink J, Vollaard P, Zwarte ND (2017) Stadsnatuur maken = Making urban nature. Nai010 Publishers
Wells M, Yeang K (2010) Biodiversity targets as the basis for green design. Archit Des 80(2):130–133
Wentworth J (2006) Ecosystems services. Parliamentary Office of Science and Technology, London
WHO (2013) WHO traditional medicine strategy: 2014–2023. World Health Organization, Geneva
WHO (2015) Sustainable cities: health at the Heart of Urban Development, World Health Organization, viewed 21 February 2019. https://www.who.int/sustainable-development/cities/Factsheet-Cities-sustainable-health.pdf?ua=1
Wilkinson JM, Hipwell M, Ryan T, Cavanagh HMA (2003) Bioactivity of Backhousia citriodora: antibacterial and antifungal activity. J Agric Food Chem 51(1):76–81
WWF (2018) Living planet report – 2018: aiming higher. World Wildlife Fund, Gland

Chapter 12
Exploring New Urban Futures Through Sydney's Hidden Grids

Mark Tyrrell

Abstract There are multiple grids that make up the modern city, these are networks of different types. Some, like the power or transport grids exist and are accepted, utilized each day and only questioned when they fail. Some, like the Green Grid are tangible and close to what people know, but are as yet not regarded as essential like power or water. Others like the Ochre Grid or Ecological Grid are barely legible, containing culture and life of the landscape within their deep memory. There is an opportunity to draw the unseen grids into the tangible experience of place, remaking a deeper connection to country for citizens of modern Sydney. Beyond a simple link or spatial connection, a revived culture of ecological system stewardship and Caring for Country can and should be inspired by the design of the public face of the city. This idea can be realized only through precise strategic system connectivity at a city scale, linked seamlessly to seductive composition and form at a site scale. This proposition outlines a design process TYRRELLSTUDIO are developing. This process organizes city scale grids with clarity through sieving and thinning of mega spatial data, whilst at the same time thickening the layers of the meaning of landscape through collage, overlay and citizen engagement. Surfaced through this process are possible connection points between site and system to catalyse major urban and cultural transformation.

Keywords Green · Blue · Ochre · Grid · Aboriginal · Landscape · Architecture · Urban · Design · Tyrrellstudio · Mark · Tyrrell · Mapping · Ecosystem · Process · Indigenous · Australia · Sydney · Future · Vision

M. Tyrrell (✉)
TYRRELLSTUDIO, Manly, NSW, Australia
e-mail: mark@tyrrellstudio.com

R. Roggema (ed.), *Nature Driven Urbanism*, Contemporary Urban Design Thinking, https://doi.org/10.1007/978-3-030-26717-9_12

12.1 Introduction

Sydney has evolved to be a shattered mosaic. There are fragments of landscape such as the drainage corridors of the city that are undeveloped, largely intact and hold deep memory. Much of this network is lost, weed infested and disconnected, through space, but also through time. The tens of thousands of years of Aboriginal care of the landscape is hidden in plain sight, in the paths we use each day, in the patterns of urbanization we develop. As citizens, our predicament is akin to the patients who suffer from amnesia and must painstakingly develop retrieval paths to find and piece back together their fragmented memories. In the same way, at the scale of the city, our hydro-ecological system as well as our collective memory has been damaged. Therefore we need to develop new urban retrieval paths, projective scaffolds or ‘grids’ to reconnect people to the environmental values and memories held in the fragments of landscape that remain.

12.2 Green and Ochre grids

The Greater Sydney Green Grid (TYRRELLSTUDIO and GANSW 2016) and The Greater Sydney Ochre Grid (under development) are two strategies that, if combined, have the potential to restructure the greater Sydney region and our relationship with place. The Green Grid began as a mapping project looking for more open space opportunities in a growing city, and to reconnect the fragments of open space in a utilitarian sense. The Ochre Grid is taking shape as a means to re-establish a framework to connect and relate the citizens of the city to the deep memory of Aboriginal Country which surround us, but remains unknown to most. Both ideas are reacting to and attempting to revolutionise an outdated planning system which gives precedence to the recording, management and development of objects rather than definition, care and management of complex natural and cultural systems.

To understand the unique potential for fragment connectivity in Sydney, one must first appreciate the value of the fragments that are still with us, and the depth of the memories they contain. These fragments still exist, due largely to the geological formation of the Sydney Basin, which has always been the major morphogenetic scaffold for the city structure, its DNA. Fragments still exist intact in the overgrown and weed infested valleys, partly due to the ferocious speed of urbanization which was required to build a city bigger in area than London in only 220 years (Citymayorsstatistics 2018). This development speed has obliterated anything in the way of the city, and the current planning resource and record of indigenous habitation of the landscape, the AHIMS database (NSW Office of Environment and Heritage undated), is less a history of Aboriginal occupation and more a map of deep wounds, sites destroyed where they lay in the path of development. The first Australians of the Sydney region were hit hardest and quickest by the wave of change and disease, and much has been lost (SBS 2009).

This speed however, has meant that land not easy to develop was left, and so, unlike London, fragments of a time before urbanization are everywhere. This deep network of green, blue and ochre and much of the undeveloped land of the city is the potential open space and cultural network of the future city.

To form the Green Grid project, Barbara Schaffer and her team at the NSW Government Architect Office translated the concepts of the London Green Grid to the context of Sydney's Planning department and became the champion of the project. Data was subsequently collected from across government and a TYRRELLSTUDIO team of Dan Sharp and Sarah Yates, with the assistance of Dr. Scott Hawken sieved these sets out into legible ecological, hydrological and recreational collections for the city. The resulting mosaic maps (Fig. 12.1) show a broken, fragmented ecosystem, made moreso by the arbitrary boundaries of the council areas which are not aligned to hydrological catchment boundaries. Immediately apparent in the gestalt of the data are characteristics in the greater Sydney landscape that make these parts of the city so different (Fig. 12.2) and it is these particularities that have been elevated by Environment commissioner Rod Simpson to define Sydneys vision of a Metropolis of Three Cities through landscape. Sydney's CBD, the easternmost 'Harbour City' has the quality of magnificent sandstone ridges and valleys in between. These heroic geologies create the internationally recognizable deep harbor. Parramatta, the 'River City', is flatter and the visible surface trickle of water whose ephemerality so disappointed the first white settlers, overlays a vast sand sheet containing the traces and memories of

Fig. 12.1 Sydney's collated Green Grid layers showing for the first time the Hydrological Grid, the Ecological Grid and the Recreational Open Space Grid (TYRRELLSTUDIO and GANSW 2016)

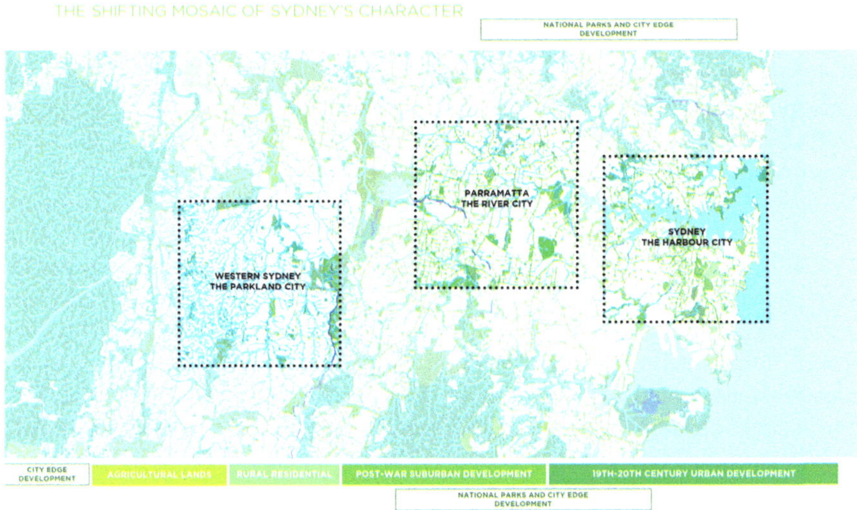

Fig. 12.2 The Green Grid helped to identify a strong identity for Sydney's Metropoloitain Plan called The Metropolis of Three Cities (TYRRELLSTUDIO and GANSW 2016)

indigenous relationship with the river. Further west, the Cumberland Plain landscapes, currently urbanizing under the Aerotropolis 'Parkland City' model, are characterized by a much flatter rolling shale landscape with wide, ephemeral rivers and hundreds of farm dams holding water in the hot, dry landscape.

The first step in the Green Grid project was to map and the second step was to listen. This process must be inverted for the Ochre Grid, where connecting with indigenous people is the only way to begin to connect with country. The two grids meet in an extremely fertile geospatial ground where future physical armatures of green infrastructure, become the public gathering lines for people and the best places to engage with an indigenous connection to the land. As our team have begun to test the potential combination of 'grids', we are beginning to revive from the land itself, a new vision for greater Sydney, where the remnant landscapes that constitute the underutilized 'infrastructure lands' of the city become the key organizational frame of city form. In its leading role as the city frame, this 'hyper – grid' must be beautifully designed to hold, protect and communicate the deep history of Country and environment. Perhaps from a combination of Green and Ochre (Fig. 12.3), emerges an Olive meta-lattice rather than a third grid, because combined, these frames are no longer government-led blueprints landing from above. Rather, they become inherently projective conceptual scaffolding devices for all to engage with. By this, I mean that we can establish a frame for new public culture to emerge, it is a designed structure, a hyper-grid embedded and charged by the specifics of place based on the values both indigenous and environmental.

If public spaces and cities are designed as open ended scaffolds rather than closed narratives, as yet unknown cultural and environmental futures can be unlocked and projected forth. In an age where the role of biofilia in preventative

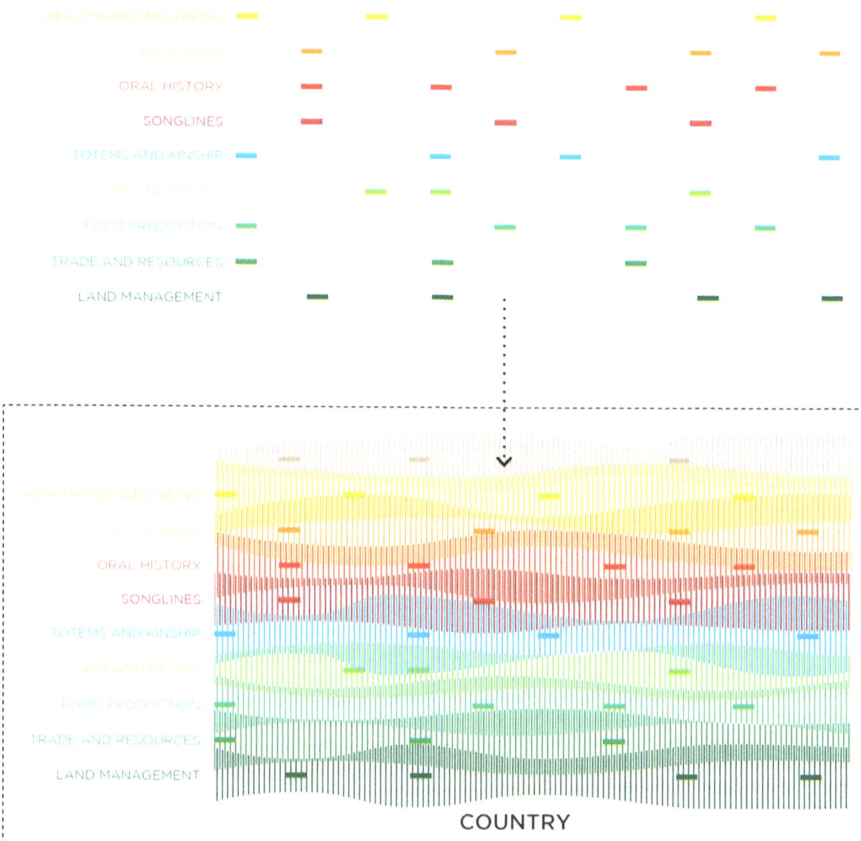

Fig. 12.3 The potential of the Ochre Grid as conceptualised by TYRRELLSTUDIO. This shows the existing object and point focus of history vs the potential reality of an interwoven position in space and time (TYRRELLSTUDIO and GANSW 2018)

health is becoming more widely appreciated, there should be even greater opportunities to value a deep connection to landscape. Culturally, the designed landscape of the city has potential to becomes a 'hyper object', defined by Professor Timothy Morton as "things that are massively distributed in time and space relative to humans" (Morton, p. 1). Acknowledgement and creation of hyper-grids to restructure Sydney will allow us to rig up accessible retrieval paths for memories and systems broken or lost and receive and process feedback loops as we learn again how as urban citizens we might collectively care for Country. To reconnect the hydro-ecological systems and to redefine a collective memory of indigenous place, would indeed redefine the spatial culture of greater Sydney.

Progressive leaps in GIS mapping technology have allowed the TYRRELLSTUDIO team to immediately project the opportunities we can see to reconnect Sydneys fractured mosaic. As part of the Green Grid process, there were many workshops held, with council staff who proposed their projects that would contribute to a connected network of green infrastructure. As soon as the mapping and consultation process began for the Green Grid it became apparent that there was a new potential in the maps produced. Simultaneously, in NSW there has been a groundswell led by the indigenous community, heritage experts and by indigenous architect Dillon Kombumerri around the need for a more sophisticated strategy to maintain and engage with living Aboriginal culture. There is clear potential for the Ochre Grid to begin to develop a co-recording of geospatial culture of indigenous place, to build and layer upon place, the intertwined meaning of Country. The Green Grid focus was on connector projects rather than parks or places and most therefore had a walking or active transport focus. The geology of Sydney has given us a pattern of roads on ridges, often directly following Aboriginal walking paths and sometimes related to songlines, whilst many steep sandstone valleys and creeks were not developed, and where creeks widened into marshes there were gathering places and hunting grounds, most incrementally filled in to became Sydneys waterfront parks.

Mapping the city in detail confirms that in general, the less developed hydrological system of the city as well as the major transport routes are where the big opportunities lie for remaking a city with a functional framework of Green and Ochre Infrastructure. If you look at the city for long enough, the opportunity emerges from the gaps, and revealed, is a vision for a connected city, a cooler city and a more equitable city, a projection of Country reconnected through space and time in a way that creates new settings for deep daily engagement between white Australians, indigenous Australians and the power of deep memory held within the landscape. Whilst the layers are there, neatly categorized in all government datasets, the vision, of the hyper grid has been missed as have the opportunities to unlock the potential of the landscape through discussion, shared experience and ultimately I believe, design composition.

Sieve mapping the landscape has underpinned much of the GIS system. However, as geospatial data is filtered into ever smaller grains of meaning, the reductionist realities of the method appear. After thinning the data to this degree through sieve mapping, one way we explore other meanings of the land is to assemble surfaces through collage (Fig. 12.4). These processes are not so much leading to visualisations

Fig. 12.4 The potential of collage to thicken the reality of place: The River City of Parramatta (Tyrrell and Griffin, 2011)

Fig. 12.5 The subtle and endless potential of vegetation to weave people and systems together, creating places of living memory. University of Canberra (Tyrrell and Wright 2009)

in the form of renders, but rather are attempts to contend with the multitudes of dimensions present in every fragment of place (Fig. 12.5). A landscape has endless depth and meaning, often hidden amongst surfaces that intersect and overlay each other (Fig. 12.6). The process of overlaying leads to new possibilities in the emergent texture. As a green grid project evolves from being a connecting line at 1:100000, it becomes a polygon at 1:10000 (Fig. 12.7). After that it becomes a texture, in which you are at first lost, yet through layers of drawing, superimposition and discussion

Fig. 12.6 The layers of vegetation structure against landform in large scale composition of place. University of Canberra (Tyrrell and Wright 2009)

emerge new possibilities of the project beyond its utility as infrastructure. How to create an appropriate communication of indigenous culture for the Ochre Grid is still in progress, but my view is that the public domain of the city needs to be composed and constructed as a scaffold for cultural exchange and also a true entry point into vast cultural and environmental systems beyond the site.

The most fertile ground for the designer building a cohesive and culturally charged scaffold occurs in the overlap between two typical modes of practice and

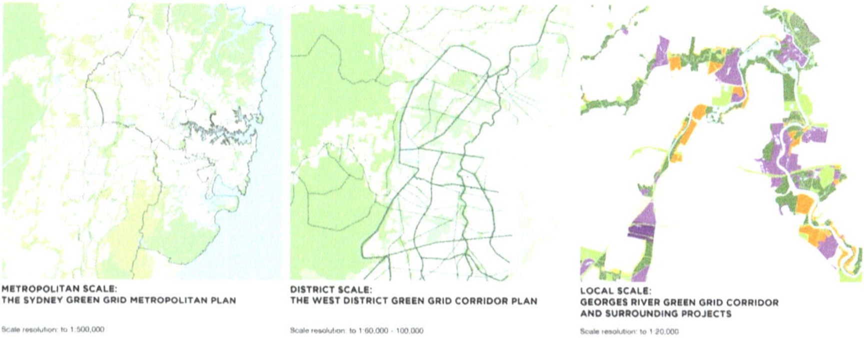

Fig. 12.7 The three resolutions of the Green Grid as a GIS dataset prior to becoming an immersive texture (TYRRELLSTUDIO and GANSW 2019)

speeds of development change. There has been a short circuit in Sydney between the polarities of strategic planning and the boomtime delivery of change. Projects leaping from intentions of liveability and equity to zones of density splashed in concentric colours from transport nodes. Sydney's brisk evolution has been a story of fortunes made through property transactions at lightning speed, and what has been missed is now clear, as we scramble for 'placemakers' to enrich our often vapid urban constructs. This veneer does not help us, but rather closes off the narrative in the realms of the urban brand. In a slower sense, the landscape is a space of flux, and there is an obvious dislocation between the deeper rhythms present in the natural environment and memory, contrasted to the speed of market and the 'fast-planning' which responds.

It is however impossible to slow down the city's growth and also mature its structure. Design and evolution of city form is progressed mainly through cycles of growth. The functionalities of city form are always triggered by the pressures of the present day attitude of 'the project' just as the designer in practice is always driven to 'get the job'. City growth is connected by necessity to the upswings of the market and the downswings of a politicians popularity yet must also be connected in all dimensions to past and future. Therefore, to remake the city in a meaningful and useful way, dense meaning must be intelligently captured from fragments of place, then translated and embedded as cultural catalysts within a projective scaffold. This, like many design processes, is not easy to neatly define. It seeks out a middle ground where strategic projections can be made, and beautiful formal catalytic interventions attached to these strategies to spark long term change in the lived experience of the city, for millions of people. (Fig. 12.8).

Fig. 12.8 The potential for a new city in Sydneys West, The Parkland City, to be designed in a new way where a cultural landscape overlays and intersects with a fragile natural system (TYRRELLSTUDIO and INSW 2019)

12.3 Conclusions

There are multiple grids that make up the city, networks of different type. Some like the power or transport grids exist and are accepted. Some, like the Green Grid are tangible and close to what people know. Others like the Ochre Grid or Ecological Grid are hardly visible, containing memory, culture and energy of the landscape and of life itself. There is an opportunity for Sydney to recognize the potential of these layers by drawing the unseen into the tangible experience of place, remaking a deeper urban culture. Beyond a simple link or spatial connection, ecological system stewardship and caring for Country can be inspired by the beautifully composed and intelligently connected public face of the city.

References

Citymayorsstatistics (2018) The largest cities in the world by land area, population and density. http://www.citymayors.com/statistics/largest-cities-area-125.html. Accessed 15 Feb 2019

Morton T (2013) Hyperobjects philosophy and ecology after the end of the world. University of Minnesota Press, Minneapolis

NSW Office of Environment and Heritage (undated) Search for NSW heritage. https://www.environment.nsw.gov.au/heritageapp/heritagesearch.aspx. Accessed 18 Feb 2019

SBS (2009) First Australians. http://www.sbs.com.au/firstaustralians/. Accessed 21 Jan 2019

Tyrrell M, Griffin D (2011) First prize winning competition entry, ideas on edge competition, City of Parramatta, NSW

Tyrrell M, Wright S (2009) First prize winning competition entry. The University of Canberra Campus Design Competition, The University of Canberra, ACT

TYRRELLSTUDIO and GANSW (2016) The Greater Sydney Green Grid Spatial Framework. Government of NSW, Sydney

TYRRELLSTUDIO and GANSW (2018) Scoping the ochre grid. Government of NSW, Sydney

TYRRELLSTUDIO and GANSW (2019) The green grid GIS dataset methodology. Government of NSW, Sydney

TYRRELLSTUDIO and INSW (2019) The Parkland City strategic framework. Government of NSW, Sydney

Chapter 13
A Bold Vision for Sydney's Future

Dajon Veldman

Abstract In the last 5–10 years, a trend has emerged where cities around the world are being compared to a greater extent for the quality of life they offer and how they perform and compete with other similar cities. There are numerous city rankings undertaken by various agencies to measure the success of cities on a global scale, such as The Global Liveability Index by The Economist (The Global Liveability Index. URL: http://www.eiu.com/topic/liveability, 2019) and the Global Quality of Living Survey by Mercer (Vienna tops Mercer's 21st quality of living ranking. URL: https://www.mercer.com/newsroom/2019-quality-of-living-survey.html, 2019). By comparing cities around the world, we essentially value cities based on several assessment criteria only. How do we ensure we meet these criteria whilst maintaining local uniqueness and cultural differences and create cities with a point of difference?

Keywords Sydney · Urban design · Urban growth · Liveability · Vision

13.1 Introduction

In the last 5–10 years, a trend has emerged where cities around the world are being compared to a greater extent for the quality of life they offer and how they perform and compete with other similar cities. There are numerous city rankings undertaken by various agencies to measure the success of cities on a global scale, such as The Global Liveability Index by The Economist (2019) and the Global Quality of Living Survey by Mercer (2019). By comparing cities around the world, we essentially value cities based on several assessment criteria only. How do we ensure we meet these criteria whilst maintaining local uniqueness and cultural differences and create cities with a point of difference?

D. Veldman (✉)
McGregor Coxall, Sydney, NSW, Australia
e-mail: Dajon.veldman@mcgregorcoxall.com

R. Roggema (ed.), *Nature Driven Urbanism*, Contemporary Urban Design Thinking, https://doi.org/10.1007/978-3-030-26717-9_13

According to the United Nations, 68% of the world's population are projected to be living in urban areas by 2050 (UN DESA 2018), This figure is in comparison to 55% currently. Australia is a highly urbanised society with over 85% of Australians currently living in urban areas (Population Australia 2019). Australia also continues to have a strong population growth, with a population milestone projection of around 40 million by 2058/59 based on a medium assumption (ABS 2018a). In order to remain competitive whilst maintaining local uniqueness in a global playing field, it is now more important than ever to prepare and design Australian cities and urban areas for a successful future.

13.1.1 Growth

Sydney's population officially reached 5.1 million in June 2017 (ABS 2018b) and it is estimated that Sydney will grow to about eight and a half million people by 2061 (ABS 2018c). This indicates the scale and speed that Sydney needs to prepare for over the coming years. Sydney will grow at a far quicker pace over the next 20 years than it has ever done before. For that growth to be realised, about 40,000 homes need to be build every year over the next 20 years. This growth will significantly impact the design of Sydney and how it is constructed and managed. Sydney and other Australian cities are increasingly struggling to cope with the pressures that come with rapid urbanisation. This struggle manifests itself particularly in an infrastructure capacity.

Many global cities are prepared and can capitalise on urban growth by having long-term strategic plans for future sustainable and holistic growth. These strategic plans are holistic plans driven by a clear vision, or concept, underpinned by clear objectives or priorities. Cities that have these plans are able to fast–track the urban growth of their city in a holistic, qualitative and diverse way. These cities are increasingly becoming more successful in the global rankings.

One of the greatest challenges we are currently facing in Australia is how to create a point of difference at a city-level and quality urban outcome at a precinct and neighbourhood level, whilst managing the urban pressures through population increase. If other global cities are well prepared and solidifying their position on the global rankings, and Australian cities are under pressure to remain competitive and are slipping in the rankings, then the question becomes: What can Australian cities learn from other cities and what do Australian cities need to do to ensure we are creating successful long-term competitive and sustainable cities?

By understanding how Australian cities are planned, designed and managed in comparison to other successful global cities will provide some insight.

13.2 History of the Australian Planning System

The Australian Planning System can be linked back to the early days of European settlement and was predominantly focussed on enabling and servicing functional and practical needs in the early days. City shaping decisions were largely based on basic infrastructure, such as roads, rail transport, water supply and usable land for farming and industries (Troy 1995). During the nineteenth and twentieth century the rapid urbanisation of Australian cities meant that governments started to develop metropolitan spatial plans (Freestone 2007). In the second half of the twentieth century with the large scale introduction of the vehicle, spatial metropolitan plans were developed mainly to indicate new land release areas, urban centres, major infrastructure and green space systems. The County of Cumberland Plan of 1948 (Cumberland County Council 1948) is broadly regarded as the first metropolitan plan for Sydney. The plan introduced land use zoning, suburban employment zones, open space acquisitions, and the idea of a 'green belt' for greater Sydney. The Main Roads Department supplied plans for an expressway network. The 'green belt' around the existing urban footprint, was designed to restrict urban sprawl, while satellite towns beyond the belt would take care of future growth (Fig. 13.1; Dictionary of Sydney staff writer 2008).

13.2.1 Urban Development History in Sydney

Sydney has seen four major waves of urban development. Firstly, the early settlements and urban growth which responded to early population increase and economic prosperity (Fig. 13.2; Livingston Mitchell 1832).

Secondly, car-based development emerged during the second half of the twentieth century. This is the era in which the Australian dream was born. The desire to own a block of land, a detached house and a car prompted the suburban sprawl. These suburbs have left an enormous legacy on Australian cities in the way they look and feel, most importantly resulting largely with an urban structure of individual landownership.

Thirdly, urban renewal has been introduced in more recent years. Urban renewal is about making better use of land within the city to contain the physical outward sprawl of cities. This results mainly in redevelopment of existing employment lands into mixed-use urban communities with a higher density. This wave of urban development is catering for changing demographics and different expectations, where high amenity, well connected, high quality places are becoming increasingly important. The dream to afford a home is not necessarily changing, but expectations are. The size of dwellings, quality of open space and ability to interact socially are greater now than they have ever been before.

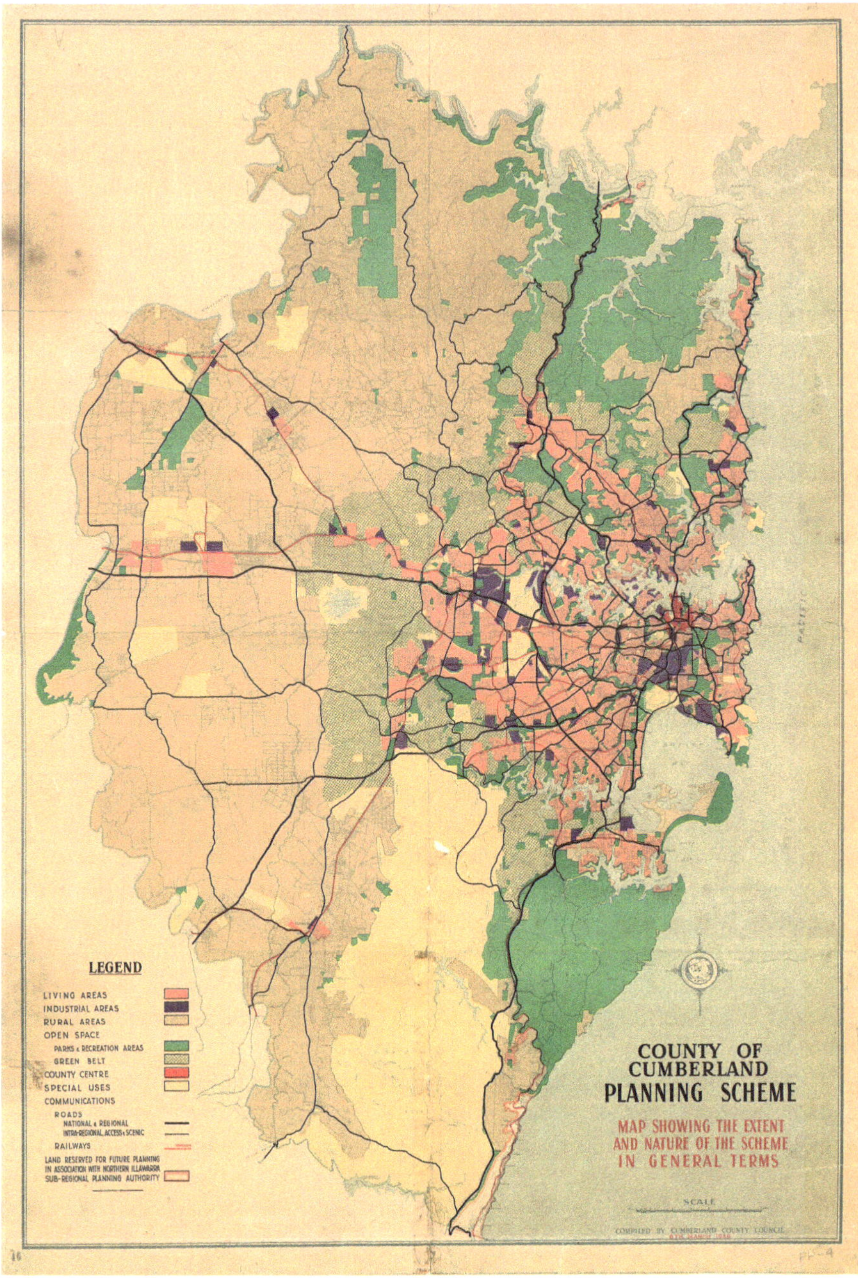

Fig. 13.1 The County of Cumberland Plan of 1948. (Source: Dictionary of Sydney Staff Writer 2008)

Fig. 13.2 Plan of the streets of the town of Sydney, 13 April 1832. (Source: Livingston Mitchell 1832)

In dealing with growth pressures more recently, the New South Wales Government has been planning and implementing major infrastructure projects, such as the upgrading of the M4 and M5 (Designinc 2017; HASSELL 2018), Sydney Metro Northwest (HASSELL 2016), Sydney Metro City & Southwest (Department of Planning and Environment 2001). This has resulted in a new type of urban development: building new rail infrastructure into existing urban and suburban areas, such as Sydney Metro Norwest. As a result, those suburban areas are being retrofitted to become Transport Oriented Developments (TOD) to support higher densities with a mixed-use urban character (Fig. 13.3).

These various waves of urban development have over time brought along different urban scales and speed of development. The Australian planning system largely remains unchanged since the 1960's and is based on a statutory system which, simply put, focusses on separating land uses and controlling the scale of developments through Floor Space Ration (FSR). FSR is the ratio of a buildings overall floor area to the size of its site, also known as FAR. The statutory system has been an efficient way to approve proposals for developments and has become the main controlling mechanism for urban growth. This in essence is a bottom up approach, designed for individual ad-hoc developments (Fig. 13.4).

Fig. 13.3 Sydenham to Bankstown urban renewal corridor, land use change map. (Source: Department of Planning and Environment 2017)

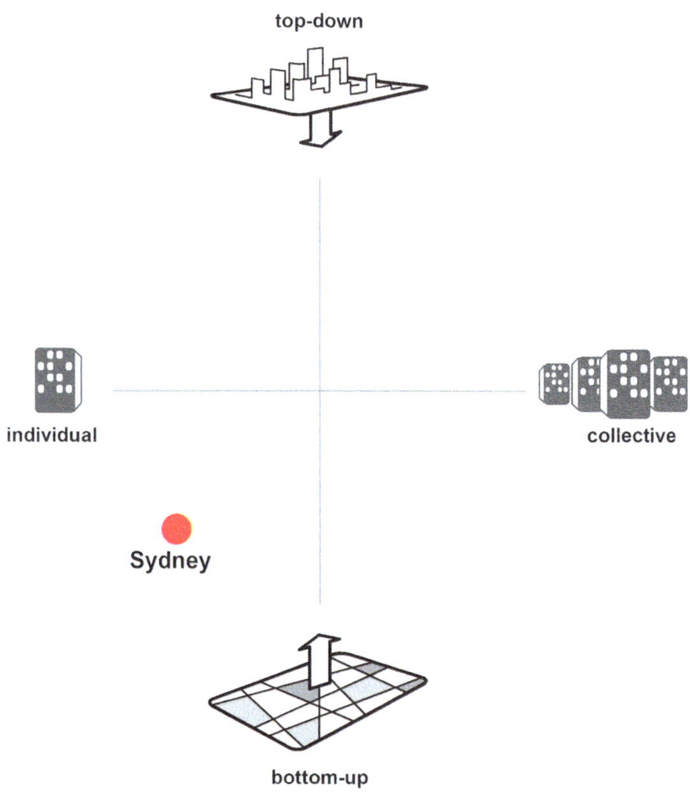

Fig. 13.4 Bottom up approach for individual developments

Without a broader strategic vision that captures both the uniqueness of areas and local neighbourhoods in site specific statutory plans, this bottom up approach leads to uniform approaches to developments of different scales and complexities. Australian cities have seen rapid urbanisation and particularly for Sydney the pressures of dealing with the urban growth through the current statutory system are now starting to impact the overall quality of the urban environment. The current issues related to housing affordability, traffic congestion and a lack of urban variety and quality are the result of this. If Australian cities want to cope with this population growth, while still maintaining their attractiveness, competitiveness and overall success, they must develop a system that delivers a variety of urban development typologies with various sizes and complexities that are specific to the context, that create quality outcomes and can be processed in a fast and efficient way. For Australian cities to deal with future growth successfully, including all pressure that comes with it, it is inevitable to adopt a more top-down approach that inspires a broader collective outcome. Successful cities around the world have been, and are, adopting this approach successfully.

To do so requires a different approach to our current planning system. One that is more visionary, strategic, inclusive, resilient and design-led to capture the complex abstract opportunity fit for a modern metropolis rather than the current plodding system of planning resulting in more of the same development outcomes. A system that anticipates future urban growth requirements and development trends and incorporates vital attributes such as environmental assets and heritage to create places with a point of difference based on contextual attributes and offer residents a choice in variety of urban environments.

13.3 Successful Global Cities

Established and continuing successful 'old' global cities such as New York, London, Paris, Amsterdam and Tokyo have been able to maintain or improve their quality standard and competitiveness. Other relatively successful 'new' global cities such as Dubai and Singapore have managed to grow at an unprecedented rate, whilst creating outstanding quality in their urban environments and developing a strong urban image. Urban precincts or neighbourhoods such as Battersea Power Station in London, Hudson Yards in New York, Downtown Dubai in Dubai and Marina Bay Sands in Singapore are now globally recognised through their urban image.

These cities have achieved a strong competitive position by working towards a shared city-wide vision and more importantly by collaborating between various government agencies, through various layers of government, the private sector and a broad range of other stakeholders, in order to develop and implement their vision. All these successful global cities will continue to attract leading companies and talent, as well as foreign investment resulting in strong economies.

13.4 Singapore as a New Global City

Since the mid-1950's Singapore has strategically moved towards a planned city-state (Chew 2016a) and undertaken a comprehensive planning process through a systematic, long-term approach to land use and transport planning. This has contributed to Singapore being a safe and attractive global city with a strong city brand. The need for a visionary plan to develop Singapore came soon after Singapore's independence in 1965. The two main priorities of a newly independent Singapore was the provision of adequate housing for all and the generation of employment opportunities for the people (Chew 2016a).

Singapore already had a statutory Master Plan in place, which was completed in 1955 and approved in 1958 (Dale 1999). The government realised that the planning strategies outlined in the statutory master plan would be inadequate to cope with the rapid social and economic changes taking place in Singapore (Chew 2016a). As a result, the first Concept Plan for Singapore was created in 1971 and was a long-term plan to guide the country's physical development for the next 20 years. (Chew 2016a). The Concept Plan captured the overall vision for Singapore and the Master Plan controlled the statutory means to implement the Concept Plan. This essentially meant that Singapore had adopted a two-tier planning system (Fig. 13.5).

The Master Plan (URA 2008) is a statutory plan, which translates the vision of the Concept Plan into detailed guidelines that steers the development over the next 10–15 years. The Master Plan details the plans for implementation by specifying the permissible land uses and densities and is reviewed once every 5 years. The Master Plan is supported by Special and Detailed Control Plans (SDCP). SDCP's are development control plans (URA 2013), which include Parks and Waterbodies, Land and Housing Areas, Street Block, Envelop Control, Building Height and Urban Design plans. Based on the planning directions set out in the Concept Plan and Master Plan, land is then released for development (Figs. 13.6 and 13.7).

Over time, revisions of the Concept Plan shifted the priories to transform Singapore into a 'tropical city of excellence'. The new priorities included; improving the quality of life, proposing a wider variety of housing, more leisure facilities and more greenery as well as aiming to make Singapore a 'thriving world class city' (Tan 2001). The development of the Singapore Green Plan (SGP) has been established in 1992 by the then Ministry of the Environment to ensure that Singapore could develop an economic growth model that does not compromise it's environment. In 2002, a second SGP known as the Singapore Green Plan 2012 (Chua 2002) was launched. By setting a series of environmental targets, the aim of SGP 2012 is to help Singapore attain environmental sustainability (Chew 2016b). A new national framework to guide Singapore's sustainable development efforts up till 2030 was launched in 2009. This framework is called the Sustainable Singapore Blueprint and has higher targets set than those in the SGP 2012 (Chew 2016b). The SGP 2012 and

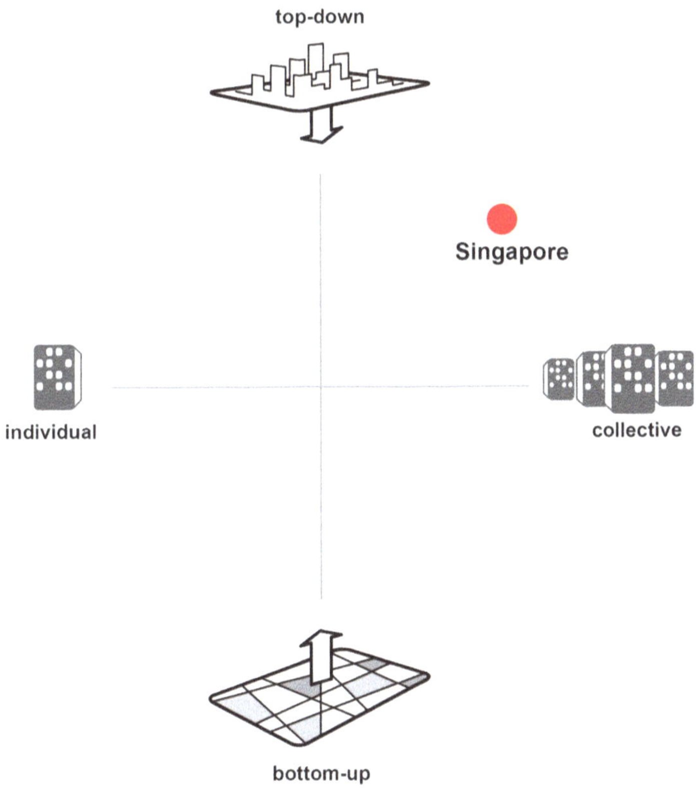

Fig. 13.5 Singapore adopted a top-down approach to collective developments

the Sustainable Singapore Blueprint will help guide The Ministry of the Environment and Water Resources and other agencies to fulfil the set targets (Chew 2016b).

One of the initiatives that directly includes the community and has developed into a major feature of Singapore today is Singapore's 'City in a Garden'. Six key areas have been identified to help implement this vision (National Parks Board 2016):

1. Engaging and inspiring communities to co-create a greener Singapore
   There are now over 1000 community gardens.
2. Enhancing competencies of the landscape and horticulture industry
   The aim is to restructure industry operations, raise industry productivity and seek new solutions in urban greening and conservation through applied research.
3. Enriching biodiversity in our urban environment

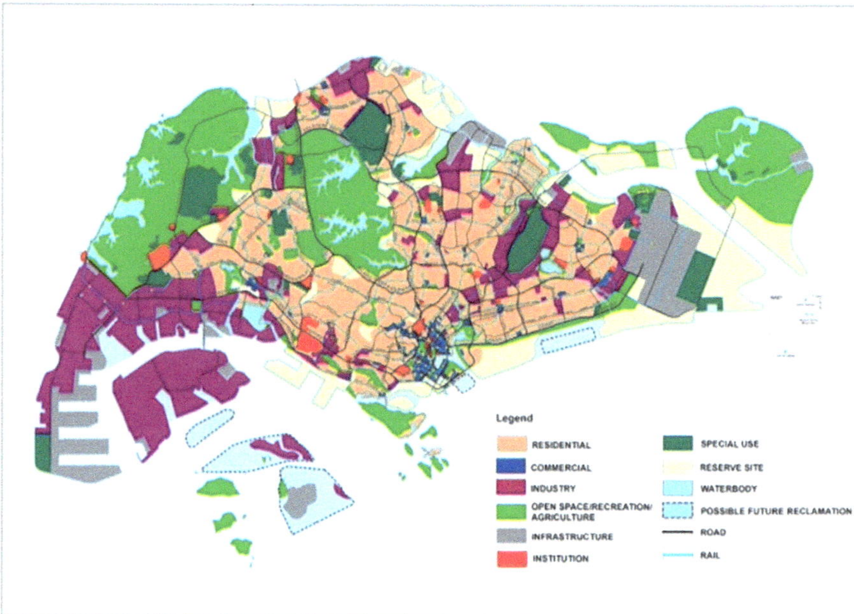

Fig. 13.6 Singapore Concept Plan, Land Use Plan 2013. (Source: URA 2013)

Fig. 13.7 Singapore Master Plan 2008. (Source: URA 2008)

By continuing to focus on; Ensuring the health of key species and ecosystems, restoring the natural areas and enhancing the nature reserves. Some native species have seen increase in numbers.

4. Establishing world-class gardens

The Singapore Botanic Gardens (SBG) is an institution for botanical research and was inscribed as a UNESCO World Heritage Site in July 2015. Opened in mid-2012, Gardens by the Bay (GB) has captured the imagination of

Fig. 13.8 Sky Garden at Pinnacle @ Duxton, Singapore

Singaporeans, and the rest of the world, by showcasing the best in horticulture and garden artistry.

5. Optimising urban spaces for greenery and recreation

   Park connectors have been developed to link Singaporeans to major parks, nature sites and housing estates. Initiative are being made to create thematic greenways that offer Singaporeans a new leisure dimension, connecting communities to nature, historical, cultural and recreational sites.

6. Rejuvenating urban parks and enliven streetscape.

There is a desire to rejuvenate key parks and develop them into leisure destinations that attract visitors from all over the island. Streetscape gardens have been developed with the support of the community.

Projects such as Gardens by the Bay, Punggol Waterway Park and the newly announced Terminal at Changi airport, which features a grand indoor garden area and a cascading waterfall, all deliver on Singapore's 'City in a Garden' Priority. Regulations and a reward system to include greenery in developments have resulted in sky gardens and green integrated into the developments to create quality and well-being. Singapore has been pioneering this and now sky gardens, rooftop gardens and greenery in developments are seen as quality additions that add value, in cities across the globe (Figs. 13.8 and 13.9).

Singapore has adapted and dealt with the pressures of urban growth by altering their statutory approach to a more visionary approach. Singapore has been able to achieve the implementation of the Concept Plan over the years and deliver on their priorities by working closely with developers, professionals, stakeholders, and public agencies to implement and integrate greenery into Singapore's urban landscape. Despite the fact that Singapore and Sydney have two very different political and cultural backgrounds, they do compare in size and scale and they are being compared as global cities on their performance.

Fig. 13.9 Gardens by the bay, Singapore

13.5 Current Sydney Urban Development

The Sydney planning system is focussed on land use allocation through Local Environmental Plans (LEP's) and controlling the scale of developments predominantly through Floor Space Ratios (FSR). The statutory system has been an efficient way to approve proposals for development and has become the main controlling mechanism for urban growth. However, this is in essence a bottom up approach, designed for individual or ad-hoc developments.

13.5.1 What Can Sydney Learn from Other Global Cities

Sydney, with it's current rapid urbanisation pressures and subsequent issues will need to look towards establishing a long-term plan to guide, coordinate and prioritise development as a city-shaping steering mechanism. This must include coordination of infrastructure provision and environmental protection and integration across regions, sectors and a range of future scenarios (demographic, environmental, technological and economic). As land-use planning is organised and mapped by statutory tools as part of a standardised process, a clear vision for cities, precincts and neighbourhoods is missing. Statutory plans can be amended to incorporate design-based outcomes at a later stage, however, this process is slow, not without risk and often expensive.

Creating design excellence in a bottom-up planning approach is, in many cases, harder to achieve than in a top-down planning approach such as Singapore applies. The statutory maps and controls lack inspiration and show little desire to create variety and unique site-specific outcomes. The plans also lack future scenario thinking and resilience. Statutory plans without vision are not conducive to design excellence. The statutory plans at best provide a convenient way for ad-hoc, short term development approvals.

In the Sydney context, a successful statutory system would need to be directed by a strong long-term vision plan that outlines key priorities, whilst protecting and enhancing key local attributes and setting out a delivery plan that will achieve long-term sustainable urban environments. A vision should not be based on statutory plans, but reversely the main goal of statutory plans should be to implement the overall vision. Like the Singapore model, the statutory plans are an efficient way of managing the planning process of implementing developments at a smaller scale, but will only be successful in creating great cities if they are informed by and part of a greater metropolitan vision.

Some of the most recent sub-divisions in Sydney's South-West are a result of the existing pragmatic approach with the aim to produce statutory ready outcomes to fast track the subdivision process due to the pressures of urbanisation. The Indicative Layout Plans for Austral and Leppington North is a plan that seems to be developed without a strong vision and as a result is a plan that does not reflect the ecological and contextual characteristics of the area, but rather produces a layout that can be found anywhere in Sydney.

In reality, this will be an area for thousands of people with multiple communities to live, work and play. The area is of natural significance and is part of a great farming heritage. Current practice however shows that the subdivision structure is based only on land ownership patterns, instead of taking advantage of the significant ecological corridors and environmental attributes as a driver for site specific outcomes. A systematic grid of roads follows the ownership pattern, with a hierarchy of centres rolled out over the top without integrating local environmental attributes in the design, or relationship between the hierarchy of roads and types of centres. Ones these plans are approved and adopted, they showcase two fundamental problems:

1. There is a lack of variety and missed opportunity to integrate local, often environmental and ecological characteristics to create unique, localised and outstanding residential environments;
2. Once the land is sold off it is very hard to implement or make structural changes. If in the event a metro line is to be implemented at a later stage, we are back at retro fitting the suburbs, which is costly, slow, divides communities and not without risk.

The same system of providing new land release is applied for the new future city expansion in the west, also known as the Western Parkland City. A long-term vision is lacking in these plans as well as a desire to create unique quality urban outcomes that add value and contribute to a holistic vision for the city (Fig. 13.10).

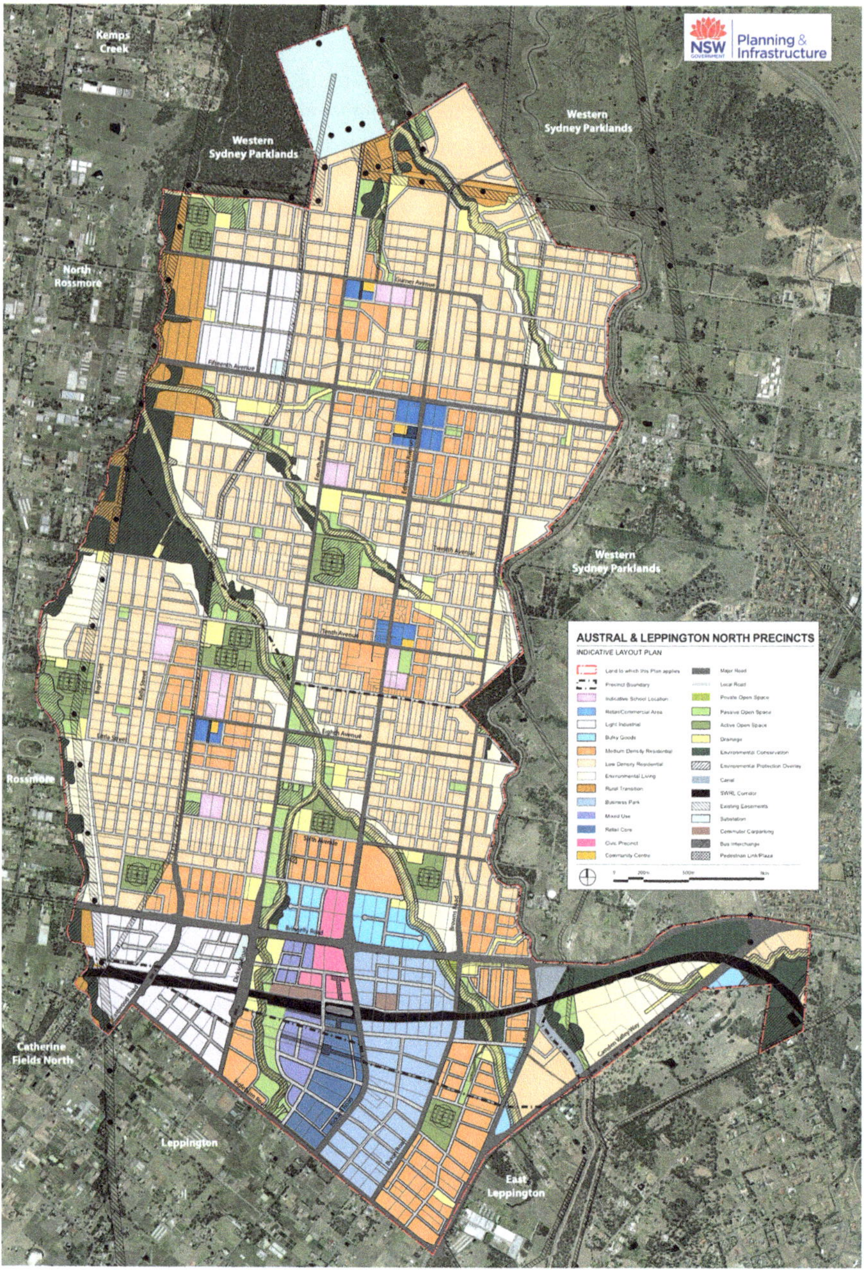

Fig. 13.10 Austral & Leppington North Precincts. (Source: NSW Planning and Infrastructure)

As a result, newly developed areas that are only approached from a statutory angle are often quickly outdated in terms of growth and infrastructure pressures and lack overall design excellence. The State Government and Local Councils in Australia are pushing hard to get better urban outcomes and places in what is called better design excellence. With the statutory focussed planning system as a bottom-up approach without the direction of an overarching vision, design excellence is forced to find resolution within an up-front controlled system, which has not necessarily been tested or crafted for a specific context. Fundamentally, the statutory system is not about design excellence but rather efficiency and since this is the system by which plans are assessed, designers are facing a tough challenge of creating design excellence in a pre-worked statutory environment.

13.6 Capitalising on Sydney's Brand

Sydney enjoys a strong brand and is well known with international audiences. The image of the Opera House and the Harbour Bridge against the backdrop of the Central Business District (CBD), combined with the beaches of Bondi and Manly, portray an image of a bustling waterfront metropolis with endless natural beauty and a great climate, especially in the perception by global audiences. The on-the-ground performance however is rated lower and Australia's cities fail to live up to their brand promise (Property Council of Australia 2018). This discrepancy is a risk that will only increase if we not adopt a different and long-term way of planning and designing Australian cities. Making sure people can relate to the brand and safeguarding the brand in the future, as part of major urban expansion will be a key challenge for Sydney. As an Australian city Sydney appears regularly in global rankings as one of the most liveable cities in the world, The Global Liveability Index 2018 prepared by The Economist Intelligence Unit places Sydney at number five, whilst the Quality of Living Ranking produced by Mercer in 2018 places Sydney at number 10. Sydney is also identified as a global Alpha-plus city by the Globalization and World Cities Research Network (GaWC) in their 2018 city classification (lboro 2018). Alpha-plus and Alpha-plus-plus cities make up the top ten cities in the world. Sydney is the only city that appears high on both the global and liveability rankings and as such it is fair to say that Sydney is the only true Global Liveable city in Australia according to these rankings.

Sydney's natural beauty and access to natural environments is a major contributor to Sydney's liveability and a major part of Sydney's identity. With the rapid urban growth fuelled by the population increase, the natural environment is now under more pressure to be retained and enjoyed. This rapid growth in Sydney is most noticeable in the Western Parkland City, including the new international Western Sydney Airport and Badgerys Creek Aerotropolis and established centres such as Campbelltown-Macathur, Greater Penrith and Liverpool. The population of the Western Parkland City is projected to grow from 740,000 in 2016 to well over 1.5 million by 2056 (Greater Sydney Commission 2018a). Maintaining and strengthening the character of these centres, as well as integrating the natural environment into the future development is crucial for the long-term success of this region and Sydney overall. A visionary plan is required for the Southwest Growth Area (SWGA) to determine a long-term holistic outcome. A visionary plan that can inform a Concept or Structure Plan, which in return can inform the localised statutory plans.

13.7 The Greater Sydney Commission

Sydney is currently seeing a major trend in comprehensive long-term holistic planning, mainly through the works of the Greater Sydney Commission (GSC). One of the major structural directions the GSC has undertaken is the organisation of Sydney into three cities (Greater Sydney Commission 2018a). Sydney is to become a Metropolis of Three Cities, made up of the Eastern Harbour City, the Central River City and the Western Parkland City (Fig. 13.11).

Furthermore, the Greater Sydney Commission championed the vision of Sydney as three 30-min cities where most residents will live within 30 min of their jobs, education and health facilities. This vision seeks to rebalance the economic and social opportunities and deliver a more equal and equitable Sydney Metropolitan Region (Greater Sydney Commission 2018b).

The Eastern City comprises the City of Sydney with it's globally recognised Central Business District next to the Sydney Harbour. In 2018 the City of Sydney resembled over 20% of the entire Gross Domestic Product (GDP) for NSW (City of Sydney 2018). Currently Sydney is largely focused on the CBD as the premier centre for employment. The Greater Sydney Commission aims to decentralise Sydney into three interconnected cities and by doing so making metropolitan Sydney less dependent on the CBD only. The Central City has Parramatta as a strong centre. Parramatta is undergoing a major urban transformation in order to attract investment and population, as well as to strengthen its position as one of Greater Sydney's metropolitan city centres (Greater Sydney Commission 2018a). It is often referred to as Sydney's second CBD. Currently, the Western Parkland City does not have a strong centre, but rather a cluster of regional centres with strong characteristics such as Liverpool, Camden, Penrith and Campbelltown. The new economic force is destined to be the new Western Sydney Airport (WSA) at Bagery's Creek. Newly generated Western Sydney jobs as a direct result from the WSA is expected to range between 12,645 and 19,982 (Kasarda 2015).

In short, the Eastern City is the most established, the Central City has a strong development focus and structure, and can increase its connectivity and economic activity, and the Western City is the new city with a strong focus around the airport.

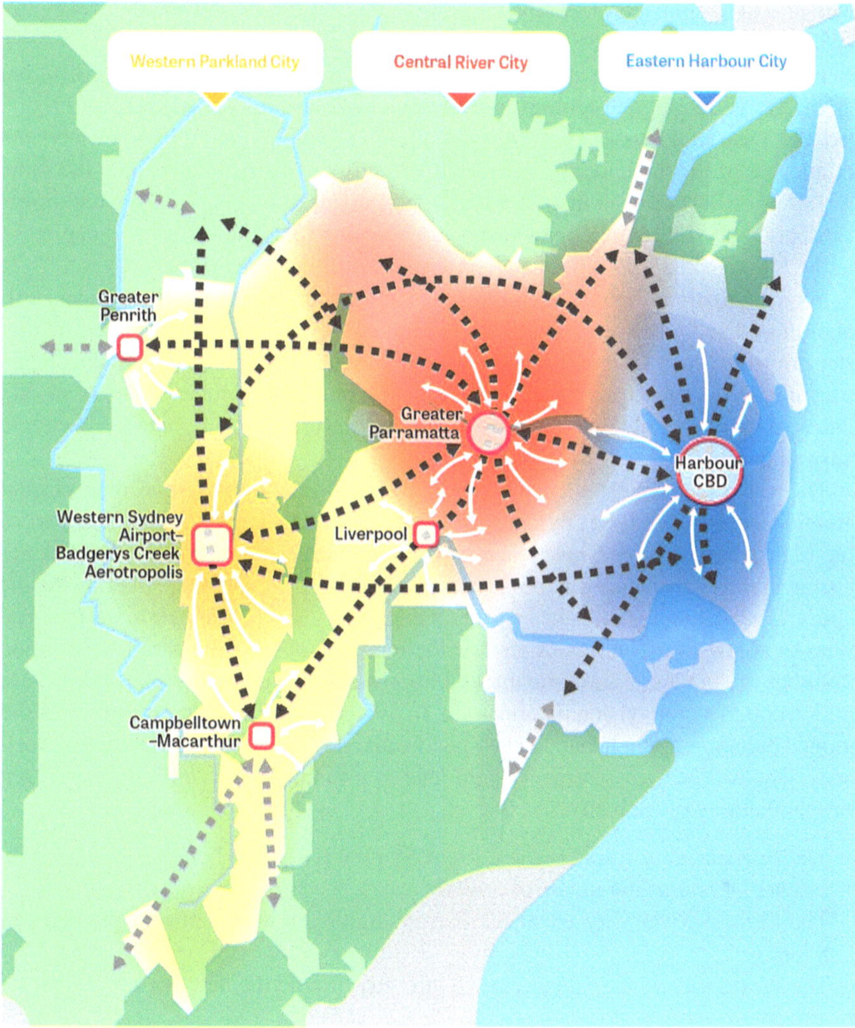

Fig. 13.11 Metropolis of three cities, Sydney. (Source: Greater Sydney Commission)

Since the introduction of the 'Metropolis of Three Cities,' the Greater Sydney Commission has developed five District Plans to guide the implementing of the Metropolis of Three Cities (Greater Sydney Commission 2018b). The key purpose of the District Plans is to inform the preparation of planning proposals. The Priorities form the criteria of assessing how proposals respond to the District Plan. These 20-year plans are a bridge between regional and local planning. They inform local environmental plans, community strategic plans and the assessment of planning proposals. The District Plans also help councils to plan and deliver for growth and change, and to align their local planning strategies to place-based outcomes (Greater Sydney Commission 2018b).

The nomination of the Three Cities is a great way of articulating the different characters, economic and ecological environments. The Greater Sydney Commission and the District Plans do however not utilise the opportunities for variety and uniqueness of the three identified cities. Currently the district plans apply the same overall priorities for each district with slightly nuanced objectives. This will direct the districts to similar outcomes instead of refusing to acknowledge and capitalise on the fundamental differences and opportunities of the three cities.

For each district, the same priorities are outlined, namely infrastructure and collaboration, liveability, productivity and sustainability. By establishing specific district priorities, each City could be approached differently and over time developed in a unique way with distinct neighbourhoods and iconic places. This is in fact placemaking on a precinct or neighbourhood level, which remains a rarity in Australian cities today.

There is a real opportunity to view each city differently based on their current situation, however holistically form a diverse overall Sydney metropolis as part of the 'Metropolis of Three Cities'. These differences can and should be translated into Vision Plans, which inform the statutory plans to give specific direction to the Cities. These Cities can then be broken up in growth areas or strategic areas to determine localised site-specific design development outcomes. An abstract description outlining the potential and fundamental approach to planning and design of these cities through a more visionary approach. In doing so it creates variety and localised attuned outcomes, which lead to site-specific and design outcomes that foster excellence. The following assessment could lead to a fundamentally different approach for the planning of each city:

- The Harbour City respects the current Sydney CBD and harbour foreshore. In essence, it retains the image of Sydney;
- The Central River City enhances Parramatta as the new centre in the heart of Sydney;
- The Western Parkland City creates a new city around the new Badgerys Creek's airport.

A strategic planning system tailored for each city based on the respect, enhance and create principles will lead to vastly different planning outcomes that can strengthen the diversity and uniqueness of the three cities. Especially the Western Parkland City is an opportunity to design and develop an entirely new city. It is here where a new planning approach should be formulated and where we should step away from our current bottom-up planning system (Fig. 13.12).

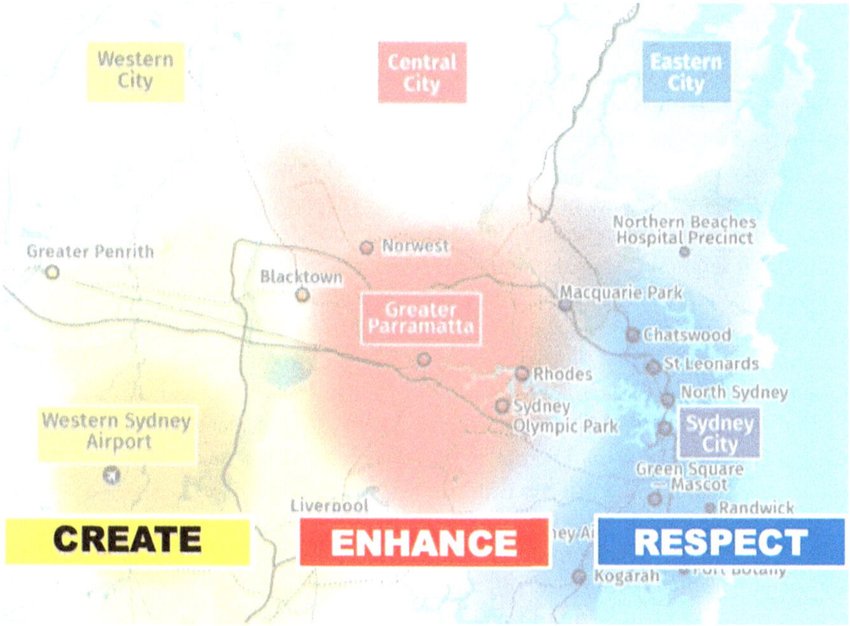

Fig. 13.12 Tailored planning system for each city

13.8 Western Parkland City Opportunity

There is a once in a lifetime opportunity to create a new city, the Western Parkland City, as part of the larger Sydney metropolis. A city where resilience, connectivity, technology and natural environments are the main priorities and where a truly health-conscious city can create a unique way of Australian living. A place where people decide to live because of it's unique character, lifestyle and opportunities nowhere else found in Sydney or Australia. This opportunity lends itself well to adopt a new holistic visionary planning approach and create a more top-down strategy that will provide a blueprint to inform long-term sustainable developments and from where regional and precinct plans can be drawn up. In doing so it is important to establish the fundamentals that will guide developing the specific priorities for the Western Parkland City. If we get the fundamentals wrong or simply fail to identify them, it will ultimately cost the city dearly in the long run by having to structurally retro-fit and it will jeopardise Sydney's position as a Global Liveable city. Without an overall vision and specific priorities the overall development of the Western Parkland City is in danger of being developed on an ad-hoc basis.

The Western Parkland City should be a thriving city in it's natural ecological environment. A city which is different to the other two Sydneys, but together form a stronger image of Sydney overall, ensuring Sydney remains the most Global Liveable City in Australia and in the world. Drawing from other successful cities around the world, a vision should be developed and plans drawn up for the entire

Western Parkland City – A proactive approach in establishing, testing and staging a desired outcome. Similarly sized cities and urban areas such as Dubai, the greater metropolitan area of Amsterdam and Singapore all draw concepts plans for the entire region with infrastructure, environment and land-use planning as an integrated outcome, staged over time.

The Western Parkland City should also outline growth aspirations and reservations for the environment, infrastructure and the increase of densities upfront so that the city does not need to be retro-fitted to accommodate growth as is the case at the moment. It requires a staged plan of developing the city in order to anticipate government spending and minimise the need for political interference over time. The large infrastructure and environmental development stages outlined in a holistic and integrated manner will define development priorities to establish a successful long-term city.

For the Western Parkland City, structure plans should identify areas where the natural environment should be protected and enhanced, not as a minimum but as a robust and fundamental priority to form the backbone of the city. A city where no deal can be made with regards to development damaging the environment alone, but rather compensate developments for introducing and restoring greening. Regional and neighbourhood centres should capture the opportunity of the broader natural setting by being planned and located next to or part of larger green systems. This way the centres can also become lifestyle centres and when connected by a truly integrated bicycle network can start to foster a much healthier way of living. The area also lends itself well for the introduction of Australia's first regional bicycle network with a hierarchy of major bicycle routes (highways) and localised routes (streets). A bicycle network that is integrated in the urban developments from the start, providing safe, comfortable and convenient transport alternative.

By celebrating and capatilising on the natural assets of the area and introducing alternatives for active living, the Western Parkland City can be a showcase of healthier urban environments resulting in improved mental and physical health.

13.9 Conclusion

Sydney is developing and becoming familiar with a more holistic planning approach. However, each City of the 'Metropolis of Three Cities,' is largely guided by the same priorities, which will ultimately result in similar urban outcomes. The three cities, whilst part of the same city of Sydney, all have fundamental differences that are neglected in the current District Plans by the lack of visionary structure plans.

Creating quality urban outcomes is a major contributor to a competitive and liveable city. Variety and site specific uniqueness is fundamental to a pleasant, comfortable and attractive city. The current District Plans are an evolution of the statutory approach of land-use allocation, especially focussed on residential and employment areas. Whilst good planning of these land uses are vital to the success of cities, the current District Plans neglect enough variety and site-specific urban outcomes by not articulating site uniqueness.

Sydney should do more to acknowledge, safeguard and strengthen the natural qualitative environment, which is such a great contributor to Sydney being the only Global Liveable city. A stronger vision to capture and implement this is required to ensure Sydney remains a Global Liveable city.

A bold vision, which captures shared priorities for all government agencies, private sectors and communities needs to be developed. Emphasis on an overarching strategy is required for the Western Parkland City to design and develop a city that values history, place and identity. Cities now and for our next generations should be formed around integration, climate resilience, adaptability, environmental inclusiveness and overall variety and livability.

The current system of urban planning is embedded in a statutory bottom up approach, which does not capture aspirational and desired outcomes. We have an opportunity to develop a new city in the Western Parkland City and we should make sure we are not restricted by our current planning system, but rather prepare for holistic growth by developing and drawing a bold vision and structure plans supported by statutory tools for implementation.

It would be a missed opportunity if the Western Parkland City is developed in the traditional bottom up planning approach. A bold vision should be developed, tested, staged and drawn up and then supported and cristalised by localised statutory plans.

References

ABS (2018a) The Australian Bureau of Statistics. November 2018

ABS (2018b) The Australian Bureau of Statistics. April 2018

ABS (2018c) Australia's population to reach 30 million in 11 to 15 years. Australian Bureau of Statistics, November 2018

Chew V (2016a) History of urban planning in Singapore. National Library Board Singapore. URL: http://eresources.nlb.gov.sg/infopedia/articles/SIP_1564_2009-09-08.html

Chew V (2016b) Singapore Green Plan. National Library Board Singapore. URL: http://eresources.nlb.gov.sg/infopedia/articles/SIP_1370_2008-11-22.html

Chua LH (2002) The Singapore Green Plan 2012: beyond clean and green towards environmental sustainability. Ministry of the Environment, Singapore

City of Sydney (2018) Sydney: City of Sydney. City of Sydney, Sydney

Cumberland County Council (1948) County of Cumberland Plan. Adopted by the Government of New South Wales in 1951

Dale OJ (1999) Urban planning Singapore: the transformation of a city. Oxford University Press, New York

Department of Planning and Environment (2001) Sydenham to Bankstown urban renewal corridor strategy. Government of New South Wales, Sydney. URL: https://www.planning.nsw.gov.au/-/media/Files/DPE/Plans-and-policies/sydenham-to-bankstown-urban-renewal-corridor-strategy-2017-06-part-1.pdf

Department of Planning and Environment (2017) Sydneyham to Bankstown urban renewal corridor, corridor strategy. Government of New South Wales, Sydney. URL: https://www.planning.nsw.gov.au/-/media/Files/DPE/Brochures/sydenham-to-bankstown-corridor-brochure-2017-06.pdf

Designinc Sydney (2017) WestConnexM4 widening urban design and landscape plan, Sydney, CPB Dragados Samsung Joint Venture. URL: https://www.westconnex.com.au/sites/default/files/Urban%20Design%20and%20Landscape%20Plan%20-%20issue%2016%20final_1.pdf
Dictionary of Sydney Staff Writer (2008) County of Cumberland Planning Scheme. URL: https://dictionaryofsydney.org/entry/county_of_cumberland_planning_scheme
Freestone R (2007) A history of planning. In: Thompson S (ed) Planning in Australia: an overview of urban and regional planning. Cambridge University Press, Cambridge
Greater Sydney Commission (2018a) Greater Sydney regional plan, a metropolis of three cities. The Government of NSW, Sydney
Greater Sydney Commission (2018b) District plans. URL: https://www.greater.sydney/district-plans
HASSELL (2016) Sydney metro northwest. System wide design. Urban design and corridor landscape plan. Transport for New South Wales, Sydney. URL: https://www.sydneymetro.info/sites/default/files/document-library/1.SM_NW_UDCLP-Introduction.pdf
HASSELL (2018) WestConnex new M5 urban design and landscape plan. CPB Dragados Samsung Joint Venture, Sydney. URL: https://www.westconnex.com.au/sites/default/files/WCXSTAG2-CDSJV-WCX-GEN-003191-180219_MSN-HSL-MNP-100-110-TR-1970_E_Urban_Design_and_Landscape_Plan_%5bLR%5d_COMBINED.pdf
Kasarda J (2015) A Western Sydney aerotropolis, maximising the benefits of Badgery's Creek. NSW Business Chamber, North Sydney. URL: https://www.nswbusinesschamber.com.au/NSWBC/media/Policy/Thinking%20Business%20Reports/A-Western-Sydney-Aerotropolis.pdf
lboro (2018) The world according to GaWC. URL: https://www.lboro.ac.uk/gawc/world2018t.html
Livingston Mitchell T (1832) Plan of the streets of the town of Sydney, 13 April 1832. State Archives & Records New South Wales [CGS 13859, [Map 5470]]. URL: https://dictionaryofsydney.org/contributor/state_archives_records_new_south_wales
Mercer (2019) Vienna tops Mercer's 21st quality of living ranking. URL: https://www.mercer.com/newsroom/2019-quality-of-living-survey.html
National Parks Board (2016) City in a garden. Singapore: Singapore Government. URL: https://www.nparks.gov.sg/about-us/city-in-a-garden
Population Australia (2019) Australia population 2019. URL: www.population.net.au
Property Council of Australia (2018) Creating great Australian cities. Property Council of Australia. Published 18 May 2018. URL: https://www.propertycouncil.com.au/Web/Content/Media_Release/National/2018/Creating_Great_Australian_Cities__Time_for_Action.aspx
Tan MB (2001) The concept plan 2001. URA, Singapore. URL: https://www.ura.gov.sg/-/media/User%20Defined/URA%20Online/publications/research-resources/plans-reports/concept_plan_2001.pdf
The Economist (2019) The Global Liveability Index. URL: http://www.eiu.com/topic/liveability
Troy P (1995) Technological change and the city. Federation Press, Sydney
UN DESA (2018) Revision of world urbanization prospects. Population Division of the UN Department of Economic and Social Affairs, UN, New York. URL: https://population.un.org/wup/
URA (2008) The master plan. Urban Redevelopment Authority, Singapore. URL: https://www.ura.gov.sg/maps/?service=mp&year=2008#
URA (2013) Special and detailed control plans. Urban Redevelopment Authority, Singapore. URL: https://www.ura.gov.sg/Corporate/Planning/Master-Plan/Control-Plans

Chapter 14
A Contemporary Approach to the Design of Road Transport Infrastructure in Balance with the Landscape

Gareth Paul Collins

Abstract Our current technological approach to design and construction is removing an important connection between humans and nature. Nature can teach us many things about design, as it represents a tried and tested response to the forces acting in our dynamic environment.

Before modern technology we had to observe and design with nature and there were few cultures that did this as well as Aboriginal peoples of Australia. Modern society can learn (or re-learn) from nature and listen to our ancient cultures and produce design outcomes that are unique, context sensitive and durable. For example, road cuttings and embankments that match natural slopes – in gradient and vegetation cover – will be more stable, less maintenance and help the road fit in to the landscape. Bridges that include arched forms respond to nature's principles regarding gravity and compressive strength and can last thousands of years. A road alignment that skirts valleys, travels through saddles and works with contours rather than against, will follow the landscape, reflect its movement and form, minimise the impact on the natural and cultural heritage and need less effort to maintain. Using stone, understanding the rock, fitting it together in a skilled way will produce dry stone walls that will look beautiful and will last for centuries. In designing long lasting artefacts and projects with a timeless elegance and beauty, we should look around, understand and work with nature, learn from and value the practices of our ancient cultures and not just rely on technology.

Keywords Nature · Geology · Landscape · Design · Aboriginal

G. P. Collins (✉)
Urban Design, Roads and Maritime Services, NSW State Government, Sydney, NSW, Australia

Australian Institute of Landscape Architects, Canberra, ACT, Australia
e-mail: Gareth.collins@rms.nsw.gov.au

R. Roggema (ed.), *Nature Driven Urbanism*, Contemporary Urban Design Thinking, https://doi.org/10.1007/978-3-030-26717-9_14

14.1 Introduction

Many of us have memories from our childhoods of drives or rides along country roads, seeing the sea appear, travelling through a tree tunnel, over a mountain pass, seeing the city on the horizon. These experiences are powerful because they are enhanced immersions in the landscape – like a film real. In a short space of time and in our bubble of a vehicle we are experiencing the grandeur of topography, the scale of forests, the vastness of the ocean and the city in its context on earth. Astronauts have similar experiences from the International Space Station but more life changing and with an 'orbital perspective' rather than a vehicular one. These experiences are significant because they were often undertaken on a piece of infrastructure that responded to and flowed with the landscape, properties that in the late 20th and early 21st centuries have lost some of their relevance due to technology and the scale of infrastructure (Fig. 14.1).

14.2 The Limitations of Modern Road Building Technology

The design and Implementation of road infrastructure is more technologically sophisticated than it has ever been. For example, drones can survey project sites quickly and efficiently using Lidar, computer models of projects can be built on these pin point accurate surveys, and the models can be sent to GPS guided earthworks graders. From inception to completion we have technological control over the building process.

The Pacific Highway Upgrades at Bulahdelah and around Byron Bay are good examples of this, where the urban designers helped design the landform on computer which was then translated to the site machinery to guide the operators. This saves time and provides surety of outcome, but if these processes lack design quality or a context sensitive approach, then the built outcome will too. Also, there has become less scope for craftsmanship in implementation. For example, in the twentieth century and the beginning of this century, projects regularly employed many skilled craftspeople. One such team in NSW, now retired, were responsible for the stone pitching on cuttings and bridge abutments and executed their role with such pride that they signed off their work in the mortar (Fig. 14.2).

Another example is the dry stone walling techniques that still exist in NSW but being used less and less. A technique that produces walls of great longevity using only rock (Fig. 14.3).

Technology in road building machinery has also become very sophisticated. Vast earthmoving machinery saves considerable time. The contemporary tunnel boring machinery – up to 100 m in length consisting of cutting heads, navigational equipment, panel storage and placing equipment, and engines have made tunnels viable in comparison to the costs of acquiring valuable land and building on the surface of cities. Modular design has also become very cost-effective in the fast

Fig. 14.1 The Sea Cliff Bridge is a modern example of a road, which for those that use it, will become a strong memory of the journey and movement through the landscape. (Photo: RMS)

implementation of projects. For example, noise wall panels come in predetermined sizes which require additional effort (texture artwork) to make them unique and respond to context. The Super-T bridge girder is a popular and vital part of the road building toolkit. It comes in predetermined sizes and is made in factories.

Fig. 14.2 Stone pitching on the Bangor Bypass, Sydney signed by 'George 04'. (Photo Gareth Collins)

Fig. 14.3 Dry stone walling at Tweed Heads hospital to minimise views and noise from the helicopter landing pad. (Photo Gareth Collins)

Fig. 14.4 Super T bridges on the Pacific Highway. (Photo Gareth Collins)

Furthermore, while the natural physical forces that apply to these girders under loads and gravity follow curved paths along the girders, it is far simpler to make the girders straight (Fig. 14.4).

We have a global market today; therefore, a material that is best for a particular task is readily available. This improves durability and safety but there is no difference to the use of that material anywhere in the world. The idea of using a locally sourced material has lost its appeal. For example, concrete over stone or a particularly successful plant species over a local native species. We also have a global market in terms of information and ideas. Skills are transferable. The Internet provides ideas that anyone can access. A solution to a particular problem which appears successful is now a strong contender as a solution to a particular problem anywhere in the world, despite there being differences in context. The cable stay bridge, reinforced earth panel or concrete road pavement are three such examples. All these factors while providing significant cost and programming benefits can have an impact on the achievement of context sensitive design or infrastructure in balance with the landscape and our cities. It is important to retain these benefits but also becoming increasingly important to design with nature. To use the inspiration and guidance provided by context, geology, landscapes and indigenous cultures to give us the guidance for unique design outcomes that suit our lives and environments and that are more cost effective in the long run.

14.3 The Rise of Appreciation of Nature in our Modern World

There is nothing new about designing with nature. We used to do it naturally and probably unthinkingly when our technologies weren't sufficient to achieve the global benefits discussed above. Cobbled streets were mined from the local quarry. Stone blocks were cut from the bedrock and used for bridges (Fig. 14.5) and walls. Local soils were used which contained local seeds, and plants were grown from the progeny of local trees. Bridges were the only solution to crossing major waterbodies before tunnel technology was readily available. Roads flowed with the landscape largely because of the costs and challenges of carving through the landscape.

The increase in populations from urbanisation, the rise of technology spurred by the World Wars, and the following austerity changed all this (reference https://architectureau.com/articles/020-la144_war_landscapearch/). Road building inspired by the effectiveness of the German Autobahns was embraced by many countries not the least the United Kingdom ('On Roads' Joe Moran A Hidden History) and the pace of change outran our capacity to understand the effect of such vast scale projects and our loss of contact with nature (Crowe 1960; McHarg 1969). Thinkers like Christopher Smith and Edward Relph who wrote 'Place and Placelessness' (Smith and Relph 1978) and Ian Nairn who wrote 'Outrage' (Nairn 1959) both expressed the lack of sense of place or nature in our towns and landscapes. Great landscape architects like Dame Sylvia Crowe, Ian McHarg and Peter Spooner understood this

Fig. 14.5 Stone bridge, Brugge. Local materials, local ways of working and craftsmanship produced infrastructure in balance with nature. (Photo: Gareth Collins)

Fig. 14.6 Sydney to Newcastle Freeway. Peter Spooner's influence helped create a road in harmony with the landscape. (Photo Michael Pratt)

and strove to set down guidance and inspiration to change the trajectory of industrial, technology-based road building. And they were successful.

Ian McHarg's 'Design With Nature' gave us a philosophy and methodology to understand landscape and respond to it (McHarg 1969). Sylvia Crowe's 'The Landscape of Roads' (Crowe 1955) gave us the principles and practices regarding the sculpting of earthworks and setting out of vegetation (responding to nature's broader vegetation patterns rather than squeezing out a strip of landscape, like toothpaste, along a road). Peter Spooner implemented these ideas (The Roadmakers, DMR) and, working with the Department of Main Roads in NSW (the first landscape architect to do so), gave us the Sydney-Newcastle freeway (Fig. 14.6) with its iconic combination of curves, deep sandstone cuts, outcrops and vistas of the landscape.

More recently we see an ever-increasing need for nature in our lives for example 'A Metropolis of Three Cities', a plan for the Greater Sydney area, talks extensively about green space, public parks and names the three cities of Sydney after the landscape – The eastern Harbour City, the central River City and the western Parkland City (Greater Sydney Commission 2018). Nature can soothe our urban souls and provide us with shade and protection from the weather and climate change. People are demanding 'greenery' in our cities. Open space, tree canopy, parks and gardens are now understood to be vital for our well-being as well as our economy by creating a liveable, attractive, vibrant and productive city. It is estimated a 10% increase in tree cover could add $50,000 to the value of properties, although not necessarily good for buyers! (Swinbourne and Rosenwax 2017).

14.4 Principles to Consider in Planning Designing and Building Infrastructure

Taking all this into account there are a number of principles that should be considered in designing infrastructure.

14.4.1 Collaborate with Landscape Architects

Ian McHarg offers the analogy of plumbers to help in deciding who should design roads. 'A plumber is the most important member of our society – our civilisation could not endure long without his services: but we do not ask plumbers to design cities or buildings' (McHarg 1969). McHarg had a disdain for engineering as a tool for the holistic design of roads, but his point was made clearly. Landscape architects need to work closely with engineers and be meaningfully engaged, not simply healing over the scars of projects.

14.4.2 Work with the Landscape Not Against It

Good design makes use of the natural processes, principles and characteristics. For example, a lack of understanding of geology can lead to the removal of rock to install concrete bridge piers. Solid substantial rock can be stronger than concrete so in this instance the durability of natural materials is not being utilised.

14.4.3 Make Use of all the Natural Resources Available

Local natural materials can in the right circumstances be a better choice than imported materials. For example, using local naturally occurring soils and seed banks to revegetate the road reserve. Provided they are relatively weed free the soils collected from site contain a natural seed bank from the surrounding trees and shrubs. This seed grows vigorously and is essentially cost free. The alternative of growing, supplying and planting trees and shrubs is a relatively slow and expensive way to revegetate an area, although necessary in the urban environment.

14.4.4 Create Opportunity for Nature to Recolonise the Project

Very often with road projects the best surface treatment – low maintenance, attractive, erosion proof – is vegetation. In the case of rock cuttings, leaving a smooth rock surface with little capability of supporting soils or plants hampers the revegetation process. Providing a roughened surface with crevices, cracks and areas where soils can accumulate can help nature recolonise an area and accelerate the revegetation process.

14.4.5 Provide a Connection to Country

Engage with local Aboriginal communities in the design of projects. Aboriginal peoples have a strong understanding of the Australian landscape both in a spiritual sense and a very practical sense. They can provide great knowledge regarding the soils, geology, landform, plants as well as the ancient yet living traditions and stories of the dreamtime. They provide us with an invaluable appreciation of Australian landscape values and principles. This principle is also applicable to all nations and cultures around the world. 'As built environment professionals who engage with land, places, cultures, history, people, natural systems and built context, landscape architects seek to recognise Aboriginal and Torres Strait Islander cultures and advocate for a 'Connection to Country' approach to landscape planning, design and management on all our projects, in varying contexts and across many scales (Corkery 2018).

14.4.6 Use the Project to Add Value or Repair the Landscape

Major infrastructure projects provide an opportunity to repair the landscape. Large scale projects can afford to deal with urban problems that would swamp many smaller ones, such as managing contamination and repairing the vegetation cover (Fig. 14.7).

Fig. 14.7 The St Peters Tip site undergoing work to reshape the landform, cap the waste, plant with native trees and shrubs and create both a vegetated interchange and a public active transport corridor. (Photo WestConnex)

14.4.7 Leave a Natural System in Place Not a Cosmetic Outcome

All projects are part of the landscape and consequently of the water, carbon and ecological systems that operate in the environment. Creating a design outcome that looks good yet has little relevance to flooding, pollination, fauna habitat, weed spread or climate extremes is unlikely to thrive and will try and revert to a natural state. Integrating with natural landform, using natural slope angles and natural forms and using native plant species are all ways to ensure the new landscape is a part of the environmental system and also looks good.

14.5 A Corridor Approach to Road Design

McHarg, Crowe, Spooner and others set the scene, but many others have continued to develop these ideas. The Pacific Highway is a case in point. A major road program of nearly 700 km of duplicated highway was needed to save lives and help connect the eastern seaboard of Australia. The program was urgent with so many traumatic crashes on the undivided road, so it needed the technological advantages of modular systems, global best practice and machinery, shared information and global materials,

Fig. 14.8 Pacific Highway Framework 2004 and the 2013 update (Roads and Maritime Services 2004, 2013)

but it has been designed and built with an underpinning of context sensitive design (Roads and Maritime Services 2004). The Pacific Highway Urban Design Framework (Fig. 14.8) was published in 2004, updated in 2013 (Roads and Maritime Services 2013) and contained two critical requirements – work with urban designers and landscape architects in the planning design and implementation of all projects and deliver the following urban design vision. 'The upgrade should be a sweeping, green highway providing panoramic views to the Great Dividing Range and the forests, farmlands and coastline of the Pacific Ocean; sensitively designed to fit into the landscape and be unobtrusive; and characterised by simple and refined road infrastructure.'

The highway has emerged relatively quickly for such a vast program of work, and the achievement in delivering a highway in balance with nature has been significant. For the majority of its length the curvilinear nature of the highway has been emphasised. The alignment, where possible, skirts valleys, avoids ridges and aims for saddles (Fig. 14.9). This minimises earthworks, opens up views of the landscape and thus provides a less monotonous journey.

The 12 m wide median (Fig. 14.10) which can accommodate potential future widening and a vegetated corridor has, when multiplied across 700 km, provided a strong landscape element to the highway its curvilinear nature and continuity breaking down of the scale of the road surface. At times the median spreads further to accommodate the retention of substantial groups of trees, at other times it narrows to reduce the footprint in sensitive ecological areas.

Fig. 14.9 Ballina Bypass alignment flows and responds to the landscape providing views and an experience of the Australian east coast. (Photo Gareth Collins)

Fig. 14.10 Wide median retaining trees, Bonville Bypass. (Photo Brett Boardman)

Earthworks have been laid back to reflect the surrounding topography and blended into the natural landform, (Fig. 14.11) at the tops and bottoms of cuttings and embankments and also at the ends. At times the cuttings have been steepened to reduce the footprint near properties or other sensitive areas. An independently graded dual carriageway has been adopted on some sections which means each carriageway can be layered or terraced down the hillsides providing a more responsive less intrusive alignment.

Bridges over the highway have been designed in include spill through abutments (Fig. 14.12). This means the bridge is longer, more slender, sits more lightly on the ground and allows better views of the landscape – all factors in creating a more context sensitive response in balance with the landscape.

Bridges over rivers have wider spans avoiding impacts on vegetation and habitats. In places large arched or haunched bridges have been used to provide wider spans (Fig. 14.13).

Fig. 14.11 New landform laid back and feathered into the existing landform. (Photo Gareth Collins)

Fig. 14.12 Spill through abutments are the standard bridge type for the Pacific Highway. Used repeatedly along the whole highway they have a powerful landscape effect opening up views. (Photo Brett Boardman)

There are two tunnels on the Pacific Highway program. The first, at the Cudgen Road ridge in the Northern Rivers area, was built to retain the forested ecological ridge line. Terracing was designed at the portal to provide a planted margin. The second was built on the Ewingsdale ridgeline to avoid a deep cutting into the ridge visible from Byron Bay and improve motorway gradients. Both tunnels retained important landscape and cultural ridgelines on the highway. However the tunnel design response to the landscape had advanced in the decade between the two tunnels. The Ewingsdale tunnel includes concrete tube structures at either end which although make the tunnel marginally longer allows the soils and landscape to wrap around the portals, avoiding retaining walls and keeping more of the ridge topography. (Fig. 14.14).

Fig. 14.13 Haunched and widely spaced girders over the Brunswick River ensuring impacts on the rare sea grass habitat is minimised and also providing an elegant durable bridge form (photo Gareth Collins)

Fig. 14.14 The ridge retained, the hillside wrapping around the portals. (Photo Gareth Collins)

Fig. 14.15 Glenugie upgrade reused the natural soils and quickly developed a native vegetation cover from the seedbank in the soils. (Photo Leigh Trevitt)

Fig. 14.16 Bonville Bypass fauna overbridge. 1m deep soils and native seeding has created a successful fauna corridor and landscape over road space. (Photo Gareth Collins)

The Highway passes through areas of forest and native vegetation (Fig. 14.15). In these areas the existing soils were carefully removed and stored in low height stock piles to avoid damage to the natural seedbank within the soils. When the earthworks were complete the soils were carefully respread on the cuttings and embankments and within a matter of months seedlings were growing fast in the light, rapidly stabilising and revegetating the corridor with local provenance native species.

Fig. 14.17 Land-bridge for open space at Banora Point – restitching the built environment. (Photo Gareth Collins)

Finally, where possible an over-highway landscape has been recreated in the form of land-bridges for fauna connections and for open space (Fig. 14.16). Fauna overpasses located for maximum effect and with deep soils for resilience and irrigation have quickly become vegetated. At Banora Point a land-bridge has been provided to continue the opens space provision in the area and retain Wilson Park (Fig. 14.17).

14.6 Conclusion

Designing modern road infrastructure in balance with the landscape is possible but requires the input of landscape architects and environmental experts, a commitment from the whole project team to achieve context sensitive design and a vision and framework to drive the project and integrate the many and varied disciplines. It also importantly requires listening to Aboriginal communities and involving them in the design process and outcomes. The Pacific Highway has shown the way on a vast scale but is only the first chapter in this contemporary approach to transport infrastructure.

References

Corkery L (2018) Australian Institute of Landscape Architects Reconciliation Action Plan. AILA, Canberra

Crowe (1955) The Landscape of Roads by Sylvia Crowe The Architectural Press: London

Crowe S (1960) The landscape of roads. The Architectural Press, London

Greater Sydney Commission (2018) A Metropolis of three cities. Greater Sydney Region Plan – connecting people. State Government of New South Wales, Sydney
McHarg IL (1969) Design with nature. Natural History Press, New York
Moran J (2009) On roads. A hidden history. Profile Books, London
Nairn I (1959) Outrage: on the disfigurement of town and countryside. Architectural Review 117(702)
Roads and Maritime Services (2004) Pacific highway urban design framework. RMS, Sydney
Roads and Maritime Services (2013) Pacific highway urban design framework – update. RMS, Sydney
Smith C, Relph E (1978) Place and Placelessness. Geogr Rev 68:116. https://doi.org/10.2307/213523
Swinbourne R, Rosenwax J (2017) Green infrastructure. A vital step to brilliant Australian cities. AECOM, Sydney

Chapter 15
Bio-inspiration: Merging Nature and Technology

Chris Bosse

Abstract Designs that merge man, nature and technology are the focus of work by innovative German/Australian firm LAVA (Laboratory for Visionary Architecture) in this chapter. The way we live, work and communicate is changing radically but the architecture around us is widely based on out-dated technologies. And yet the answer is there right before us – nature. Nature has optimised its systems and processes. Its principles don't change, but it can adapt quickly when environments change.

Using examples, both conceptual and realised, this chapter looks at how merging future technologies with the patterns of organisation found in nature forges a smarter, friendlier, more socially and environmentally responsible future. A forest of sunflowers in an oasis of the future in the desert, a crater-shaped football stadium, an eco-home based on cells and a hotel resort designed as a topological structure of valleys and canyons merge 'man, nature and technology'. Nature's forms – cells, trees, plants, dunes, membranes and craters – connect structure with landscape for people in the twenty-first century.

Keywords Bio mimicry · Responsive architecture · Minimal surface · Adaptability · Sustainability

15.1 Introduction

A forest of sunflowers in an oasis of the future in the desert, a crater-shaped football stadium, an eco-home based on cells and a hotel resort designed as a topological structure of valleys and canyons merge 'man, nature and technology' (Fig. 15.1). The understanding of nature and technology is the driver for all these designs.

Nature has been around for millions of years and has optimised its systems and processes (Fig. 15.2). Its principles don't change, but it can adapt quickly when environments change.

C. Bosse (✉)
Laboratory for Visionary Architecture, Sydney, NSW, Australia
e-mail: bosse@l-a-v-a.net

R. Roggema (ed.), *Nature Driven Urbanism*, Contemporary Urban Design Thinking, https://doi.org/10.1007/978-3-030-26717-9_15

Fig. 15.1 Examples of design that merge man, nature and technology (LAVA 2015)

Fig. 15.2 Nature as inspiration (LAVA 2017)

Technology advances constantly to a point where we can't even use the same telephone anymore after a software update: the way we communicate, commute, transport, workflow is changing continuously and rapidly. Society has changed too. The smartphone is our new workplace. Communication occurs through social media. In short, the way we live, work and communicate is transformed radically: yet the architecture around us is still the same and is widely based on old and outdated technologies.

15.2 Responsive Architecture

The beauty of architecture is that buildings are built to last forever! The reason for this is the capital investment put in at the start of the building project. Then, there is the challenge that the process from design to construction of a large-scale building is, on average, five years. During this period our telephones are redesigned five times, our computers are written off twice, and chances are our job specification has transformed three times. Architecture needs to adapt and respond in order to keep up with technology, culture and society.

How do we, can we, accommodate these constantly changing demands into a constantly updateable architecture? How do we allow for current activities and adapt for functions that may be required in the future? It may be that flexible structures are the answer, but the twenty-first century is also about mass customisation, individual expression and reactivity. While the industrial revolution was about mass producing a singular solution the twenty-first century is about mass producing mass customisable solutions that are each individually different and relevant to the specific needs of the end consumer (Fig. 15.3).

The city is a technological replacement of nature (Wallisser 2010). Water from a waterfall or spring becomes water available on tap (literally) and energy comes from power sockets instead of the complex process in nature that produces energy. In the west, these services become readily available, but it's based on individual consumption and we haven't learned to share resources and systems.

Our answer to these questions is that the future is not about how buildings look, their shape, but rather their performance and interactivity, how they connect with each other and how they adapt (Chua 2013). The city of the future should be like a coral reef – thousands of species thrive in coexistence of each other and the elements, air, water and sun. These reefs are like the cities of the future. The intelligence of the smallest unit results in the intelligence of the overall organism (Lynch 2013). Or the city should be seen as a tree that filters the air, filters water, produces oxygen, and is self-generating. It is carrying leaves and fruit, a multiple of its own structural weight (Figs. 15.4 and 15.5).

This future world has buildings that are intelligent and responsive to external influences like air pressure, temperature, humidity, air pollution and solar radiation (CUSP undated). They are not singular structural entities (designed, serviced and accessed as isolated units), but rather they are part of large networked systems, where the whole is greater than the sum of its parts (Fig. 15.6).

Fig. 15.3 City of the Future. (LAVA 2015)

Fig. 15.4 Ecosystems (LAVA 2013)

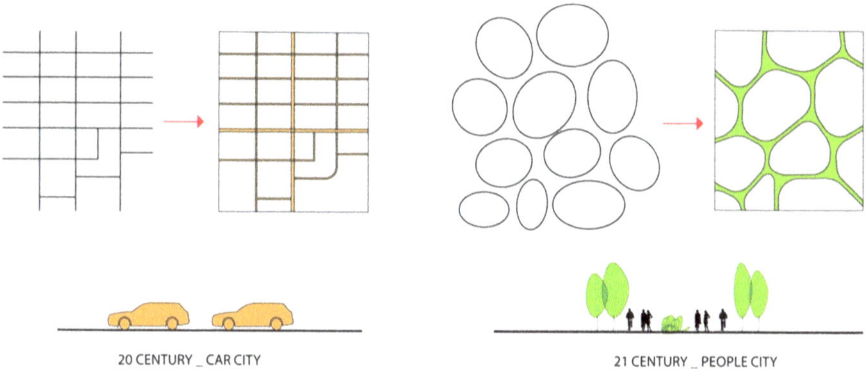

Fig. 15.5 From car city to people city (LAVA 2017)

Fig. 15.6 Bionic Tower (LAVA 2007)

An example of this is the Kacare Master Plan (LAVA 2010) (Fig. 15.7), which transforms a dry riverbed into a thriving city in a post oil world. This 'City of Clouds' uses concepts of water and growth. Kacare is a new city in Saudi Arabia to be developed with renewable energy as its engine. Located as part of a wadi system, a network of natural valleys and riverbeds, the form of the system is in direct dialogue with the topology in which it is located. The wadi system is enclosed under an artificial 'cloud' surface that mimics the action of the atmospheric process (Fig. 15.8). This Cloudscape naturally reduces the reliance on energy to cool the city as well as acting as a tool for capturing and collecting sunlight and rainwater, both precious commodities in a post oil arid nation. The Cloudscape makes it possible to convert extreme environmental conditions into a comfortable environment for human habitation.

Fig. 15.7 Kacare Masterplan (LAVA 2010)

Fig. 15.8 Kacare Cloudscape (LAVA 2010)

15.3 Biomimicry

At the heart of our work is biomimicry – the examination of nature's models, systems, processes, and elements in order to solve human problems (Armstrong 2013). Nature and its symbiotic processes figure strongly in a vision of the architecture of the future. Nature holds all the answers (Fig. 15.9).

The potential for these naturally evolving systems, such as bubbles, spider webs and corals, to create new building typologies and structures informs all our work (LAVA undated-a) – the geometries in nature create both efficiency and beauty. The concepts of structure, space and architectural expression are unified to create a unique experience. Computation allows the simulation of this natural behaviour (Höltgen and Fischer 2018). It is often misunderstood as superficial mimicry, but the potential lies in understanding the principles behind nature, such as growth and adaptation of species, not only their appearance (Höltgen and Fischer 2018) (Fig. 15.10).

Utilising the latest available technology, these new forms harmonise and integrate with nature. Urban structures become landscape, landscape becomes buildings. We have explored this natural phenomenology through the use of membranes, minimal surfaces, inflation/pneumatic structures, branching, adaptive structures and cellular and evolutionary structures (Krzykowski 2008) (Fig. 15.11).

Fig. 15.9 Beijing Watercube, detail (Chris Bosse 2008)

Fig. 15.10 Minimal surface (LAVA 2008)

Fig. 15.11 Origami structures (Ian Barnes 2010)

15.4 Minimal Surface

The organisational principle of a minimal surface (Pritzker Architecture Prize, 2 undated) gleaned from the experiments of German architect Frei Otto's soap bubble experiments for the Munich Olympic Stadium in the early 1970s (Glancey 2004), is critical to many of these projects. Our Green Void project renounced the application of a structure in the traditional sense. Instead, a space was filled with a three-dimensional lightweight-sculpture, solely based on minimal surface tension, freely stretching between wall and ceiling and floor (Pohl 2008) (Fig. 15.12).

An example of minimal surface is the Beijing Olympic Watercube swimming pavilion (Wikepedia undated-b) by PTW and CSCEC with ARUP which associates water as a

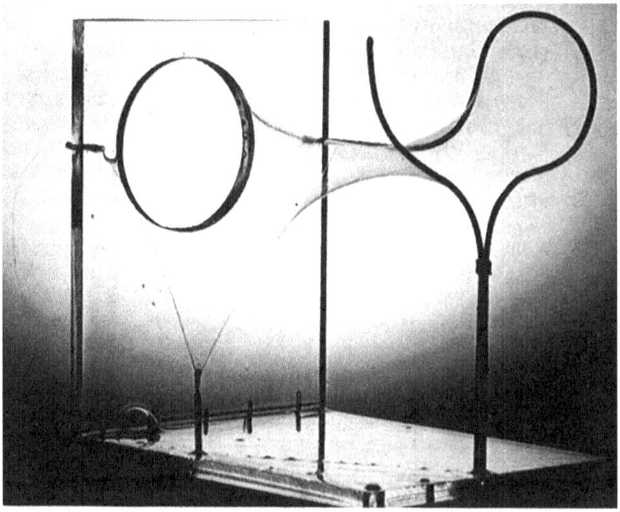

Fig. 15.12 Frei Otto's soap bubble experiments (Otto 1985)

structural and thematical "leitmotiv" with the square, the primal shape of the house in Chinese tradition and mythology (Wikepedia undated-a). Whilst associate architect at PTW I was a key designer of the project. The entire structure is based on a unique lightweight construction derived from the structure of water in the state of aggregation of foam. The aquatic centre is based on the geometrical arrangement of soap bubbles, the most efficient way of filling space with structure. The skin weighs 1/15 of the weight of glass. A strict geometry can be found in natural systems like crystals, cells and molecular structures (Wikepedia undated-b). By applying this technology, the transparency and apparent randomness of bubbles is transposed into the inner and outer skins of ethylene tetrafluoroethylene (ETFE) cushions. Unlike traditional stadium structures with gigantic columns and beams, to which a facade system is applied, the space, structure and façade are one and the same element (Fig. 15.13).

The Bionic Tower in the United Arab Emirates (Grozdanic 2012), moves beyond the superseded modernist concept of extruded footprint and applied curtain wall to create a fully integrated intelligent façade that harvests its surrounding environment to create maximum energy efficiency and user comfort. No building skin today approaches the performance of the biological world (Wikepedia undated-c). The traditional curtain wall is passive, lacking the power to adjust to the fluctuating external environment (Wikepedia undated-d). It should be able to intervene actively in the buildings struggle to maintain its internal stability. By parametric modelling of the 'behavioural logic' the façade has been constantly optimised throughout the design process to create a 'whole' that is greater than the sum of its parts. Instead of an array of individual elements the building behaves like an organism or ecosystem, with a skin that controls air pressure, temperature, humidity, air pollution and solar radiation. New materials and technologies enable an adaptability, responsiveness, environmental awareness and strength not seen in conventional architectural design (Fig. 15.14).

Fig. 15.13 Beijing Watercube (Chris Bosse 2008)

Fig. 15.14 Bionic Tower (LAVA 2008)

15.5 Adaptability

Apart from sustainability, adaptability is critical. How do we revamp existing buildings, without having to pull them down and waste all the money and energy invested in them in the first place? And how do we design buildings that are open, versatile, adaptable and reactive to external forces, contemporary times and changing needs? Buildings of the future have to be robust and flexible, technology has to be accessible and interchangeable, services have to be retractable and serviceable. Building components have to be recyclable, and whenever a new "update" looms, able to be implemented in real time (Inhabitat undated-a). These environmental considerations are an integral part of the design process, in fact a central part (Figs. 15.15 and 15.16).

The idea of 'reskinning', which was proposed for the UTS-tower in Sydney, involves a new high-tech, lightweight-material to transform these outdated buildings (The Independent 2010). Like a skin of a snake or a spacesuit for a new environment, the new skin addresses contemporary needs for flexibility, light, air and views. A new skin can react to the environment, to temperature, humidity and air pressure, and can have embedded layers of technology and sustainability, saving water, producing energy and communicating information to occupants inside the building as well as to the outside world (Norrie 2011) (Fig. 15.17).

Another application is an existing cylindrical-shaped storage centre in Heidelberg, Germany, which is being transformed into a knowledge centre, tourist attraction and city icon with a new multi-layered facade inspired by the geometries of nature (Stevens 2017).

Fig. 15.15 Reskin UTS Tower (LAVA 2009)

Fig. 15.16 Day view, Reskin UTS Tower (LAVA 2009)

Fig. 15.17 Night view, Reskin UTS Tower (LAVA 2009)

An inner shell coloured in different shades of blue wraps the building. Tilted elliptical rings positioned around the cylinder continue in the outer façade with "energy loops" circling the structure and rising dramatically to the top. A cable network between the steel rings forms the outer façade layer (Fig. 15.18).

Around 11,000 diamond-shaped plates of thin stainless steel are hooked with an ingenious connection to this steel network allowing them to twist up to 45 degrees in the wind. The complex interplay of movement, light and shadow is generated by sun and wind, with no additional energy or complicated technology required (Writer 2017) (Figs. 15.19 and 15.20).

Meanwhile a youth hostel in Bayreuth, Germany (Wu 2018) can transform into an aged care facility and other uses in the future. It features innovative spatial configurations, sustainability at environmental, and structural and social levels, and integrated sporting facilities. We chose a 'Y' shape because it generates a connective and beautiful space offering expansive views and multiple openings to the sport fields and gardens. The central atrium is a hub for offline and online entertainment, interaction and communication. A skylight provides natural daylight to a central amphitheatre that connects the different levels, whilst horizontal and diagonal sightlines direct guests to different building functions. There are flexible room walls with contemporary modular 'built-in furniture' elements, and rooms, grounds and sports fields are all wheelchair accessible, enabling it to be future proof (Fig. 15.21).

Fig. 15.18 Energy Storage Centre (LAVA 2017)

Fig. 15.19 Energy Storage Centre (LAVA 2017)

Fig. 15.20 Bayreuth Youth Hostel (LAVA/Häfele, Studio Huber 2017)

Fig. 15.21 Bayreuth Youth Hostel (LAVA/Häfele, Studio Huber 2017)

15.6 Sustainability

And of course, the problem facing designers today is how to create more with less, create more architecture and more amenities, with less material, a lower energy footprint and where possible, less money (Fig. 15.22).

Masdar City is a carbon neutral, car free, solar powered city being built in Abu Dhabi (Dürheimer 2009) which is rethinking the city from scratch. LAVA won the international competition to design the centre of the 50,000-person city, a plaza comprising a hotel, shopping centre, cinema complex and a conference centre (Fairs 2009). LAVA designed an 'oasis of the 21st century', a green heart of the city where giant solar powered sunflower umbrellas create the first mediated outdoor plaza in the Middle East, which in normal circumstances would be uninhabitable throughout the year. These adaptive branch-like structures (Fig. 15.20) shade the space, move with the sun, store heat and release it at night. Conceived as an open spatial experience the plaza, just like an oasis, becomes the city's social epicentre with 24 hours access enabled through interactive, heat sensitive technology that activates low intensity lighting in response to pedestrian traffic and mobile phone usage (Fig. 15.23).

Evolutionary structures such as cells act as a metaphor for an architecture where the individual components interact in symbiosis to create an environment. In urban terms, the smallest homes, the spaces they create, the energy they use, the heat and moisture they absorb, multiply into a bigger organisational system, whose sustainability depends on their intelligence.

Fig. 15.22 Masdar City Centre (LAVA 2008)

Fig. 15.23 The Masdar Plaza at night (LAVA 2008)

This concept of evolutionary structures lies behind the design for the Home of the Future (Fig. 15.24), a showcase for future living, in which nature, technology and man cohabit in a new harmony (Chatterjee 2011). Its fluid design and organisational strategy is based on cells and can be easily modified to suit specific requirements. An ETFE geodesic sky dome provides a year-round microclimate that opens up the home to a garden filled with sun, light and fresh air, away from the pollution and noise of the city. Visitors experience 15 different living spaces, from internal/external bathroom zones to kitchens flowing to veggie patches and barbecues to sunken bedrooms with dream inducing lighting. It integrates the latest improvements in comfort and instantaneous information technology. Smart floors and walls with lighting and heating sensors, intelligent surfaces, fully integrated media displays combine with air and water purification, fully sustainable food generation, water recycling, passive systems, self-sufficient power generation and recycled waste (Fig. 15.25).

The vision for international climate protection was realised in the design for Germany's bid to host the Secretariat of the Green Climate Fund [GCF] in Bonn (Schaefers 2012). The design for the proposed headquarters is 'an ecological model project' and demonstrates the motto: 'green is the new black'. Facades are articulated according to building orientation, surfaces integrate various means of regenerative energy production, photovoltaics and bioreactors demonstrate the application of the latest technologies. The building responds to the site to maximise the experience of its natural qualities (Fig. 15.26). With a design inspired by the beautiful setting in the Rhine valley, and with curvilinear forms, natural light wells, roof top gardens and a large sunken terrace for the restaurant, the three-level structure complies with the latest

Fig. 15.24 Home of the Future (LAVA 2011)

Fig. 15.25 Home of the Future (LAVA 2011)

Fig. 15.26 Green Climate Fund (LAVA 2012)

Fig. 15.27 Green Climate Fund (LAVA 2012)

energy and building ecology standards (Richelle 2012). It features 'environmentally friendly functional spaces that foster working productivity' (Frearson 2012), state-of-the-art technology, security controls, a visitor reception, auditorium, conference room, a canteen and underground car parking (Figs. 15.27 and 15.28).

15.7 In Harmony with Nature

Architectural design must be brought to live in harmony with nature. The design for a National Stadium and Sport Village in Addis Ababa, Ethiopia (Passarivaki 2012), combines geological-inspired forms and local culture with new technology. With partners JDAW and DESIGNSPORT, LAVA went back to the very origin of stadium design with a sunken arena surrounded by grandstands formed from excavated material.

The man-made crater is a clever remodelling of the existing terrain and generates efficient spaces (Fig. 15.30), optimises environmental performance, minimises construction costs and integrates facilities within the existing landscape. Tectonic structures and movement are the underlying concept for the Master Plan (Staff 2012). The breathtaking beauty of the surrounding Entoto Hills is the backdrop to a design that responds to the volcanic geology of the region. The roof of the stadium, an intelligent membrane, appears like a cloud on the horizon of the vast Ethiopian sky, a lightweight tensile structure floating over the formed-earth landscape (Fig. 15.29). Gently

Fig. 15.28 Green Climate Fund (LAVA 2012)

Fig. 15.29 National Stadium and Sport Village (LAVA 2012)

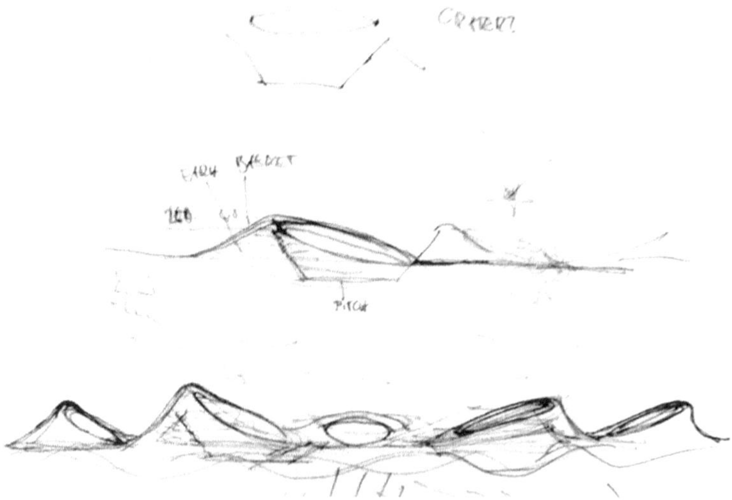

Fig. 15.30 National Stadium and Sport Village (LAVA 2012)

undulating urban parkland follows the lines of the crater and is conceived as a continuous spatial experience strategically activated to balance people movement, climate, experience and efficiency (Fig. 15.30). Giant solar powered umbrellas provide shade and shelter whilst pedestrian activated light and water features appear as fissures in the ground surface, providing way-finding and creating animated art works.

The design for a hotel on Jeju, a mountainous volcanic island, resembles the topological structure of valleys and canyons. Located off South Korea the island is home to the world heritage Hallasan volcano and lava tubes, a national icon and major tourist destination (Bojovic 2017). The hotel is designed as another landform. The 'valleys and canyons' generate 2000 rooms around a terracing façade (Fig. 15.31). Interconnected open atrium spaces form a continuous green landscape throughout the interior of the hotel creating a natural atmosphere. This landscape continues to the canyon between the two sinuous residential zones, accentuating the connection with the beachfront forest over which they look, creating an uninterrupted landscape, within and out (Fig. 15.32).

The concept for 'Forest City', a new 20 square kilometre green smart city on reclaimed land in southern Malaysia (Inhabitat undated-b) sees the city defined not as one iconic building, nor as a skyline, but as a central public space, a real forest with nature as the icon. It's an inverse city skyline where the icon of the city is a public space, not an object/ building. This central space demonstrates the equation: PEOPLE = CITY. The city is not a (series of) object(s) but a place for people (Figs. 15.33 and 15.34).

Fig. 15.31 Jeju Hills Hotel Resort (LAVA 2011)

Fig. 15.32 Interior, Jeju Hills Hotel Resort (LAVA 2011)

Fig. 15.33 Forest City (LAVA 2017)

Fig. 15.34 Forest City (LAVA 2017)

Future city notions are embedded in the overall concepts of a

- 'Public city', a central public space surrounded by buildings;
- 'Layered city', where people, railways and traffic are separated with vehicles underground;
- 'Loop city', a closed loop system reusing its resources and controlling the out-flow;
- 'Sponge city', with recycling processes hidden underground. (Fig. 15.5)

A group of buildings step down towards this green public centre and are an architectural interpretation of the rainforest, with various layers of program and vertical stratification (Fig. 15.35).

Green shapes a new residential quarter based on nature's principles in the concept for Garden Island in Sydney Harbour (McCarthy 2017), turning a previously inaccessible headland into a cultural, leisure, and community destination. The former dry-dock could be used for floating markets, harbour baths, theatre performances and boat shows, whilst a new residential precinct is inspired by the sweeping curves of Sydney Harbour with all its wonderful bays, beaches and sandstone headlands. The new buildings make a departure from the traditional vertical apartment box model, instead a green and sustainable, airy tower landscape, with roof terraces, balconies, swimming pools and community facilities (Fig. 15.36).

Natural and intelligent light is used for information, visualisation, emotion and enabling in the Philips Lighting Headquarters Eindhoven, The Netherlands (Foges 2017; Philips undated). Light is the tool to create a volume of space. Visitors are

Fig. 15.35 Garden Island concept (LAVA 2017)

Fig. 15.36 Garden Island concept (LAVA 2017)

greeted by a huge parametrically designed interactive light 'tree' (Fig. 15.38), a sculpture of light containing 1500 'leaves', panels suspended from the atrium ceiling, an experiential welcome that fosters creativity. Reflective surfaces create a play of light and shadow. The concept is of golden light filtering through trees. The latest lighting technology is showcased and the sculpture gives shape and visibility to light. Intelligent light control generates different scenarios that activate or relax users – every panel is programmed for the whole calendar year, a bit like an ecosystem, with light effects turning golden, for example, when meetings are in progress (Figs. 15.37 and 15.38).

A twisting tree-like tower with ETFE curtain walls, smart surfaces and green vertical gardens is the centre of a 'five-finger' Master Plan for the 2017 Astana Expo Kazakhstan (Kannfinch undated). This Master Plan in collaboration with Kann Finch is inspired by nature. With its trees and branches the five finger-pads house a cluster of facilities linked through green fingers, connecting visually to Astana's city icons. This paradigm for global city planning transforms the twentieth century 'car city' into the twenty-first century 'people city' (Fig. 15.39). It fuses the traditional city grid with fluid organic shapes providing a human scale, resolving the conflict between built and natural forms (Fig. 15.39). The design enables future organic growth, whilst the Expo buildings are all 'powerplants': they collect, generate and share local renewable energy sources. Other sustainable features include bio-energy production using the facade, (hyper)localised food production, a bio filtration system, wind energy generation, facade airflow and ventilation, local water harvesting and reuse, photovoltaic integrated membrane shading and a buffer of winter gardens.

Fig. 15.37 Philips Lighting Headquarters (Jonathan Andrew 2016)

Fig. 15.38 Philips Lighting Headquarters (Jonathan Andrew 2016)

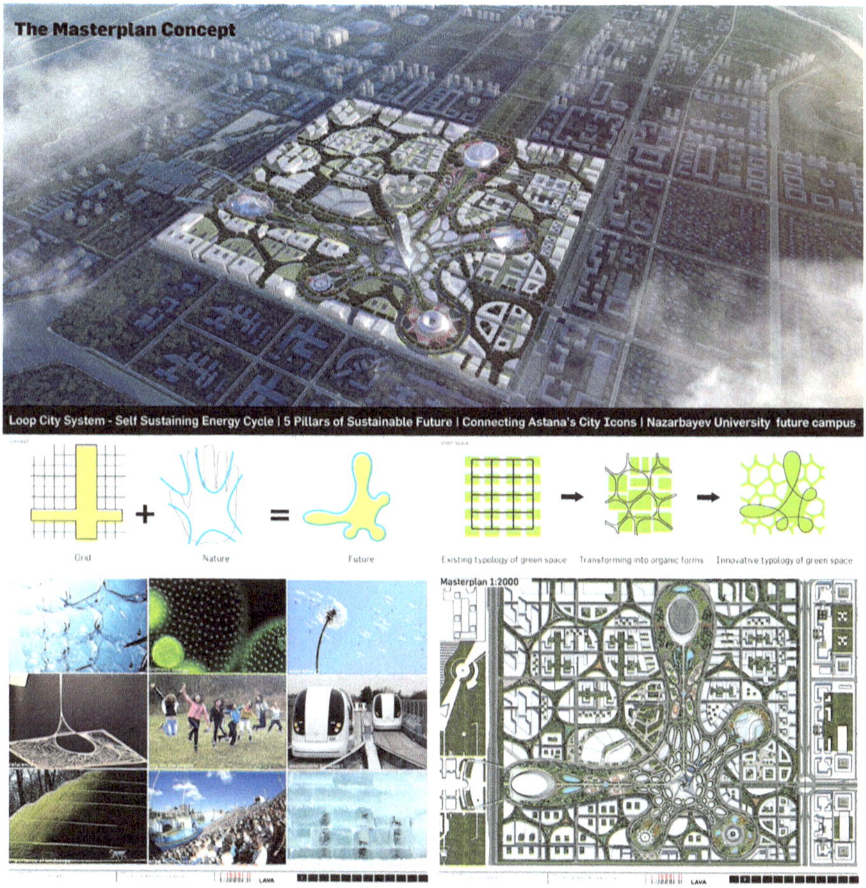

Fig. 15.39 Astana Expo (LAVA 2013)

15.8 Design Principles

LAVA's design principles are Man. Nature. Technology. These underpin all of LAVA's work. It starts with naturally evolving structural systems, such as snowflakes, spider webs and soap bubbles. The geometries in nature generate both efficiency and beauty. These are merged with future technologies. But above all the human is the centre of every project.

15.9 Conclusion

The examples shown in this chapter, such as Masdar City, a green heart of the city oasis of the future, Astana Expo, a future sustainability masterplan, Forest City, nature as the icon and Garden Island, shaping a new residential quarter based on nature's principles, exemplify the fusion of nature and technology.

Nature's forms – cells, trees, plants, dunes, membranes and craters – connect structure with landscape for people in the twenty-first century. By merging future technologies with the patterns of organisation found in nature a smarter, friendlier, more socially and environmentally responsible future lies ahead.

15.10 About LAVA

LAVA was founded in 2007 by directors Chris Bosse, Tobias Wallisser, and Alexander Rieck and was the 2016 European Laureate for Architecture, Europe's highest award for architecture presented by the European Centre and The Chicago Athenaeum.

LAVA merges future technologies with the patterns of organisation found in nature. Digital workflow, nature's structural principles and the latest digital fabrication technologies are combined to build MORE WITH LESS: more (architecture) with less (material/energy/time/cost) (Fig. 15.40).

Fig. 15.40 LAVA designs and directors (LAVA 2018)

LAVA designs everything from pop up installations to master-plans and urban centres, from homes made out of PET bottles to 'reskinning' aging 60s icons, from furniture to hotels, houses and airports of the future.

www.l-a-v-a.net

References

Armstrong R (2013) Rachel Armstrong on Biomimicry as parametric snake oil. The Architectural Review. Published 11 July 2013. URL: https://www.architectural-review.com/8650000.article?fbclid=IwAR0ahbpeQobgYCUd5i-Wc59mgDZmXpvR_PWmHgcdE1-iZ5wZJFb_S5XHdOw

Bojovic M (2017) Jeju Hills hotel resort mimics surrounding landscape – LAVA. eVolo. URL: http://www.evolo.us/jeju-hills-hotel-resort-mimics-surrounding-landscape-lava/. Published 1 June 2017

Chatterjee A (2011) LAVA: home of the future. Studio International. URL: https://www.studiointernational.com/index.php/lava-home-of-the-future. Published 9 May 2011

Chua G (2013) CUSP exhibition to present ideas on architecture and design around Australia. Architecture & Design. URL: https://www.architectureanddesign.com.au/news/cusp-exhibition-travels-australia-presenting-ideas. Published 11 July 2013

CUSP (undated) Chris Bosse. URL: https://cusp-design.com/designer/chris-bosse/

Dürheimer A (2009) LAVA design for the centre of fosters Masdar City. Detail. URL: https://www.detail-online.com/article/lava-design-for-the-centre-of-fosters-masdar-city-13762/. Published 28 Aug 2009

Fairs, M. (2009) Masdar City centre by LAVA. Dezeen. URL: https://www.dezeen.com/2009/08/28/masdar-city-centre-by-lava/. Published 28 Aug 2009

Foges C (2017) Philips Lighting headquarters by LAVA. Architectural Record. URL: https://www.architecturalrecord.com/articles/12208-philips-lighting-headquarters-by-lava?v=preview. Published 1 Feb 2017

Frearson A (2012) Green Climate Fund headquarters by LAVA. Dezeen. URL: https://www.dezeen.com/2012/06/06/green-climate-fund-headquarters-by-lava/. Published 6 June 2012

Glancey J (2004) The lightweight champion of the world. The Guardian. URL: https://www.theguardian.com/artanddesign/2004/oct/04/architecture. Published 4 Oct 2004

Grozdanic J (2012) Bionic tower combines structure and ornament/LAVA. eVolo. URL: http://www.evolo.us/bionic-tower-combines-structure-and-ornament-lava/. Published 2 Jan 2012

Höltgen S, Fischer T (2018) Medientechnisches Wissen, Band 2: informatik, Programmieren, Kybernetik. Taschenbuch. De Gruyter Oldenbourg, Munich

Inhabitat (undated-a) Smart Facades. URL: https://inhabitat.com/tag/smart-facade/

Inhabitat (undated-b) Futuristic green city design runs like a real rainforest in Malaysia. URL: https://inhabitat.com/futuristic-green-city-design-runs-like-a-real-rainforest-in-malaysia/forest-city-by-lava-4/

Kannfinch (undated) Atana Expo 2017 – LAVA. KANNFINCH. URL: http://www.kannfinch.com/projects/astana-expo-2017-astana-kazakhstan

Krzykowski M (2008) Michael Schumacher World Champion Tower. Dezeen. URL: https://www.dezeen.com/2008/10/07/mswct-tower-by-lava/. Published 7 October 2008

LAVA (2010) Kacare City of Clouds, Final Design Report

LAVA (undated-a) About LAVA. URL: http://www.l-a-v-a.net/about-lava/

Lynch O (2013) CUSP: designing into the next decade. Indesignlive. URL: https://www.indesignlive.com/projects/cusp-designing-into-the-next-decade. Published 1 Aug 2013

McCarthy A (2017) Sydney's Garden Island cold be transformed into a cultural district. Lonely Planet. URL: https://www.lonelyplanet.com/news/2017/10/24/sydney-garden-island-new-plans/. Published 25 Oct 2017

Norrie, J. (2011) How to transform an ugly university: just don't call it The Condom-inium. Sydney Morning Herald. URL: https://www.smh.com.au/national/nsw/how-to-transform-an-ugly-university-just-dont-call-it-the-condom-inium-20110618-1g8un.html. Published 18 June 2011

Passarivaki M (2012) Sports complex in Ethiopia-LAVA. Design Exchange. URL: http://www.demagazine.co.uk/architecture/sports-complex-in-ethiopia. Published 5 July 2012

Philips (undated) Discussing with light. A showcase for light. Philips Lighting. URL: http://www.lighting.philips.com/main/cases/luminous-magazine/luminous-14/luminous-14-showcase-for-light

Pohl EB (2008) Green Void/LAVA. Architecture Daily. URL: https://www.archdaily.com/10233/green-void-lava. Published 16 Dec 2008

Pritzker Architecture Prize (undated) Laureates/Otto Frei. URL: https://www.pritzkerprize.com/laureates/frei-otto

Richelle (2012) LAVA: green climate fund headquarters, Bonn. Designboom. URL: https://www.designboom.com/architecture/lava-green-climate-fund-headquarters-bonn/. Published 21 May 2012

Schaefers S (2012) Green Climate Fund Bonn by LAVA. Frame. URL: https://www.frameweb.com/news/green-climate-fund-bonn-by-lava. Published 22 May 2012

Staff CW (2012) Qatari, Oz stadium design a winner in Addis Ababa. Construction Week. Online. URL: http://www.constructionweekonline.com/article-18021-qatari-oz-stadium-design-a-winner-in-addis-ababa/. Published 12 Aug 2012

Stevens P (2017) LAVA's sculptural and sustainable energy tower breaks ground in Heidelberg. Designboom. URL: https://www.designboom.com/architecture/lava-energy-storage-tower-stadtwerke-heidelberg-germany-08-04-2017/. Published 8 April 2017

The Independent (2010) Condom architecture. URL: https://www.independent.co.uk/arts-entertainment/architecture/condom-architecture-1949391.html. Published 21 April 2010

Wallisser, T. (2010) Future Cities: towards a synergy of nature and technology. TEDX-talk. URL: https://www.youtube.com/watch?v=q71qGavyxQU

Wikepedia (undated-a) Chinese architecture. URL: https://en.wikipedia.org/wiki/Chinese_architecture

Wikepedia (undated-b) Weaire Phelan structure. https://en.wikipedia.org/wiki/Weaire%E2%80%93Phelan_structure

Wikepedia (undated-c) Eden Project. URL: https://en.wikipedia.org/wiki/Eden_Project

Wikepedia (undated-d) Seagram building. URL: https://en.wikipedia.org/wiki/Seagram_Building

Writer S (2017) Germany builds revolutionary water tank for energy storage. The Urban Developer. URL: https://theurbandeveloper.com/articles/energy-storage-centre-beacon-sustainable-energy-germany. Published 7 Aug 2017

Wu D (2018) Lava youth hostel hotel – Bayreuth, Germany. Wallpaper. URL: https://www.wallpaper.com/travel/germany/bayreuth/hostels/lava-youth-hostel. Published 17 Mar 2018

Chapter 16
The Future of Nature-driven Urbanism

Rob Roggema

Abstract Current urban design and urban planning aim to facilitate global, regional and local urbanization programs. This implies most of the planning documents give room to the types of land use that seem to require space 'here and now'. The amount of new housing, office and other industrial and commercial space, accompanying amounts of parking lots and the necessity of new transportation routes, infrastructure and corridors are the main topics in the majority of future oriented plans. This is what is called 'fast urbanism' ((Roggema, R., Special Issue Urban Planning 6:946-956, October 2015)). It is the natural preferred habit of planners, decision-makers and politicians, and many developers, economists and municipal land departments. It seems as if this way of future planning brings the highest revenues, and this may be true, on the short term and for only a limited part of involved groups in the city. The impact of this way of planning the city has negative consequences for our health in general (see Roggema, this volume, Chap. 5; Han and Keeffe, this volume, Chap. 4; Monti, this volume, Chap. 11), and more specifically the quality of nature and biodiversity in our urban and natural environments (Birtles, this volume, Chap. 10; Tillie, this volume, Chap. 6; Monti, this volume, Chap. 11; Backes et al., this volume, Chap. 3; Sijmons, this volume, Chap. 2). One way of coping with the effects is to 'repair' the damage after the city has been built. Aiming to increase the quality of small green spaces (Veldman, this volume, Chap. 13; Casagrande, this volume, Chap. 7), add temporary nature (Backes et al., this volume, Chap. 3), or greening buildings (Bosse, this volume, Chap. 15), could help to prevent the largest impacts of fast urbanism. However, this will always be a solution that repairs, or greenwashes urbanization that has neglected the natural systems in the first place.

Keywords Nature driven urbanism · Future · Natural system · Landscape

R. Roggema (✉)
Research Centre for the Built Environment NoorderRuimte,
Hanze University of Applied Sciences, Groningen, The Netherlands

CITTA IDEALE, Office for Adaptive Planning, Wageningen, The Netherlands
e-mail: r.e.roggema@pl.hanze.nl; rob@cittaideale.eu

R. Roggema (ed.), *Nature Driven Urbanism*, Contemporary Urban Design Thinking, https://doi.org/10.1007/978-3-030-26717-9_16

Current urban design and urban planning aim to facilitate global, regional and local urbanization programs. This implies most of the planning documents give room to the types of land use that seem to require space 'here and now'. The amount of new housing, office and other industrial and commercial space, accompanying amounts of parking lots and the necessity of new transportation routes, infrastructure and corridors are the main topics in the majority of future oriented plans. This is what is called 'fast urbanism' (Roggema 2015). It is the natural preferred habit of planners, decision-makers and politicians, and many developers, economists and municipal land departments. It seems as if this way of future planning brings the highest revenues, and this may be true, on the short term and for only a limited part of involved groups in the city. The impact of this way of planning the city has negative consequences for our health in general (see Roggema this volume-a; Han and Keeffe this volume; Monti this volume), and more specifically the quality of nature and biodiversity in our urban and natural environments (Birtles this volume; Tillie this volume; Monti this volume; Backes et al. this volume; Sijmons this volume). One way of coping with the effects is to 'repair' the damage after the city has been built. Aiming to increase the quality of small green spaces (Veldman this volume; Casagrande this volume), add temporary nature (Backes et al. this volume), or greening buildings (Bosse this volume), could help to prevent the largest impacts of fast urbanism. However, this will always be a solution that repairs, or greenwashes urbanization that has neglected the natural systems in the first place.

Instead, urban design and planning could and should make space for vulnerable land-uses from the beginning of the planning process, at every scale, and every time again. These vulnerable uses in the city comprise not only nature, but also sensitive water systems, urban agriculture and culture. These require to be watered every now and then, and succinct being given attention throughout the planning, and lifetime of their existence. The space needed for flourishing ecological, cultural, aquatic and agricultural systems must be found as an integrated part within the city, moreover as the basis for every planning or design step undertaken. If the landscape with all its essential functions and systems is taken as the first step in the planning process (Roggema this volume-a, this volume-b; Tyrrell this volume; Sijmons this volume), slow urbanism (Roggema 2015) can find the basis for a structural and qualitative development within dynamic and urban constraints. To design literally space for the water and ecological systems, to grow food and allow for cultural events and encounters, for instance in public spaces, the city gains value. This may not be of direct financial benefits, but as mentioned in several of the chapters in this book, having these 'slow' elements in the vicinity has profound positive implications on health, well-being, stress relief, not only for human beings but also for animals and plants.

Additionally, current times hold many uncertainties. The pace, dynamic and impact of climate change, new geo-political relationships, or the increasing unrest in many parts of the world, leading to migration and political instability, all, and more have profound impact on the success of planning. Events occur suddenly, without announcement, are unprecedented and have unforeseen consequences for political, social, physical and special systems around the world. These impacts will have a stronger effect on societies if the flexibility in those societies is low. When planners plan the urban environments as if they remain unchanged and reflect the ultimate final version of the city, flexibility, or the capability to easily deal with change, is not incorporated in the urban fabric hence impacts will do more damage than needed. On the contrary, planners need to prepare the city so it can cater for 'suddenism' (Roggema 2015), dealing with sudden events in a way the impact is not felt rather turned around in a positive, antifragile, manner (Roggema 2018). This implies designing the urban environments in a way spaces are created that 'offer help' when sudden events occur. Instead of fixing the buildings and public spaces till the max, redundancy should be embraced, and voids and potentials should become part of the plan. A conscious inclusion of spaces that can change use, are flexible enough to take up sudden spatial demands, could help the city being more resilient. The landscape points us at the places where these voids can be found, where we need to create emptiness so its potential can be capitalized whenever necessary.

The three types of urbanism (fast, slow and sudden), play a role at every moment in every design process. Giving priority to the short-term economically driven uses will only bring a short time gain for a selected group of people. When other paces of urbanism are given a serious position, in every stage of the design, build and maintenance process, and at every scale, the benefits for the society as a whole will be bigger, and more people will profit from it, for instance in the form of living in a safer place or feeling healthier and better. The reduced costs for repairing and rebuilding after damage of a disaster, or reduction of the collective costs for care and hospitalization should be weighted while planning the city of the future.

Fast urbanism can be planning with our eyes closed, but when we open them, we will need to create space for slow urbanism and suddenism. Coincidentally, these two forms of urbanization require green, open spaces, and the landscape is just offering this. Nature based solutions offer solutions literally based on what nature requires. Necessary but not enough as if this effort, which in itself is not easy, will lead to more than patches of nature scattered in the city. No, instead of basing the city on demands from nature, we need to shift to nature-driven solutions. These solutions are different because the city, the entire city is driven by the requirements and desires nature poses on human life. Not only is nature realized for their own good, it will contribute to the overall quality of urban life when nature can drive urbanization. If in the future urban design and planning are driven by the powers of nature, the processes that shape ecosystems and the constructing mechanisms found in nature, the city will be embedded in nature, green space and the landscape and not the other way around. This will add the quality, the anticipative capacity, the adaptive capability, learned from nature to the urban society, the infrastructure and the people living in the city. The future city has to be nature-driven, as our lives depend on it, and even better our lives gain quality and longevity. Not the least important reason for a change.

References

Backes C, Krefeld A, Schoukens H (this volume) Chapter 3: Temporary Nature - a win-win for nature and developers: tinkering with the law in order to combat biodiversity loss. In: Roggema R (ed) Nature-driven urbanism, Contemporary urban design thinking, vol 2. Springer, Dordrecht

Birtles P (this volume) Chapter 10: South Creek in Far Western Sydney: opportunities for a new waterway focused city. In: Roggema R (ed) Nature-driven urbanism, Contemporary urban design thinking, vol 2. Springer, Dordrecht

Bosse C (this volume) Chapter 15: Bio-inspiration: merging nature and technology. In: Roggema R (ed) Nature-driven urbanism, Contemporary urban design thinking, vol 2. Springer, Dordrecht

Casagrande M (this volume) Chapter 7: From urban acupuncture to the third generation city. In: Roggema R (ed) Nature-driven urbanism, Contemporary urban design thinking, vol 2. Springer, Dordrecht

Han Q, Keeffe G (this volume) Chapter 4: Stepping-stone city: process-oriented infrastructures to aid forest migration in a changing climate. In: Roggema R (ed) Nature-driven urbanism, Contemporary urban design thinking, vol 2. Springer, Dordrecht

Monti S (this volume) Chapter 11: Nature-inclusive cities: concepts and considerations. In: Roggema R (ed) Nature-driven urbanism, Contemporary urban design thinking, vol 2. Springer, Dordrecht

Roggema R (2015) Three urbanisms in one city: accommodating the paces of change. J Environ Protect Special Issue Urban Plann 6:946–956. https://doi.org/10.4236/jep.2015.69084

Roggema R (2018) Design with voids. How inverted urbanism increases urban resilience. Architectural Sci Rev (ASR) Special issue: Time, Place and Architecture 61:349–357. https://doi.org/10.1080/00038628.2018.1502153

Roggema R (this volume-a) Chapter 5: Landscape first! Nature-based design for Sydney's third city. In: Roggema R (ed) Nature-driven urbanism, Contemporary urban design thinking, vol 2. Springer, Dordrecht

Roggema R (this volume-b) Chapter 8: Urbanism on water and ecology: the early example of Westerpark, Breda. In: Roggema R (ed) Nature-driven urbanism, Contemporary urban design thinking, vol 2. Springer, Dordrecht

Sijmons D (this volume) Chapter 2: Contrast, contact & contract, pathways to pacify urbanization and nature. In: Roggema R (ed) Nature-driven urbanism, Contemporary urban design thinking, vol 2. Springer, Dordrecht

Tillie N (this volume) Chapter 6: From urban green structure to tidal river in Rotterdam: testing grounds for Urban Ecology. In: Roggema R (ed) Nature-driven urbanism, Contemporary urban design thinking, vol 2. Springer, Dordrecht

Tyrrell M (this volume) Chapter 12: Exploring new urban futures through Sydney's hidden grids. In: Roggema R (ed) Nature-driven urbanism, Contemporary urban design thinking, vol 2. Springer, Dordrecht

Veldman D (this volume) Chapter 13: A bold Vision for Sydney's future. In: Roggema R (ed) Nature-driven urbanism, Contemporary urban design thinking, vol 2. Springer, Dordrecht

Index

R. Roggema (ed.), *Nature Driven Urbanism*, Contemporary Urban Design Thinking, https://doi.org/10.1007/978-3-030-26717-9

H

I

L

M

N

O

P

T

U

V

W

Printed by Printforce, the Netherlands